MODERN CERTIFICATE MATHEMATICS

7157 1356 – 6

MODERN CERTIFICATE MATHEMATICS

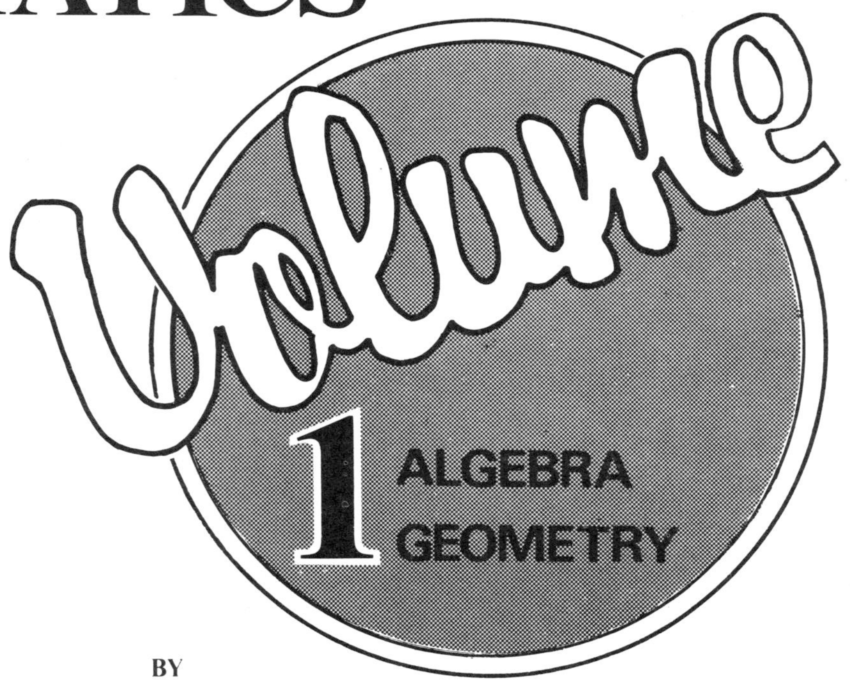

BY

F. CAPANNI M.A., B.Sc.

Head of the Mathematics Department
Holyrood Secondary School, Glasgow.

Designed by ROGER SADLER D.A. (Edin.)

HOLMES McDOUGALL, EDINBURGH

CONTENTS

ALGEBRA

GEOMETRY

ANSWERS

PREFACE

This book originated from a series of notes which I issued to my certificate pupils each year. They formed the nucleus of the work required for the Scottish Certificate of Education Examination at the higher grade. Since most pupils do not wade through large chunks of mathematics in order to find an explanation of some particular difficulty I have kept explanations as concise as possible but have reinforced them with many worked examples. It should be noted, however, that all the proofs required for the S.C.E. examination are included in the text.

Volume 1 contains two sections, algebra and geometry, and Volume 2 three sections, coordinate geometry, trigonometry, and calculus. Each section is divided into several units. As each part of each unit is completed an assignment on that particular part has been inserted and to meet modern examination trends an extra assignment has been included at the end of each unit containing 12 objective items, these items being of three different types. Thus the two volumes contain over 300 objective type questions. Over and above this there are supplementary examples at the end of each section.

These two books cover the syllabus required for the Scottish Certificate Examination at the higher grade. They may be found useful either as standard text or as supplementary material for the ordinary and advanced levels of the various Associated Examining Boards or Joint Matriculation Boards of the English Universities and Schools.

I wish to thank my wife Anne for her help in checking part of the work in the algebra section, Mr. Iain McLean of Holmes McDougall for also checking part of the work and for his patience with my many mistakes in the original draft and lastly Mr. Rodger Sadler for the excellence of the art-work.

Holyrood Secondary School,
Glasgow,
May, 1977.

F.C.

NOTATION

Different writers use different letters to denote various sets of numbers. This does not matter, provided the notation is stated clearly at some point in the work and not contradicted later in the same work.

In this book the following notation for sets of numbers is used.

N the set of natural numbers or positive integers $\{1, 2, 3, \ldots\}$.

W the set of whole numbers $\{0, 1, 2, 3, \ldots\}$.

Z the set of integers $\{\ldots -2, -1, 0, 1, 2, \ldots\}$.

Q the set of rational numbers which includes Z together with the positive and negative fractions.

R the set of real numbers which includes numbers such as $\sqrt{2}, -\sqrt{3}, \pi, e, \ldots$.

Hence $N \subset W \subset Z \subset Q \subset R$.

The following are also used:

R^+ the set of positive real numbers which does not include zero.

$\varnothing$ the empty set.

E the universal set.

ALGEBRA

UNIT 1: SEQUENCES AND SERIES

INTRODUCTION

1.1

A **sequence** is a set of numbers written down *in accordance with some given rule or law*. Each member of the set is called a **term** of the sequence. A **series** (or **progression**) is used for an expression consisting of a sum of terms.

Example: 1, 3, 5, 7, 9, . . . is a sequence of numbers.
$1+3+5+7+9+\ldots$ is a series or progression.

Examples of sequences

(i) 1, 3, 5, 7, . . .

(ii) 1, 1, 2, 3, 5, 8, . . .

(iii) 1, 2, 6, 24, 120, . . .

(iv) 2, 4, 8, 16, 32, . . .

A sequence is a mapping from the set of natural numbers into the set of real numbers and it is usual to define the mapping by means of a **formula**.

Example 1

For the sequence 1, 3, 5, 7, . . . the associated function is $u: n \rightarrow 2n-1,\ n \in N$, i.e. the formula is $u(n) = 2n-1$. This is usually written $u_n = 2n-1$ where $n \in N$ and gives the ***n*th term** of the sequence.

Domain	*Range*
1	$2\times 1-1 = 1$
2	$2\times 2-1 = 3$
3	$2\times 3-1 = 5$
. . . .	. . . .
n	$2\times n-1 = 2n-1$

The 1st term, denoted by u_1 is $u_1 = 2\times 1-1 = 1$
the 2nd term, is $u_2 = 2\times 2-1 = 3$
the 3rd term, is $u_3 = 2\times 3-1 = 5$
....
the nth term is $u_n = 2\times n-1 = 2n-1$

Note: If we know the nth term, we can find all the other terms of the sequence by replacing n by 1, 2, 3 and so on to obtain the 1st, 2nd, 3rd etc. terms of the sequence.

Example 2

Give the first five terms of the sequence defined by $u_n = 3n+2$. Which term of the sequence is 197?

$$u_1 = 3\times 1+2 = 5$$
$$u_2 = 3\times 2+2 = 8$$
$$u_3 = 3\times 3+2 = 11$$
$$u_4 = 3\times 4+2 = 14$$
$$u_5 = 3\times 5+2 = 17$$

Hence the first five terms of the sequence are 5, 8, 11, 14, 17.
If 197 is the nth term then $u_n = 197$, but $u_n = 3n+2$

$$\Rightarrow 3n+2 = 197$$
$$\Rightarrow 3n = 195$$
$$\Rightarrow n = 65$$

i.e. 197 is the 65th term of the sequence.

ASSIGNMENT 1.1

1. Write down the first four terms and the 10th term of the sequences given by the formulae:
   (i) $u(n) = 3n-1$ (ii) $u(n) = 4n^2$
   (iii) $u_n = 11-5n$ (iv) $u_n = 2^n$
   (v) nth term $= n(n-1)$ (vi) nth term $= \dfrac{n}{n+1}$
   (vii) $u_n = (-1)^{n-1}$ (viii) $u_n = (-1)^n$
2. A sequence is defined by $u_n = 4n-18$. How many terms of the sequence are negative?
   Show that $u_9 - u_8 = 4$.
3. $u_n = (n-1)x^{n-1}$. Write down the first three terms of the sequence.
4. If the nth term of a sequence is defined by $u_n = 3(2n-1)$, write down $u_2, u_{20}, u_k, u_{2k}, u_{n-1}$.
5. If $u_n = 16.(\frac{1}{2})^{n-1}$, how many terms of the sequence are greater than 1?
   Which term is equal to 1?
6. Write down the simplest formula which defines the nth term of each of the following sequences:
   (i) $1, 2, 3, 4, 5, \ldots$ (ii) $2, 4, 6, 8, 10, \ldots$
   (iii) $5, 10, 15, 20, \ldots$ (iv) $1, 4, 9, 16, 25, \ldots$
   (v) $2, 4, 8, 16, 32, \ldots$ (vi) $\frac{1}{3}, \frac{1}{9}, \frac{1}{27}, \frac{1}{81}, \ldots$
   (vii) $4, 7, 10, 13, \ldots$ (viii) $11, 7, 3, -1, -5, \ldots$
   (ix) $-1, 1, -1, 1, \ldots$ (x) $1, -\frac{1}{2}, \frac{1}{4}, -\frac{1}{8}, \ldots$
7. Write down the first three terms of the sequence defined by $u_n = n^2-7n$. Which term of the sequence is 18?
8. Which term of the sequence defined by $u_n = n^2-5n$ is the same as the 5th term of the sequence defined by $u_n = 3n-1$?
9. The eighth term of the sequence $u_n = n^2-2kn$ is 100. Find the value of k.
10. A sequence is defined by $u_n = 7n-25$.
    (i) How many terms of the sequence are negative?
    (ii) Is 80 a term of the sequence? If so which one?

11. A series is given by $1+\frac{1}{2}+\frac{1}{4}+\frac{1}{8}+\ldots$
 (i) Write down u_n, u_{n-1} and u_{n+1}.
 (ii) Simplify as far as possible $\frac{u_{n+1}}{u_n}$ and $\frac{u_{n-1}}{u_n}$.

12. A sequence is defined by $u_n = a+(n-1)d$, where $a, d \in R$. If $u_5 = 14$ and $u_8 = 23$ find a and d. With these values of a and d rewrite u_n and find the value of $u_n - u_{n-1}$.

13. List the first five members of the sequence defined by $u_n = 4+\frac{u_{n-1}}{10}$ where $u_1 = 4$ and $n = (2, 3, 4, \ldots)$.

14. Write down the first four terms of the sequences whose nth term is defined by

 (i) $u_n = (-\frac{1}{4})^{n-1}$ (ii) $u_n = \frac{x^{n-1}}{n}$

15. u_k is the kth term of a sequence satisfying the recurrence relation $u_k = \frac{1}{2}(4-u_{k-1}^2)$ with $u_1 = 0$. Find the sequence.

16. u_k is the kth term of a sequence satisfying the recurrence relation $u_{k+1} = 2u_k - 3$ with $u_1 = 1$. Find the first five terms of the sequence and hence or otherwise show that $u_k - u_{k+1} = 2^k$.

ARITHMETIC SEQUENCES AND SERIES

Each term of this sequence is obtained by adding the same quantity to the preceding term.

The sequence can start at any number which is called the **first term** of the sequence and the constant quantity is called the **common difference** (c.d.).

Examples of arithmetic sequences

(i) $2, 4, 6, 8, \ldots, 2n$ (c.d. $= 2$)
(ii) $7, 3, -1, -5, \ldots, -4n+11$ (c.d. $= -4$)
(iii) $2\frac{1}{4}, 1\frac{1}{2}, \frac{3}{4}, 0, -\frac{3}{4}, \ldots, -\frac{3}{4}n+3$ (c.d. $= -\frac{3}{4}$)
(iv) $a, a+d, a+2d, a+3d, \ldots, a+(n-1)d$ (c.d. $= d$)

(iv) is called the **Standard Arithmetic Sequence** and $a+(n-1)d$ is often referred to as the nth or last term of the sequence or series, denoted by,

$$u_n = a+(n-1)d$$

Note: For an arithmetic sequence

$$u_n - u_{n-1} = u_{n+1} - u_n$$

Check this for the above sequences, when $n = 3$.

Example 1

Find the arithmetic sequence whose 8th term is 19 and whose 14th term is 37. Let a be the 1st term and d the common difference.

The nth term $u_n = a+(n-1)d$.

When $n = 8$, $u_n = 19 \Rightarrow a+7d = 19$
When $n = 14$, $u_n = 37 \Rightarrow a+13d = 37$
$\Rightarrow 6d = 18$
$\Rightarrow d = 3$

When $d = 3$, $a = 19-7d = 19-21$
$= -2$.

Hence sequence is $-2, 1, 4, 7, 10, \ldots$

Example 2

Is 209 a term of the sequence 3, 7, 11, 15, ...? Here $a = 3$ and $d = 4$. Let $u_n = 209$, where $u_n = a+(n-1)d$, $n \in N$. Hence 209 is a term of the sequence if

$a+(n-1)d = 209$ gives a value of $n \in N$.

But $3+(n-1)4 = 209$

$$3+4n-4=209$$
$$4n-1=209$$
$$4n=210$$
$$\Rightarrow n=52\tfrac{1}{2}$$

Hence 209 is not a term of the sequence since $n=52\frac{1}{2}\notin N$.

FORMULA FOR THE SUM TO n TERMS OF AN ARITHMETIC SERIES

Example 3

Find the sum of all the even numbers from 2 to 50. Let S denote the sum,

$$\begin{aligned}
\text{then}\quad S &= 2+4+6+8+\ldots+46+48+50\\
\text{and}\quad S &= 50+48+46+44+\ldots+6+8+2\\
\text{Adding}\quad 2S &= 52+52+52+52+\ldots+52+52+52\\
&= 25\times 52\\
\text{Hence}\quad S &= \frac{25\times 52}{2}=25\times 26\\
&= 650
\end{aligned}$$

This gives us a way of finding the sum to n terms of any arithmetic series.

Let a be the first term, d the common difference and let S_n denote the sum of the first n terms of the series, then
$S_n=a+(a+d)+\ldots+[a+(n-2)d]+[a+(n-1)d]$
also
$S_n=[a+(n-1)d]+[a+(n-2)d]+\ldots+(a+d)+a$

$$\begin{aligned}
\text{Adding}\quad 2S_n &= [2a+(n-1)d]+[2a+(n-1)d]+\ldots\\
&\quad +[2a+(n-1)d]+[2a+(n-1)d]\\
&= n[2a+(n-1)d]
\end{aligned}$$

$$\text{Hence}\quad S_n=\frac{n}{2}[2a+(n-1)d] \qquad (1)$$

This sum can also be written

$$S_n=\frac{n}{2}[a+\ a+(n-1)d].$$

If l is the last term of the series then $l=a+(n-1)d$ and the sum becomes

$$S_n=\frac{n}{2}(a+l) \qquad (2)$$

Example 4

Find the sum of all the even numbers from 1 to 100. Here $n=50$, $a=2$ and $d=2$. Substituting these values in the formula

$$S_n=\frac{n}{2}[2a+(n-1)d],\text{ gives}$$

$$\begin{aligned}
S_{50}=\tfrac{50}{2}[4+(50-1)2] &= 25[4+98]\\
&= 25\times 102\\
&= 2550
\end{aligned}$$

Or since we know the last term, i.e. $l=100$

$$\begin{aligned}
S_{50}=\frac{n}{2}(a+l)=\tfrac{50}{2}(2+100) &= 25\times 102\\
&= 2550
\end{aligned}$$

Example 5

Find the sum of all the numbers between 1 and 100 which are divisible by 7. Since the first number is 7 and the last is $98=7\times 14$ then $a=7$, $d=7$, $n=14$ and $l=98$.

Hence $\quad S_{14}=\frac{14}{2}(7+98)=7\times 105=735$

Example 6

In a certain arithmetic sequence the sum of the first n terms is given by $S_n=6n-3n^2$. Find the sequence.

Given $S_n = 6n - 3n^2$ hence

$$S_1 = 6 - 3 = 3$$
$$S_2 = 12 - 12 = 0$$
$$S_3 = 18 - 27 = -9$$
$$S_4 = 24 - 48 = -24$$

S_1 = sum to one term $= u_1 = 3$

S_2 = sum of first two terms, $\therefore u_2 = S_2 - S_1 = 0 - 3 = -3$

S_3 = sum of first three terms,

$\therefore u_3 = S_3 - S_2 = -9 - 0 = -9$

Hence a (*first term*) $= 3$ *and* d (common difference) $= -6$

$\therefore$ Sequence is $3, -3, -9, -15, \ldots$

ARITHMETIC MEAN

The arithmetic mean of n numbers is found by adding the numbers and dividing by n.

Example 7

Find the arithmetic mean of 4, 10, 16, 14.

$$\text{Arithmetic Mean} = \frac{4+10+16+14}{4} = \frac{44}{4} = 11$$

ASSIGNMENT 1.2

1. In the following arithmetic sequences with given nth term u_n, show that $u_n - u_{n-1} = u_{n+1} - u_n$ = common difference.

   (i) $1, 3, 5, 7, \ldots, (2n-1)$
   (ii) $3, 1, -1, \ldots, (5-2n)$
   (iii) $6, 11, 16, \ldots, (5n+1)$
   (iv) $k, 2k, 3k, \ldots, kn$

2. The following are the nth terms of certain sequences. Show that each is an arithmetic sequence and find the first term and the common difference of each.

   (i) $n+2$ (ii) $4n-1$ (iii) $a+3n$
   (iv) $3-2n$ (v) $bn+2$.

3. Find the sum of the following arithmetic series,
   (i) $1+3+5+7+\ldots$ to 30 terms.
   (ii) $5+7+9+11+\ldots$ to 11 terms.
   (iii) $3+1\frac{1}{2}+0-1\frac{1}{2}-\ldots$ to 12 terms.
   (iv) $-4-1\frac{1}{2}+1+\ldots$ to 12 terms.
   (v) $20+17+14+11+\ldots$ to 18 terms.
   (vi) $1+2+3+4+5+\ldots$ to n terms.
   (vii) $2+4+6+8+\ldots$ to k terms.
   (viii) $1+3+5+7+9+\ldots$ to $2n$ terms.

4. Find the sum of all the odd numbers from 1 to 99.

5. Find the sum of all the numbers between 1 and 101 which are multiples of 5.

6. Find the sum of all the natural numbers between 1 and 1000 which are divisible by 9.

7. An arithmetic sequence has $u_1 = -4$ and $u_6 = 21$, find the common difference.

8. How many terms of the sequence $3, 7, 11, 15, \ldots$ must be taken to give a sum of 465? What would be the last term of this sequence giving this sum?

9. An arithmetic sequence has first term 24 and common difference -2. Write down the first 3 terms. How many terms are required to give a sum of 84?

10. The fifth term of an arithmetic sequence is 23 and the twelfth term is 72. Find the first term, common difference and hence the first five terms of the sequence.

11. A shop sells various sizes of pots and the prices of successive sizes rises by equal amounts. The smallest costs 30p and the largest £1·14. It costs £5·76 to buy one of each. How many kinds does the shop sell and what is the price of each pot?

12. In an arithmetic series whose first term is 30 the sum to four terms is the same as the sum to 9 terms. Find the common difference of the series.

13. In a certain arithmetic series, the sum of the first n terms is given by $S_n = 3n^2 - 8n$. Calculate S_1, S_2, S_3, S_4 and hence find the first four terms of the series.
    Write down an expression for u_n.

14. Find the sum of the first 10 terms of the sequence 4, 9, 14, 19, ... Find also the sum of the second 10 terms.

15. The sum to n terms of a series is $kn^2 + n$, where k is a constant. Show that the series is arithmetic.

16. The sum of the first three terms of an arithmetic series is 12. If the sum of their squares is 66 find the arithmetic series.

17. The sum of the first n terms of an arithmetic series is given by $S_n = 2n + 3n^2$. Find the first four terms of the series and an expression for u_n.

18. The same quantity is added to each term of an arithmetic sequence. Show that the resulting sequence is also arithmetic with the same common difference.

19. -5 is added to each term of the arithmetic sequence whose first term is p and whose common difference is k. Is the resulting sequence arithmetic? If so state its first term and common difference.

20. If all the terms of an arithmetic sequence be multiplied by the same quantity k, show that the resulting sequence is also arithmetic. Do both sequences have the same common difference?

21. Every year I save £20 more than I saved the year before. The first year I saved £100. How many years will it take me to save £3600?

22. u_k is the kth term of a sequence defined by the recurrence relation $u_{k+1} = u_k - 3$ with $u_1 = 1$. Find the sum of the first ten terms.

GEOMETRIC SEQUENCES AND SERIES

Each term of this sequence is obtained by multiplying the previous term by the same factor. This constant factor is called the **common ratio** (c.r.).

Examples of geometric sequences

(i) $4, 8, 16, 32, \ldots, 2^{n+1}$ (c.r. $= 2$)

(ii) $1, \frac{1}{4}, \frac{1}{16}, \frac{1}{64}, \ldots, (\frac{1}{4})^{n-1}$ (c.r. $= \frac{1}{4}$)

(iii) $9, -27, 81, \ldots, 9(-3)^{n-1}$ (c.r. $= -3$)

(iv) $a, ar, ar^2, ar^3, \ldots, ar^{n-1}$ (c.r. $= r$)

(iv) is called the **standard geometric sequence** and the nth term is $u_n = ar^{n-1}$ where a is the 1st term and

$$r = \frac{u_n}{u_{n-1}} \text{ the common ratio.}$$

Note: For a Geometric Sequence

$$\frac{u_{n+1}}{u_n} = \frac{u_n}{u_{n-1}}$$

Check this for the above sequences when $n = 4$.

Example 1

Find the 8th term of the geometric sequence

$$9, 9.\tfrac{2}{3}, 9.(\tfrac{2}{3})^2, 9.(\tfrac{2}{3})^3, \ldots$$

Here $a = 9$, $r = \frac{2}{3}$ and $u_8 = ar^{8-1}$

Hence $\quad u_8 = 9.\left(\dfrac{2}{3}\right)^7 = 9.\dfrac{2^7}{3^7} = \dfrac{2^7}{3^5} = \dfrac{128}{243}$

Example 2

Find the geometric sequence whose 4th term is -6 and whose 7th term is 162.

Let a be the 1st term and r the common ratio.

Then $\quad u_n = ar^{n-1} \Rightarrow u_4 = -6 = ar^3$

and $\quad u_7 = 162 = ar^6$

Hence $\quad \dfrac{ar^6}{ar^3} = \dfrac{162}{-6}$

$$\Rightarrow r^3 = -27$$

$$\Rightarrow r = -3$$

Also $-6 = ar^3$, and when $r = -3$, $-6 = a(-27)$

$$\Rightarrow a = \tfrac{2}{9}$$

Hence sequence is $\frac{2}{9}, -\frac{2}{3}, 2, -6, 18, \ldots$

FORMULA FOR THE SUM TO n TERMS OF A GEOMETRIC SERIES

Example 3

Find the sum of the first n terms of the series,

$$3 + \tfrac{3}{2} + \tfrac{3}{4} + \tfrac{3}{8} + \tfrac{3}{16} + \ldots$$

Let S_n denote the sum of the first n terms of the series which has c.r. $= \frac{1}{2}$.

Hence $\quad S_n = 3 + \dfrac{3}{2} + \dfrac{3}{2^2} + \dfrac{3}{2^3} + \ldots + \dfrac{3}{2^{n-2}} + \dfrac{3}{2^{n-1}}$

Multiply each side by the c.r. and shift each term one place to the right.

Hence $\quad \dfrac{1}{2}S_n = \dfrac{3}{2} + \dfrac{3}{2^2} + \dfrac{3}{2^3} + \ldots + \dfrac{3}{2^{n-2}} + \dfrac{3}{2^{n-1}} + \dfrac{3}{2^n}$

By subtraction, $S_n - \dfrac{1}{2}S_n = 3 - \dfrac{3}{2^n}$

and so $\quad \dfrac{1}{2}S_n = 3 - \dfrac{3}{2^n}$

Hence $\quad S_n = 6\left(1 - \dfrac{1}{2^n}\right)$

This gives us a way of finding the sum to n terms of a geometric series.

Let S_n denote the sum to n terms of a geometric series whose first term is a and whose common ratio is r.

Then $\quad S_n = a + ar + ar^2 + ar^3 + \ldots + ar^{n-1}$

and $\quad rS_n = ar + ar^2 + ar^3 + \ldots + ar^{n-1} + ar^n$

By subtraction $(1-r)S_n = a - ar^n$

$$= a(1-r^n)$$

Hence $\quad S_n = \dfrac{a(1-r^n)}{1-r} \ldots \text{(A)}$

Or $\quad S_n = \dfrac{a(r^n-1)}{r-1} \ldots \text{(B)}$

for $r \neq 1$

Use formula (A) when r is negative or a proper fraction, i.e. $r < 1$.

Use formula (B) otherwise, i.e. when $r > 1$.

Example 4

Find the sum of the geometric series,

$$2 + 2\sqrt{2} + 4 + 4\sqrt{2} + \ldots \text{ to 14 terms.}$$

Here $a = 2$, $r = \sqrt{2}$ and $n = 14$.

$$S_n = \frac{a(r^n - 1)}{r - 1}$$

$$\therefore S_{14} = \frac{2[(\sqrt{2})^{14} - 1]}{\sqrt{2} - 1} = \frac{2(2^7 - 1)}{\sqrt{2} - 1}$$

$$= \frac{2 \times 127}{\sqrt{2} - 1} = \frac{254(\sqrt{2} + 1)}{(\sqrt{2} - 1)(\sqrt{2} + 1)}$$

$$= \frac{254(\sqrt{2} + 1)}{2 - 1} = 254(\sqrt{2} + 1)$$

THE GEOMETRIC MEAN

The geometric mean of n numbers is found by multiplying the numbers and taking the nth root of the product.

Example 5

Find the geometric mean of 3, 9, 27.

$$\text{Geometric Mean} = (3 \times 9 \times 27)^{\frac{1}{3}} = (27 \times 27)^{\frac{1}{3}} = 9$$

ASSIGNMENT 1.3

1. In each of the following geometric sequences with given nth term u_n write down u_{n+1} and u_{n-1} and show that

$$\frac{u_{n+1}}{u_n} = \frac{u_n}{u_{n-1}} = \text{common ratio.}$$

   (i) $1, 3, 9, 27, \ldots, 3^{n-1}$

   (ii) $5, 10, 20, 40, \ldots, 5.2^{n-1}$

   (iii) $1, \frac{1}{4}, \frac{1}{16}, \frac{1}{64}, \ldots, \frac{1}{4^{n-1}}$

   (iv) $3, \frac{9}{2}, \frac{27}{4}, \ldots, \frac{3^n}{2^{n-1}}$

2. Write down the first four terms and the common ratio of the sequences whose nth term is given by

   (i) $u_n = 3^{1-n}$ (ii) $u_n = 2(-3)^n$

   (iii) $u_n = 8 \times (-\frac{1}{4})^{n-1}$ (iv) $u_n = \frac{3}{4}(\frac{2}{3})^{n-1}$

3. Write down the nth term of the following geometric sequences.

   (i) $1, 2, 4, 8, 16, \ldots$

   (ii) $1, -3, 9, -27, \ldots$

   (iii) $1, \frac{1}{2}, \frac{1}{4}, \frac{1}{8}, \ldots$

   (iv) $25, 5, 1, \ldots$

   (v) $b^2, -b, 1, -\frac{1}{b}, \ldots$

   (vi) $a, ar, ar^2, ar^3, \ldots$

4. Which term of the sequence $3, 6, 12, 24, \ldots$ is 384?

5. Which term of the sequence $\frac{1}{2}, \frac{1}{8}, \frac{1}{32}, \ldots$ is $\frac{1}{512}$?

6. Find the sums of the following geometric series. Your answer may be left in index form.

   (i) $1 + \frac{1}{2} + \frac{1}{4} + \frac{1}{8} + \ldots$ to 8 terms.

   (ii) $1 - 3 + 9 - 27 + \ldots$ to 8 terms.

   (iii) $6 + 18 + 54 + \ldots$ to 7 terms.

   (iv) $2 - 4 + 8 - 16 + \ldots$ to 10 terms.

   (v) $1 - \frac{1}{3} + \frac{1}{9} - \frac{1}{27} + \ldots$ to 8 terms.

   (vi) $18 + 6 + 2 + \ldots$ to 10 terms.

   (vii) $1 + a + a^2 + a^3 + \ldots$ to n terms.

   (viii) $1 - \frac{1}{3}x + \frac{1}{9}x^2 - \frac{1}{27}x^3 + \ldots$ to 12 terms.

7. How many terms of the sequence $1, 3, 9, 27, \ldots$ must be taken for their sum to be 364?

8. The series $2 + 2^2 + 2^3 + 2^4 + \ldots$ has a sum of 254. How many terms are in the series?

9. If the nth term of the series $S = 1+x+x^2+\ldots$ to n terms is denoted by s_n and the nth term of

$$R = 1+\frac{1}{x}+\frac{1}{x^2}+\ldots$$

to n terms is denoted by r_n, show that $s_n . r_n = 1$.

10. Find the greatest value of n for which

$$\frac{1}{2}+\frac{1}{2^2}+\frac{1}{2^3}+\ldots+\frac{1}{2^n}$$

is less than $\frac{99}{100}$.

11. Find the value of n for which the series $1+\frac{1}{3}+\frac{1}{9}+\ldots$ is equal to $\frac{40}{27}$.

12. Find the geometric sequence whose 2nd term is -12 and whose 7th term is $\frac{729}{8}$.

13. Which term of the geometric sequence $144, 108, 81, \ldots$ is $\frac{729}{16}$?

14. The formula for the sum to n terms of a geometric series is given by $4^n - 1$ for $n \in N$. Find the first three terms.

15. A geometric series has first term 12 and the sum of the first three terms is 21. Show that there are two possible series and find them.
 Find two other geometric series whose first term is 8 and the sum of whose first three terms is 14.

THE LIMIT OF A SEQUENCE

1.4

Consider the sequence whose nth term is defined by $u_n = 1 + \dfrac{1}{n^2}$.

Then

$$u_1 = 1+\tfrac{1}{1}$$
$$u_2 = 1+\tfrac{1}{4}$$
$$u_5 = 1+\tfrac{1}{25}$$
$$u_{100} = 1+\frac{1}{100^2}.$$

This suggests that by taking n sufficiently large u_n can be made as close as we wish to 1. We say that 1 is the limit of the sequence defined by $u_n = 1 + \dfrac{1}{n^2}$.

In general: A sequence with nth term u_n is said to have a limit l, if by taking n sufficiently large u_n can be made as close as we wish to l.

Example 1

If $u_n = \dfrac{2n^2+1}{n^2}$, find u_1, u_{10}, u_{2n} and the limit of the sequence for large n.

$$u_n = \frac{2n^2+1}{n^2}$$

Hence

$$u_1 = \frac{2+1}{1} = 3$$
$$u_{10} = \frac{200+1}{100} = 2\tfrac{1}{100}$$
$$u_{2n} = \frac{8n^2+1}{4n^2}$$

If n is large, we write $u_n = 2 + \dfrac{1}{n^2}$

Hence by taking n sufficiently large, we can make $\dfrac{1}{n^2}$ as small as we please.

Thus the limit of the sequence is 2.

THE SUM TO INFINITY OR LIMITING SUM OF A GEOMETRIC SERIES

Consider the geometric series, $S_n = 1+\frac{1}{2}+\frac{1}{4}+\frac{1}{8}+\ldots+\frac{1}{2^{n-1}}$

$$S_1 = 1$$
$$S_2 = 1+\tfrac{1}{2} = 1\tfrac{1}{2}$$
$$S_3 = 1+\tfrac{1}{2}+\tfrac{1}{4} = 1\tfrac{3}{4}$$
$$S_4 = 1+\tfrac{1}{2}+\tfrac{1}{4}+\tfrac{1}{8} = 1\tfrac{7}{8}$$
$$S_5 = 1+\tfrac{1}{2}+\tfrac{1}{4}+\tfrac{1}{8}+\tfrac{1}{16} = 1\tfrac{15}{16}$$
$$S_6 = 1+\tfrac{1}{2}+\tfrac{1}{4}+\tfrac{1}{8}+\tfrac{1}{16}+\tfrac{1}{32} = 1\tfrac{31}{32}$$

Hence the more terms we take, the nearer the sum gets to 2, i.e. as n increases S_n approaches nearer and nearer to the value 2.

Now $S_n = \dfrac{a(1-r^n)}{1-r}$ where $r = \frac{1}{2}$.

Thus $r^2 = \frac{1}{4}$, $r^3 = \frac{1}{8}$, $r^4 = \frac{1}{16}$ and r^{100} is obviously small, i.e. if n is large r^n is extremely small. Hence by taking n sufficiently large r^n can be made as close as we wish to 0.

But S_n can be written $S_n = \dfrac{a}{1-r} - \dfrac{ar^n}{1-r}$, and so when n is made sufficiently large, r^n is as close as we wish to zero and we say S_n becomes the **sum to infinity** or the **limiting sum** of the series and is denoted by S.

Hence $S = \dfrac{a}{1-r}$ is the formula for the sum to infinity or the limiting sum for $-1 < r < 1$.

This is sometimes written "S_n tends to S as n tends to infinity".

Or $S_n \to S$ as $n \to \infty$.

Or $\lim\limits_{n\to\infty} S_n = S$, where "lim" is an abbreviation of the word "limit".

Note that if $r \geqq 1$, or $r \leqq -1$, there is no "sum to infinity".

Example 2

Find the sum to n terms and the sum to infinity of the series $\frac{3}{2}+\frac{3}{4}+\frac{3}{8}+\ldots$

How many terms of the series must be taken so that the sum to n terms is $\frac{63}{64}$ of the sum to infinity?

Here $a = \frac{3}{2}$ and $r = \frac{1}{2}$.

Thus $$S_n = \frac{a(1-r^n)}{1-r} = \frac{\frac{3}{2}[1-(\frac{1}{2})^n]}{1-\frac{1}{2}} = 3[1-(\tfrac{1}{2})^n]$$

$$S = \frac{a}{1-r} = \frac{\frac{3}{2}}{1-\frac{1}{2}} = \tfrac{3}{2}\times\tfrac{2}{1} = 3$$

Also $S_n = \frac{63}{64}S \Rightarrow 3[1-(\frac{1}{2})^n] = \frac{63}{64}\times 3$

$$\Rightarrow \quad 1-(\tfrac{1}{2})^n = \tfrac{63}{64}$$
$$\Rightarrow \quad (\tfrac{1}{2})^n = 1-\tfrac{63}{64} = \tfrac{1}{64} = (\tfrac{1}{2})^6$$
$$\Rightarrow \quad n = 6$$

ASSIGNMENT 1.4

1. Find the limits of the following sequences defined by u_n.

   (i) $u_n = \dfrac{n+1}{n}$ (ii) $u_n = \dfrac{2n^2+1}{n^2}$

   (iii) $u_n = \dfrac{n}{n^2-1}$ $n \neq 1$ (iv) $u_n = (1+2^{-n})$

   (v) $u_n = \dfrac{\frac{1}{2}n(n-1)}{n^2}$ (vi) $u_n = \dfrac{n^2+3n-4}{n(n+1)}$

2. The nth term of a sequence is $\dfrac{n}{n+1}$. Write down the first six terms of the sequence. What is the limit of the sequence? Which term of the sequence differs from the limit of the sequence by $\frac{1}{100}$?

3. Show that $\frac{3n^2+n}{n^2+1} = 3 + \frac{n-3}{n^2+1}$ and hence find the limit of the sequence defined by $u_n = \frac{3n^2+n}{n^2+1}$.

4. Which of the following geometric series has a sum to infinity, and find the sum to infinity if it exists.

   (i) $1+\frac{1}{4}+\frac{1}{16}+\ldots$ (ii) $\frac{1}{3}+\frac{1}{9}+\frac{1}{27}+\ldots$

   (iii) $4+6+9+\ldots$ (iv) $3+2+\frac{4}{3}+\ldots$

   (v) $1-3+9-27+\ldots$ (vi) $\frac{1}{2}-\frac{1}{4}+\frac{1}{8}-\ldots$

   (vii) $1-1+1-1+\ldots$ (viii) $10+1+0{\cdot}1+\ldots$

   (ix) $8-4+2-1+\ldots$ (x) $\frac{1}{16}+\frac{1}{8}+\frac{1}{4}+\ldots$

5. A geometric series has first term 2 and its sum to infinity is 6. Find the series.

6. Find the difference between the sum to infinity and the sum to 8 terms of the series $1 + \frac{1}{1{\cdot}5} + \frac{1}{1{\cdot}5^2} + \ldots$

7. Write down the conditions that the series

$$1 + \frac{1}{a} + \frac{1}{a^2} + \ldots$$

   should have a sum to infinity. If the sum to infinity is 5 find a.

8. Write down the conditions that the series

$$1+a^2+a^4+a^6+\ldots$$

   should have a sum to infinity. If the sum to infinity is $1\frac{1}{3}$ find a.

9. Write $0{\cdot}\dot{4}$ as an infinite series. [Note: $0{\cdot}\dot{4} = 0{\cdot}444444\ldots$] Hence by summing the series to infinity find the value $0{\cdot}\dot{4}$ as a fraction.

10. If $0{\cdot}\dot{6}\dot{3} = 0{\cdot}63636363\ldots$ write this recurring decimal as a series and hence find the value of $0{\cdot}\dot{6}\dot{3}$ as a proper fraction.

11. A ball is dropped from a height of 3 metres above the floor of a room. After each bounce it reaches to $\frac{2}{3}$ of the height from which it fell. Calculate the total distance travelled by the ball before it comes to rest on the floor.

12. A tree is already 5 metres in height when planted and increases $1\frac{1}{2}$ metres in the next year. If in each succeeding year the growth is $\frac{11}{12}$ of that in the previous year, find the limiting height of the tree.

13. ABC is an equiangular triangle of side 16 cm. The mid-points of the sides are joined to form another equiangular triangle $A_1B_1C_1$. Triangles $A_2B_2C_2$, $A_3B_3C_3$, ... are formed by joining the mid-points of the sides of the previous triangles. What is the sum of the perimeters of all the triangles so formed?

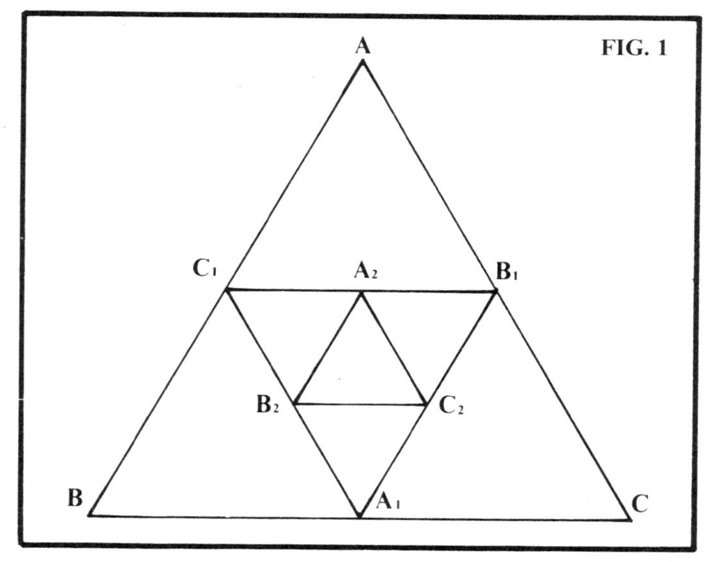

FIG. 1

14. S denotes the sum to infinity of the geometric series $a+ar+ar^2+ar^3+\ldots$. Write down S_1 and S_2 the sums to infinity of the odd and even terms of the series respectively. If the sum of the odd terms S_1 is $5\frac{2}{5}$ and the sum of the even terms S_2 is $3\frac{3}{5}$ find the series.

    State the value of S.

15. A sequence is defined by $u_n = \dfrac{n+3}{2n}$. Write down the first four terms of the sequence and the limit of the sequence.

    By how much does the limit of the sequence differ from the 200th term?

16. The first three terms of a geometric series are $3(x+2)$, $3(x-2)$ and $x+6$ respectively. Show that there are two values of x and find the two possible series.

    Show that only one of these series has a limiting sum and find this sum.

    Show that the sum to n terms of the other series takes one of two values depending on the value of n.

17. Find the sum to infinity of the series

$$1.\frac{1}{4}+3.\frac{1}{4^2}+9.\frac{1}{4^3}+27.\frac{1}{4^4}+\ldots$$

1.5 COMPOUND INTEREST OR GROWTH, DECAY OR DEPRECIATION

If £P is invested at r per cent per annum compound interest then the amount £A after n years is given by

$$A = £P\left(1+\frac{r}{100}\right)^n.$$

For Principal for the 1st year $= £P$

i.e. $u_0 = £P$

Interest for the 1st year at $r\% = £\dfrac{Pr}{100}$

Amount at the end of the 1st year

$$= £P + £P\frac{r}{100}$$

$$= £P\left(1+\frac{r}{100}\right)$$

i.e. $u_1 = \left(1+\dfrac{r}{100}\right)u_0$

Principal at the beginning of 2nd year

$$= £P\left(1+\frac{r}{100}\right)$$

Interest for the 2nd year at $r\%$

$$= £P\left(1+\frac{r}{100}\right).\frac{r}{100}$$

Amount at the end of the 2nd year

$$= £P\left(1+\frac{r}{100}\right)+£P\left(1+\frac{r}{100}\right).\frac{r}{100}$$

$$= £P\left(1+\frac{r}{100}\right)\left(1+\frac{r}{100}\right)$$

$$= £P\left(1+\frac{r}{100}\right)^2$$

i.e. $u_2 = \left(1+\dfrac{r}{100}\right)^2 u_0$ or $u_2 = \left(1+\dfrac{r}{100}\right)u_1$

Similarly,

Amount at the end of the 3rd year

$$= £P\left(1+\frac{r}{100}\right)^3$$

i.e. $u_3 = \left(1 + \frac{r}{100}\right)^3 u_0$ or $u_3 = \left(1 + \frac{r}{100}\right) u_2$

Hence,

At the end of the nth year, the amount A

$$= £P\left(1 + \frac{r}{100}\right)^n$$

i.e. $u_n = \left(1 + \frac{r}{100}\right)^n u_0$ or $u_n = \left(1 + \frac{r}{100}\right) u_{n-1}$

where u_0 = the amount or value at the beginning

u_n = the amount or value at the end of n years.

$\left(1 + \frac{r}{100}\right)$ is called the *Growth factor* for a quantity that increases in value or size.

If some quantity decreases in size or value (e.g. a car, a piece of machinery, neutron decay), then we get:

$u_n = \left(1 - \frac{r}{100}\right)^n u_0$ or $u_n = \left(1 - \frac{r}{100}\right) u_{n-1}$ after n years

In this case $\left(1 - \frac{r}{100}\right)$ is called the *Decay factor*.

Example 1

A tree 400 cm high increases its height by 2 per cent each year. What is its height after 10 years?

Here the growth factor is $\left(1+\frac{r}{100}\right) = \left(1+\frac{2}{100}\right) = 1{\cdot}02$.

Hence the height after 10 years is given by,

$u_{10} = (1{\cdot}02)u_9$ or $u_{10} = (1{\cdot}02)^{10}u_0$ where $u_0 = 400$

After 10 years the height $u_{10} = (1{\cdot}02)^{10} \times 400$

$= 492$ cm.

Example 2

A piece of machinery valued at £1,000 depreciates 15 per cent each year. What is its value after 4 years?

Here the decay factor is $\left(1 - \frac{r}{100}\right) = \left(1 - \frac{15}{100}\right) = 0{\cdot}85$.

Hence the value after 4 years is given by,

$u_4 = (0{\cdot}85)u_3$ or $u_4 = (0{\cdot}85)^4 u_0$ where $u_0 = 1000$

After 4 years the value is $u_4 = £(0{\cdot}85)^4 \times 1000$

$= £520$.

ASSIGNMENT 1.5

1. Calculate the growth factor or decay factor in each of the following.
   (i) A sum of money gaining interest at 5% per annum.
   (ii) A motor-car depreciating at 20% per annum.
   (iii) A piece of radioactive substance disintegrating by 12% every minute.
   (iv) A plant increasing its height by 3% every month.
2. £100 is invested at $4\frac{1}{2}$% compound interest. Find the amount after 8 years.
3. A car costs £1500 and depreciates 15% per annum. Calculate the decay factor. Give the value of the car at the beginning of the 2nd year and at the beginning of the 5th year.
4. A new born baby weighs $3\frac{1}{2}$ kg, and increases its weight by 2% each week. What is its weight at the end of the first week? By how much does its weight increase in 8 weeks?

5. A plant 256 cm in height increases its height by 25% each year. In how many years will it stand 625 cm high?

6. A man pays £200 at the beginning of each year into a superannuation fund at 8% per annum compound interest. Write down the growth factor for his money. What sum should he receive at the end of 10 years?

7. A tree 1 metre high increases its height each year by 3% of its previous year's height. What is the height of the tree after 10 years, 15 years, 20 years?

8. Machinery is bought for a plant and is valued at £100,000. It depreciates by 2% every half year.
   (i) Write down the decay factor.
   (ii) What is the value of the machinery at the end of the first year?
   (iii) What is the value of the machinery at the end of 10 years?

   Give your answers to 2 significant figures.

9. The population of a region is 220,000. It is estimated (due to births and deaths) that the population increase is five per cent each year and the death rate is three and a half per cent each year.
   Calculate
   (i) the growth (or birth) factor and the decay (or death) factor for the population,
   (ii) the estimated population of the region in five years time.

Objective test items

In the objective test items situated at the end of each unit, answer the items as follows (unless otherwise stated).

(i) In questions 1–8 the correct answer is given by one of the options A, B, C, D or E.

(ii) In questions 9 and 10 one or more of the three statements is/are correct.

Answer A if statement (1) only is correct.
B if statement (2) only is correct.
C if statement (3) only is correct.
D if statements (1), (2) and (3) are correct.
E if some other combination of the given statements is correct.

(iii) In questions 11 and 12 two statements are numbered (1) and (2).

Answer A if (1) implies (2) but (2) does not imply (1).
B if (2) implies (1) but (1) does not imply (2).
C if (1) implies (2) and (2) implies (1).
D if (1) denies (2) or (2) denies (1).
E if none of the above relationships hold.

ASSIGNMENT 1.6

Objective items testing sections 1.1–1.5

Instructions for answering the following items are given above.

1. Which one of the following gives the value for $u_n - u_{n-1}$ for the series $1+\frac{1}{2}+\frac{1}{4}+\frac{1}{8}+\ldots$?

   A. 2^{n-1}

   B. $-\dfrac{1}{2^{n-1}}$

   C. $\dfrac{1}{2^{n-1}}$

   D. $-\dfrac{1}{2^n}$

   E. $\dfrac{1}{2}$

2. A sequence is defined by $u_n = \dfrac{2n^2-5n}{3n^2}$. Find the limit of u_n for large n.

   A. 2

   B. $\frac{5}{3}$

   C. 0

   D. $\frac{2}{3}$

   E. It has no finite limit.

3. Find the sum to infinity of the series

$$3-\frac{3^2}{2}+\frac{3^3}{2^2}-\frac{3^4}{2^3}+\ldots$$

   A. -6

   B. $-\frac{3}{2}$

   C. $\frac{3}{2}$

   D. 6

   E. It has no sum to infinity.

4. If the value of a car at the beginning of the first three years is £1200, £1020 and £867 then the decay factor for the value of the car is

   A. 1·05

   B. 1·15

   C. 0·95

   D. 0·85

   E. 0·15

5. $1-\frac{1}{3}+\frac{1}{9}-\frac{1}{27}+\ldots$ is an infinite series. The sum to infinity is

   A. $\frac{3}{4}$

   B. $\frac{2}{3}$

   C. $\frac{3}{2}$

   D. $-\frac{3}{4}$

   E. non-existent.

6. If $u_n = 81(\frac{1}{3})^{n-1}$ defines a sequence, which one of the following is false?

   A. 4 terms of the sequence are greater than 1.

   B. $u_5 = 1$.

   C. The limit of the sequence for large n is zero.

   D. The series defined by u_n has no limiting sum.

   E. The sum of the first 4 terms of the sequence is 120.

7. The sum to infinity of the geometric series whose third and fourth terms are 12 and -6 respectively is

   A. 72

   B. 32

   C. 16

   D. 8

   E. -48

8. $(2x+5y)$, $(x+y)$, $(x-y)$ are the first three terms of a geometric series. Which of the following gives x in terms of y?

$$\text{I } x = 2y; \quad \text{II } x = -2y; \quad \text{III } x = -3y$$

A. Only I

B. Only II

C. I and III

D. II and III

E. I, II and III

9. Which of the following series have a sum to infinity?

(1) $1+p+p^2+p^3+\ldots \quad -1 < p < 1$

(2) $1+q^2+q^4+q^6+\ldots \quad -1 < q < 1$

(3) $1+pq+p^2q^2+p^3q^3+\ldots \quad -1 < pq < 1$

10. If $S_n = 4^n - 1$ is the expression for the sum of the first n terms of a series, $n \in N$, then

(1) the first 3 terms of the series are 3, 12, 48.

(2) the sum to infinity is -1.

(3) the sum to $2n$ terms is $2^{4n} - 1$.

11. (1) The nth term of a sequence is $3(-2)^{n-1}$, $n \in N$.

(2) -384 is a member of the sequence.

12. (1) The sum to n terms of a series is $S_n = \dfrac{n^2}{4}(1+n)^2$.

(2) The first three terms of the series are 1, 9, 36.

UNIT 2: MATRICES

MATRIX NOTATION: ADDITION AND SUBTRACTION OF MATRICES

2.1

A **Matrix** is an array of numbers written as rows or columns. Each number is called an **entry** or **element** of the matrix. The **order** of the matrix is described by stating first the *number of rows*, then the *number of columns*.

Example 1

(i) $P = \begin{pmatrix} 8 & 1 & 0 & -5 & 3 \\ 4 & 2 & 1 & 0 & -3 \end{pmatrix}$

is a matrix of order 2×5 i.e. 2 rows and 5 columns.

(ii) $Q = \begin{pmatrix} 3 & 8 & 14 & -2 \\ -3 & 4 & 6 & 0 \\ 0 & 2 & 1 & 1 \end{pmatrix}$ has order 3×4

In general a matrix which has m rows and n columns is called an $m \times n$ matrix. If $m = n$ the matrix is an $n \times n$ matrix and is called a **Square Matrix** of order n.

Example 2

$R = \begin{pmatrix} -1 & \frac{1}{2} \\ \frac{1}{3} & 0 \end{pmatrix}$ is a square matrix of order 2.

Sometimes we use square brackets.

Example 3

$S = [2 \quad -\frac{1}{2} \quad -4]$ has order 1×3

The element in the ith row and jth column is denoted by a_{ij}.

Example 4

In matrix P, a_{23} means the entry in the 2nd row, 3rd column. $a_{23}=1$.

EQUAL MATRICES

Two matrices A and B are equal if and only if they are of the same order and each element of A is equal to the corresponding element of B. We then write $A=B$.

Example 5

$$\begin{pmatrix} 3 & 4 & 0 \\ -1 & 2 & -3 \\ 1 & -2 & 5 \end{pmatrix} = \begin{pmatrix} 3 & 4 & a \\ b & 2 & -3 \\ 1 & c & 5 \end{pmatrix} \Leftrightarrow a=0, b=-1 \text{ and } c=-2.$$

THE TRANSPOSE OF A MATRIX A

The transpose of a matrix A is a new matrix obtained by interchanging the rows and columns. It is called the transpose of A and is written A'.

Example 6

If $A = \begin{pmatrix} 3 & -5 & 1 & 0 \\ 6 & 4 & 3 & -2 \end{pmatrix}$ is of order 2×4

Then $A' = \begin{pmatrix} 3 & 6 \\ -5 & 4 \\ 1 & 3 \\ 0 & -2 \end{pmatrix}$ is of order 4×2

A' is the transpose of A, and conversely A is the transpose of A'.

THE ZERO MATRIX

The zero matrix is denoted by O and has all its elements zero. It can be of any order.

Example 7

$O = \begin{pmatrix} 0 & 0 & 0 \\ 0 & 0 & 0 \end{pmatrix}$ is the zero matrix of order 2×3

$O = \begin{pmatrix} 0 & 0 & 0 & 0 \\ 0 & 0 & 0 & 0 \\ 0 & 0 & 0 & 0 \end{pmatrix}$ is the zero matrix of order 3×4.

ADDITION OF MATRICES

If A and B are matrices of the **same order,** then $A+B$ the *sum* of A and B, is obtained by adding corresponding elements. From the definition, $A+B$ must be of the same order as each of A and B.

If two matrices have different orders, addition is not defined.

Example 8

If $A = \begin{pmatrix} 3 & 4 & 6 \\ 3 & 1 & -1 \end{pmatrix}$ 2×3 $\quad B = \begin{pmatrix} -1 & 0 & -4 \\ 2 & 1 & 0 \end{pmatrix}$ 2×3

then $A+B = \begin{pmatrix} 3-1 & 4+0 & 6-4 \\ 3+2 & 1+1 & -1+0 \end{pmatrix}$

$= \begin{pmatrix} 2 & 4 & 2 \\ 5 & 2 & -1 \end{pmatrix}$ 2×3

Note: (i) Matrix addition is Commutative, i.e.

$$A+B=B+A$$

(ii) Matrix addition is Associative, i.e.

$$(A+B)+C = A+(B+C)$$

SUBTRACTION OF MATRICES

To subtract matrix B from matrix A, we subtract corresponding elements.

Example 9

From A and B above,

$$A-B = \begin{pmatrix} 3 & 4 & 6 \\ 3 & 1 & -1 \end{pmatrix} - \begin{pmatrix} -1 & 0 & -4 \\ 2 & 1 & 0 \end{pmatrix}$$

$$= \begin{pmatrix} 3-(-1) & 4-0 & 6-(-4) \\ 3-2 & 1-1 & -1-0 \end{pmatrix}$$

$$= \begin{pmatrix} 4 & 4 & 10 \\ 1 & 0 & -1 \end{pmatrix} \quad 2\times 3$$

ASSIGNMENT 2.1

1. Write out the following matrices and state their order.

   (i) $\begin{pmatrix} 2i & -i \\ -4i & 3i \end{pmatrix}$ given $i = -3$.

   (ii) $\begin{pmatrix} a_{11} & a_{12} & a_{13} \\ a_{21} & a_{22} & a_{23} \end{pmatrix}$

   given $a_{ij} = 4i - j$, where i, j are variables on the set $\{1, 2, 3\}$.

2. Add or subtract the following matrices where possible,

   (i) $\begin{pmatrix} 3 & 4 \\ 7 & 8 \end{pmatrix} + \begin{pmatrix} -2 & 1 \\ 1 & -3 \end{pmatrix}$

   (ii) $\begin{pmatrix} 1 & 3 & 5 \\ 7 & 9 & 11 \end{pmatrix} + \begin{pmatrix} 1 & 0 \\ 4 & -1 \end{pmatrix}$

   (iii) $\begin{pmatrix} 2 & -7 \\ 6 & -3 \end{pmatrix} - \begin{pmatrix} 5 & -8 \\ 4 & -2 \end{pmatrix}$

   (iv) $\begin{pmatrix} a & b \\ c & d \end{pmatrix} - \begin{pmatrix} p & q \\ r & s \end{pmatrix}$

3. Given $A = \begin{pmatrix} 2 & -1 \\ 0 & 3 \end{pmatrix}$ $B = \begin{pmatrix} -1 & 4 \\ -2 & 3 \end{pmatrix}$ $C = \begin{pmatrix} 2 & 0 \\ 5 & -3 \end{pmatrix}$

   (i) Calculate $A+B$ and compare with $B+A$.

   (ii) Calculate $(A+B)+C$ and compare with $A+(B+C)$.

   Which two laws do each of these results illustrate?

4. Solve the following matrix equations for the 2×2 matrix X.

   (i) $X + \begin{pmatrix} 2 & 0 \\ -1 & -2 \end{pmatrix} = \begin{pmatrix} 3 & 5 \\ 2 & -1 \end{pmatrix}$

   (ii) $\begin{pmatrix} -3 & -2 \\ 1 & 4 \end{pmatrix} + X = \begin{pmatrix} 3 & -1 \\ -1 & 0 \end{pmatrix}$

   (iii) $X + \begin{pmatrix} 3 & 0 \\ -6 & 9 \end{pmatrix} = 4X$

5. If $P = \begin{pmatrix} 5 & -4 \\ -2 & 1 \end{pmatrix}$ write down $-P$.

6. If $A = \begin{pmatrix} 3 & 1 \\ -1 & -4 \end{pmatrix}$ and $A+B = O$, where O is the 2×2 zero matrix, write down the 2×2 matrix B.

7. Two matrices A and B of the same order are equal if and only if each element of A is equal to the corresponding element of B.

Which of the following matrices are equal?

$$A = \begin{pmatrix} 2 & -5 \\ 0 & 1 \end{pmatrix} \quad B = \begin{pmatrix} 2 & -5 \\ 0 & -1 \end{pmatrix}$$

$$C = \begin{pmatrix} -2 & -5 \\ 0 & 1 \end{pmatrix} \quad D = \begin{pmatrix} 2 & -5 \\ 0 & 1 \end{pmatrix}$$

8. If $A = \begin{pmatrix} x & 3 \\ y & -1 \end{pmatrix}$ and $B = \begin{pmatrix} -2 & a \\ -1 & y \end{pmatrix}$ find x, y and a so that matrix A = matrix B.

9. Write down the transpose of the following matrices.

$$A = \begin{pmatrix} 2 & -4 \\ 0 & 1 \end{pmatrix} \quad B = \begin{pmatrix} 7 & 1 \\ -2 & 3 \\ 0 & 2 \end{pmatrix} \quad C = \begin{pmatrix} 2 & 0 & 1 \\ 1 & -1 & 0 \end{pmatrix}$$

State the order of the transpose of A, B, and C.

10. If $X = \begin{pmatrix} x & y \\ -1 & t \end{pmatrix}$ find the values of x, y and t in the following matrix equations.

(i) $X + X' = \begin{pmatrix} 4 & -3 \\ -3 & 2 \end{pmatrix}$ (ii) $X = -X'$

11. Matrices can be used to store information in the following way. Suppose Mary, John and Paul receive 40p, 45p and 60p pocket money respectively each week, then this can be written as

$$\begin{pmatrix} M \\ J \\ P \end{pmatrix} = \begin{pmatrix} 40 \\ 45 \\ 60 \end{pmatrix}$$ where M = Mary, J = John, P = Paul.

(a) What would the following matrix equation represent

$$\begin{pmatrix} M \\ J \\ P \end{pmatrix} = \begin{pmatrix} 40 \\ 45 \\ 60 \end{pmatrix}?$$

(b) After 10 weeks each is given an increase in pocket money represented in matrix form thus,

$$\begin{pmatrix} M \\ J \\ P \end{pmatrix} = \begin{pmatrix} 40 \\ 45 \\ 60 \end{pmatrix} + \begin{pmatrix} 20 \\ 25 \\ 12 \end{pmatrix}$$

How much does each now get in pocket money each week?

Represent this new pocket allowance in matrix form as in part (a).

(c) If each saves 15p every week out of his or her allowance, write down in matrix form the amount each now spends every week.

12. The matrix $X = \begin{pmatrix} y & y \\ z & 1 \end{pmatrix}$ equals its transpose. Write down the relation between y and z. Hence state which of the following matrices equals its transpose,

(i) $\begin{pmatrix} a & b \\ b & c \end{pmatrix}$ (ii) $\begin{pmatrix} a & b & 1 \\ b & a & 0 \end{pmatrix}$

(iii) $\begin{pmatrix} 3 & -4 \\ -4 & 0 \end{pmatrix}$ (iv) $\begin{pmatrix} 2 & -1 \\ 1 & 2 \end{pmatrix}$

13. Find the value of e so that matrix $A = \begin{pmatrix} g & e \\ e^2 & g \end{pmatrix}$ is equal to its transpose.

14. State the conditions so that the matrix $\begin{pmatrix} a & b & c \\ d & e & f \\ g & h & i \end{pmatrix}$ equals its transpose.

Find values of a, b, and c so that the matrix $\begin{pmatrix} 0 & 2 & b \\ a & 0 & c \\ -1 & -4 & 0 \end{pmatrix}$ equals its transpose.

15. Use matrices $A=\begin{pmatrix} a & b \\ c & d \end{pmatrix}$ and $B=\begin{pmatrix} p & q \\ r & s \end{pmatrix}$ to show that $A'+B'=(A+B)'$.

16. Find values of p, q, r and s which satisfy the matrix relationship

$$\begin{pmatrix} p+3 & 2q-8 \\ r+1 & 4p+6 \\ s-3 & 3s \end{pmatrix} = \begin{pmatrix} 0 & -6 \\ -3 & 2p \\ 2s+4 & -21 \end{pmatrix}$$

17. If $\begin{pmatrix} 2 & -3 \\ 5 & 0 \end{pmatrix} - \begin{pmatrix} x & y \\ z & t \end{pmatrix} = \begin{pmatrix} -3 & 4 \\ -1 & 4 \end{pmatrix}$ find values of x, y, z and t.

MULTIPLICATION OF A MATRIX BY A NUMBER K

If $A=\begin{pmatrix} a & b \\ c & d \end{pmatrix}$ and k a real number

then $$kA = k\begin{pmatrix} a & b \\ c & d \end{pmatrix} = \begin{pmatrix} ka & kb \\ kc & kd \end{pmatrix}$$

MULTIPLICATION OF MATRICES

If $A=\begin{pmatrix} a_1 & a_2 \\ a_3 & a_4 \end{pmatrix}\begin{matrix} \leftarrow R_1 \\ \leftarrow R_2 \end{matrix}$ $\quad B=\begin{pmatrix} b_1 & b_2 \\ b_3 & b_4 \end{pmatrix}$

$\qquad\qquad\qquad\qquad\qquad\qquad\uparrow C_1 \quad \uparrow C_2$

Where, R_1 denotes row 1

R_2 denotes row 2

C_1 denotes column 1

C_2 denotes column 2

Then $$AB = \begin{pmatrix} R_1C_1 & R_1C_2 \\ R_2C_1 & R_2C_2 \end{pmatrix} = \begin{pmatrix} a_1b_1+a_2b_3 & a_1b_2+a_2b_4 \\ a_3b_1+a_4b_3 & a_3b_2+a_4b_4 \end{pmatrix}$$

Note that:

$$\underset{2\times 2}{\begin{pmatrix} 4 & 1 \\ 0 & 3 \end{pmatrix}} \underset{3\times 3}{\begin{pmatrix} 1 & 2 & 3 \\ 4 & 5 & 4 \\ 3 & 2 & 1 \end{pmatrix}}$$

is impossible, so like addition, multiplication is not always possible

but

$$\begin{pmatrix} 1 & 3 & 5 \\ 2 & 4 & 1 \end{pmatrix}\begin{pmatrix} 1 & 2 \\ 0 & 1 \\ 3 & 4 \end{pmatrix} = \begin{pmatrix} 1+0+15 & 2+3+20 \\ 2+0+3 & 4+4+4 \end{pmatrix}$$

2×3 by 3×2 $\quad = \underset{2\times 2}{\begin{pmatrix} 16 & 25 \\ 5 & 12 \end{pmatrix}}$ is possible

2.2

Note: The product AB can be formed only if the number of columns of A is the same as the number of rows of B.

Hence, if A is of order $m\times p$, and B is of order $p\times n$ then AB is of order $m\times n$ since p, the number of columns of A, equals p, the number of rows of B.

Also if A is of order 5×3, and B is of order 3×4 then AB is of order 5×4.

When multiplication AB is possible A and B are said to be *Conformable for multiplication.*

Example 1

Let $A=\begin{pmatrix} 1 & 4 \\ 5 & 3 \end{pmatrix}$, $B=\begin{pmatrix} 1 & 6 \\ 9 & 2 \end{pmatrix}$ and $C=\begin{pmatrix} 2 \\ 3 \end{pmatrix}$

Then $AB = \begin{pmatrix}1 & 4\\5 & 3\end{pmatrix}\begin{pmatrix}1 & 6\\9 & 2\end{pmatrix} = \begin{pmatrix}37 & 14\\32 & 36\end{pmatrix}$

$BA = \begin{pmatrix}1 & 6\\9 & 2\end{pmatrix}\begin{pmatrix}1 & 4\\5 & 3\end{pmatrix} = \begin{pmatrix}31 & 22\\19 & 42\end{pmatrix}$

Note that $AB \neq BA$

Also $AC = \begin{pmatrix}1 & 4\\5 & 3\end{pmatrix}\begin{pmatrix}2\\3\end{pmatrix} = \begin{pmatrix}14\\19\end{pmatrix}$

but $CA = \begin{pmatrix}2\\3\end{pmatrix}\begin{pmatrix}1 & 4\\5 & 3\end{pmatrix}$ is not defined.

Hence even if the product AB exists, it need not be equal to the product BA, indeed the latter might not exist.

THE UNIT OR IDENTITY MATRIX I

This matrix is **square** and of any order.

$$I = \begin{pmatrix}1 & 0\\0 & 1\end{pmatrix} \quad 2\times 2, \qquad I = \begin{pmatrix}1 & 0 & 0\\0 & 1 & 0\\0 & 0 & 1\end{pmatrix} \quad 3\times 3$$

Note that in the unit matrix ones appear in the leading diagonal and that the other elements are zero.

This type of matrix is the identity element for multiplication, for if A is a square matrix then

$$AI = IA = A$$

ASSIGNMENT 2.2

1. $A = \begin{pmatrix}1 & -3\\2 & 0\end{pmatrix}$ $\quad B = \begin{pmatrix}-3 & -2\\4 & 1\end{pmatrix}$
   (i) Find $3A$ and $4B$.
   (ii) Find $3A-2B$.
   (iii) Find $2(A+B)$ and verify that this is equal to $2A+2B$.
   Hence *note* that $k(A+B) = kA+kB$, $k \in R$.

2. If $P = \begin{pmatrix}a & b\\c & d\end{pmatrix}$ and $Q = \begin{pmatrix}x & y\\z & t\end{pmatrix}$
   Verify the following results,
   (i) $(-1)P = -P$
   (ii) $0Q = 0$
   (iii) $4\times(3P) = 12P$
   (iv) $aP+aQ = a(P+Q)$

3. If $p\begin{pmatrix}4 & -1\\3 & 2\end{pmatrix} + q\begin{pmatrix}-2 & 3\\0 & 1\end{pmatrix} = \begin{pmatrix}10 & -5\\6 & 3\end{pmatrix}$ find p and q.

4. Simplify $\sqrt{3}\begin{pmatrix}\tan 60^\circ & \sin 60^\circ\\-\cos 30^\circ & -\tan 30^\circ\end{pmatrix}$

5. Multiply the following matrices where possible. In the possible products state the order of the resulting matrix.

   (i) $(7 \quad 5)\begin{pmatrix}3\\2\end{pmatrix}$
   (ii) $\begin{pmatrix}-1 & 3\\4 & 0\end{pmatrix}\begin{pmatrix}5 & 6\\2 & 1\end{pmatrix}$
   (iii) $\begin{pmatrix}1 & 2\\3 & 4\end{pmatrix}\begin{pmatrix}1 & 2 & 3\\4 & 5 & 6\end{pmatrix}$
   (iv) $(2 \quad -1 \quad 4)\begin{pmatrix}5\\-3\end{pmatrix}$
   (v) $\begin{pmatrix}-2 & -3\\4 & 1\end{pmatrix}\begin{pmatrix}2\\-1\end{pmatrix}$
   (vi) $\begin{pmatrix}2\\-1\end{pmatrix}\begin{pmatrix}-2 & -3\\4 & 1\end{pmatrix}$
   (vii) $\begin{pmatrix}-2 & 3\\0 & -1\end{pmatrix}\begin{pmatrix}2\\1\end{pmatrix}$
   (viii) $\begin{pmatrix}-1 & 4\\2 & 1\end{pmatrix}\begin{pmatrix}3 & 1\\6 & -2\\-4 & 0\end{pmatrix}$
   (ix) $\begin{pmatrix}0 & 4\\2 & -2\\-1 & 1\end{pmatrix}\begin{pmatrix}4\\-2\\1\end{pmatrix}$

(x) $\begin{pmatrix} 4 & 0 \\ 2 & -1 \\ -1 & 2 \end{pmatrix}\begin{pmatrix} 2 & 1 & 0 \\ -1 & 0 & 2 \\ 3 & -1 & -1 \end{pmatrix}$

6. $A = \begin{pmatrix} 1 & -2 \\ 3 & -4 \end{pmatrix}$ $B = \begin{pmatrix} 2 & 0 \\ -2 & 3 \end{pmatrix}$ $C = \begin{pmatrix} -1 & 2 \\ 5 & 1 \end{pmatrix}$

Calculate

(i) AB and BA. Does $AB = BA$?

(ii) $AB(C)$ and $A(BC)$. Does $(AB)C = A(BC)$?

What law appears to hold and what law does not hold for matrix multiplication?

7. Find the following matrix products.

(i) $\begin{pmatrix} 1 & 4 \\ 3 & -2 \end{pmatrix}\begin{pmatrix} 3 & \frac{3}{2} \\ \frac{1}{2} & -1 \end{pmatrix}$ (ii) $\begin{pmatrix} a & -b \\ 1 & 0 \end{pmatrix}\begin{pmatrix} a & -1 \\ b & 0 \end{pmatrix}$

(iii) $\begin{pmatrix} 1 & 3 \\ 2 & 4 \\ -1 & -1 \end{pmatrix}\begin{pmatrix} 4 & 2 \\ 0 & -1 \end{pmatrix}$ (iv) $\begin{pmatrix} a & b \\ c & d \end{pmatrix}\begin{pmatrix} k & 0 \\ 0 & 1 \end{pmatrix}$

8. $A = \begin{pmatrix} 2 & -1 \\ 5 & 0 \end{pmatrix}$ $B = \begin{pmatrix} -2 & 3 \\ 4 & 2 \end{pmatrix}$ $C = \begin{pmatrix} 1 & 2 \\ 0 & 3 \end{pmatrix}$

Calculate

(i) $AB + AC$ and $A(B+C)$

(ii) $AB - AC$ and $A(B-C)$

(iii) $(B+C)A$ and $BA + CA$

What law appears to hold?

9. $A = \begin{pmatrix} 0 & -1 \\ 4 & 2 \end{pmatrix}$ $B = \begin{pmatrix} 1 & -1 \\ -2 & 2 \end{pmatrix}$

Calculate

(i) $A.A$ and $A.A.A$. These are written A^2 and A^3 respectively.

(ii) $(A+B)^2$ and $A^2 + 2AB + B^2$.

Does $(A+B)^2 = A^2 + 2AB + B^2$? If not, write down what $(A+B)^2$ should equal. Verify that your result is correct.

(iii) Write down a similar result for $(A-B)^2$ and verify your answer.

(iv) Does $(A-B)(A+B) = A^2 - B^2$? Verify your answer.

10. If $P = \begin{pmatrix} 4 & -3 \\ 5 & 6 \end{pmatrix}$ and $Q = \begin{pmatrix} -3 & 4 & -1 \\ 5 & 6 & 2 \\ -2 & 0 & 1 \end{pmatrix}$

and I is the unit 2×2 or 3×3 matrix, verify that

$$PI = IP = P \quad \text{and} \quad QI = IQ = Q$$

11. If $A = \begin{pmatrix} -4 & -2 \\ -1 & 3 \end{pmatrix}$, $B = \begin{pmatrix} 0 & 1 \\ 2 & 3 \end{pmatrix}$ and k is any non-zero real number, verify that,

(i) $A(kB) = k(AB)$

(ii) $(kA)B = k(AB)$

12. If matrix $A = \begin{pmatrix} k & -1 \\ 0 & m \end{pmatrix}$, matrix $B = \begin{pmatrix} 2 & 1 \\ -2 & -1 \end{pmatrix}$ and the product $AB = \begin{pmatrix} 6 & 3 \\ -4 & -2 \end{pmatrix}$ find the values of k and m. With these values of k and m find the matrix product BA.

13. $A = \begin{pmatrix} 7 & 1 \\ -2 & 3 \end{pmatrix}$ and $B = \begin{pmatrix} 2 & 0 \\ 3 & 1 \end{pmatrix}$

Calculate AB and write down $(AB)'$.
Write down A', B' and calculate $B'A'$.
Does $(AB)' = A'B'$ or $B'A'$?
What would $(ABC)'$ equal in terms of A', B' and C'?

14. Given that $\begin{pmatrix} 3 & x \\ y & 3 \end{pmatrix}\begin{pmatrix} 3 & 2 \\ 4 & 3 \end{pmatrix} = \begin{pmatrix} 1 & 0 \\ 0 & 1 \end{pmatrix}$ find x and y.

15. Show that $(A-I)(A+I) = A^2 - I$ and $(A+I)(A-I) = A^2 - I$, where A is any square matrix of order 2 and I is the unit 2×2 matrix.

16. If $A = \begin{pmatrix} 2 & 1 \\ 2 & 3 \end{pmatrix}$, verify that $A^2 - 5A + 4I = O$.

17. The point $P(x, y)$ is mapped onto the point $P'(x', y')$ by the operation of the matrix $\begin{pmatrix} 1 & 0 \\ 3 & 1 \end{pmatrix}$ such that $\begin{pmatrix} 1 & 0 \\ 3 & 1 \end{pmatrix}\begin{pmatrix} x \\ y \end{pmatrix} = \begin{pmatrix} x' \\ y' \end{pmatrix}$.

Write down the equations giving x', y' in terms of x and y.

Write down the coordinates of the images A', B', C' of the points $A(2, 0)$, $B(2, 1)$, $C(4, 3)$ under this mapping.

Find the equations which give x and y in terms of x' and y' and hence the matrix of the mapping which maps the point (x', y') back onto the point (x, y).

18. The rectangle whose vertices are $A(1, 1)$, $B(3, 1)$, $C(1, 2)$ and $D(3, 2)$ is mapped by the matrix $\begin{pmatrix} 1 & 0 \\ 2 & 1 \end{pmatrix}$ onto the figure with vertices A', B', C', D' such that $A \to A'$, $B \to B'$, $C \to C'$ and $D \to D'$.

What kind of figure is $A'B'D'C'$?

19. The point $P(x, y)$ is mapped onto the point $P'(x', y')$ by a rotation through an angle θ about the origin. The matrix for this mapping is

$$\begin{pmatrix} \cos\theta & -\sin\theta \\ \sin\theta & \cos\theta \end{pmatrix}$$

and is written

$$\begin{pmatrix} x' \\ y' \end{pmatrix} = \begin{pmatrix} \cos\theta & -\sin\theta \\ \sin\theta & \cos\theta \end{pmatrix}\begin{pmatrix} x \\ y \end{pmatrix}$$

The point $P'(x', y')$ is mapped onto the point $P''(x'', y'')$ by the matrix $\begin{pmatrix} 1 & 0 \\ 0 & -1 \end{pmatrix}$. Write down a similar matrix equation for this mapping.

What single matrix maps P onto P''?

If the operations are reversed so that $P \to P'$ by the matrix $\begin{pmatrix} 1 & 0 \\ 0 & -1 \end{pmatrix}$ and $P' \to P''$ by the matrix

$$\begin{pmatrix} \cos\theta & -\sin\theta \\ \sin\theta & \cos\theta \end{pmatrix}$$

will P'' be the same point as in the first case?

20. A biscuit manufacturer makes 3 kinds of biscuits comprising flour, fat, sugar and chocolate. The requirements for each batch of biscuits is given in kilograms in Table 1.

Table 1

	flour	fat	sugar	chocolate
Plain biscuits	20	3	4	0
Butter biscuits	20	8	4	0
Chocolate biscuits	15	5	3	2

The cost in pence per kilogram for flour, fat, sugar and chocolate is given in Table 2.

Table 2

	Cost in pence per kg
flour	8p
fat	15p
sugar	6p
chocolate	20p

If A denotes the "weight" matrix as a 3×4 matrix and B denotes the "cost" matrix as a column matrix, write down Table 1 and Table 2 as matrices A and B.

Calculate the product matrix AB. What does each element in the product matrix AB represent?

Find the total cost of manufacturing a batch of each kind of biscuit.

21. X is the 2×1 matrix $\begin{pmatrix} x \\ y \end{pmatrix}$ and P the 2×2 matrix $\begin{pmatrix} p & p-1 \\ q-1 & q \end{pmatrix}$

If $PX = X$ show that either $p = 1$ and $q = 1$ or $x = -y$.

22. If $A = \begin{pmatrix} 1 & 1 \\ -2 & 1 \end{pmatrix}$ and $B = \begin{pmatrix} 2 & 1 \\ 0 & -1 \end{pmatrix}$, find

(i) AB, BA and $A^2 - B^2$.

(ii) Find the matrix X such that $ABX = BA$.

23. If $A = \begin{pmatrix} 0 & 0 \\ 1 & 0 \end{pmatrix}$ and $B = \begin{pmatrix} 1 & 0 \\ 0 & 0 \end{pmatrix}$ which of the following statements are true?

(i) $A \neq O$ and $B \neq O$

(ii) $AB \neq O$

(iii) $BA \neq O$

(iv) $AB = O \Rightarrow B = O$ or $A = O$

24. Show that for all matrices of the form $P = \begin{pmatrix} p & q \\ -q & p \end{pmatrix}$ and $Q = \begin{pmatrix} r & s \\ -s & r \end{pmatrix}$ we have $PQ = QP$.

Illustrate this by giving numerical values to p, q, r, s with $p, q, r, s \in Z$.

25. Show for all matrices of the form $A = \begin{pmatrix} ab & b^2 \\ -a^2 & -ab \end{pmatrix}$ that $A^2 = O$.

26. If $A = \begin{pmatrix} 2 & -1 \\ 1 & 0 \end{pmatrix}$ find A^2, A^3, A^4 and hence show that A^n can be written $\begin{pmatrix} n+1 & -n \\ n & -n+1 \end{pmatrix}$

2.3 THE INVERSE OF A SQUARE MATRIX A OF ORDER 2

If the product of two numbers is equal to 1 then each is said to be the multiplicative inverse of the other, 1 being the identity element for the multiplication of numbers.

Example 1

Since $3 \times \frac{1}{3} = 1$, 3 is the inverse of $\frac{1}{3}$ and $\frac{1}{3}$ is the inverse of 3.

I is the unit or identity matrix for multiplication of 2×2 matrices. If the product of two matrices is equal to I then each is said to be the inverse of the other, i.e. if $AB = I$ then B is the inverse of A and is denoted by A^{-1}.

In general if A and B are two square matrices of the same order such that $AB = BA = I$, then B is the inverse of A denoted by A^{-1}, and A is the inverse of B denoted by B^{-1}.

Given a square matrix A, is there a matrix B such that $AB = I$, where I is the unit or identity matrix for multiplication of 2×2 matrices?

Let $A = \begin{pmatrix} 3 & 4 \\ 5 & 7 \end{pmatrix}$ and suppose $B = \begin{pmatrix} a & b \\ c & d \end{pmatrix}$

If $AB = I$, then $\begin{pmatrix} 3 & 4 \\ 5 & 7 \end{pmatrix}\begin{pmatrix} a & b \\ c & d \end{pmatrix} = \begin{pmatrix} 1 & 0 \\ 0 & 1 \end{pmatrix}$

$$\Rightarrow \begin{pmatrix} 3a+4c & 3b+4d \\ 5a+7c & 5b+7d \end{pmatrix} = \begin{pmatrix} 1 & 0 \\ 0 & 1 \end{pmatrix}$$

$$\Rightarrow \begin{array}{ll} 3a+4c = 1 \quad \text{and} & 3b+4d = 0 \\ 5a+7c = 0 & 5b+7d = 1 \\ \hline a = 7, c = -5 & d = 3, b = -4 \end{array}$$

Hence, $$B = \begin{pmatrix} 7 & -4 \\ -5 & 3 \end{pmatrix}$$

Note its similarity to matrix A. Elements in the leading diagonal have been interchanged, whilst the signs of the elements in the other diagonal have been altered.

Note that $\begin{pmatrix} 3 & 4 \\ 5 & 7 \end{pmatrix}\begin{pmatrix} 7 & -4 \\ -5 & 3 \end{pmatrix} = \begin{pmatrix} 1 & 0 \\ 0 & 1 \end{pmatrix}$ i.e. $AB = I$

Similarly we find that $BA = I$.

Hence, if $A = \begin{pmatrix} 3 & 4 \\ 5 & 7 \end{pmatrix}$ then $A^{-1} = \begin{pmatrix} 7 & -4 \\ -5 & 3 \end{pmatrix}$

is the inverse of the matrix A.

Consider now the question whether every 2×2 matrix has an inverse.

If $A = \begin{pmatrix} a & b \\ c & d \end{pmatrix}$, let $A^{-1} = \begin{pmatrix} d & -b \\ -c & a \end{pmatrix}$

then $$AA^{-1} = \begin{pmatrix} a & b \\ c & d \end{pmatrix}\begin{pmatrix} d & -b \\ -c & a \end{pmatrix} = \begin{pmatrix} ad-bc & 0 \\ 0 & ad-bc \end{pmatrix}$$
$$= (ad-bc)\begin{pmatrix} 1 & 0 \\ 0 & 1 \end{pmatrix}$$

Hence, our choice of A^{-1} does not satisfy $AA^{-1} = I$.

If $$A^{-1} = \frac{1}{ad-bc}\begin{pmatrix} d & -b \\ -c & a \end{pmatrix}$$

then $$AA^{-1} = \begin{pmatrix} a & b \\ c & d \end{pmatrix} \frac{1}{ad-bc} \begin{pmatrix} d & -b \\ -c & a \end{pmatrix}$$
$$= \frac{1}{ad-bc} \begin{pmatrix} a & b \\ c & d \end{pmatrix}\begin{pmatrix} d & -b \\ -c & a \end{pmatrix}$$
$$= \frac{1}{ad-bc} (ad-bc) \begin{pmatrix} 1 & 0 \\ 0 & 1 \end{pmatrix}$$
$$= I, \quad \text{provided. } ad-bc \neq 0.$$

Similarly $A^{-1}A = I$, provided $ad-bc \neq 0$.

Therefore, the matrix $\begin{pmatrix} a & b \\ c & d \end{pmatrix}$ has inverse

$$\frac{1}{ad-bc}\begin{pmatrix} d & -b \\ -c & a \end{pmatrix}$$

provided $ad-bc \neq 0$.

Example 2

Let $A = \begin{pmatrix} 5 & 6 \\ 3 & 4 \end{pmatrix}$ then $ad-bc = 20-18 = 2$.

Hence, $$A^{-1} = \frac{1}{2}\begin{pmatrix} 4 & -6 \\ -3 & 5 \end{pmatrix} = \begin{pmatrix} 2 & -3 \\ -\frac{3}{2} & \frac{5}{2} \end{pmatrix}$$

If $ad-bc = 0$ then the procedure has no meaning, i.e. if $ad-bc = 0$ the matrix has no inverse.

A matrix which does not have an inverse is called a *singular matrix*.

$ad-bc$ is called the **determinant** of the matrix $\begin{pmatrix} a & b \\ c & d \end{pmatrix}$ and is usually denoted by det A.

Hence, a 2×2 matrix A has an inverse matrix if and only if det $A \neq 0$.

ASSIGNMENT 2.3

1. Find the inverse of each of the following matrices, if the inverse exists. [Note: $ad-bc \neq 0$ for the inverse to exist.]

   (i) $\begin{pmatrix} 6 & 4 \\ 4 & 3 \end{pmatrix}$ (ii) $\begin{pmatrix} 5 & 8 \\ 1 & 2 \end{pmatrix}$ (iii) $\begin{pmatrix} 6 & 9 \\ 4 & 6 \end{pmatrix}$

   (iv) $\begin{pmatrix} 4 & 5 \\ 2 & 1 \end{pmatrix}$ (v) $\begin{pmatrix} 3 & 6 \\ 4 & 8 \end{pmatrix}$ (vi) $\begin{pmatrix} 2 & -1 \\ 0 & 1 \end{pmatrix}$

   (vii) $\begin{pmatrix} -2 & -1 \\ 3 & 2 \end{pmatrix}$ (viii) $\begin{pmatrix} 5 & 10 \\ 3 & 6 \end{pmatrix}$ (ix) $\begin{pmatrix} p & -2q \\ p & -q \end{pmatrix}$

   (x) $\begin{pmatrix} a & b \\ -b & a \end{pmatrix}$

2. If $A = \begin{pmatrix} 3 & 2 \\ 0 & 1 \end{pmatrix}$ and $B = \begin{pmatrix} -2 & 1 \\ -5 & 3 \end{pmatrix}$, calculate,

   (i) A^{-1} (ii) B^{-1} (iii) $A^{-1}B^{-1}$

   (iv) $B^{-1}A^{-1}$ (v) $(AB)^{-1}$

   Note: $(AB)^{-1} = B^{-1}A^{-1}$ from parts (v) and (iv). This is called the **"reversal rule for inverses".**

3. $A = \begin{pmatrix} 4 & 5 \\ 2 & 3 \end{pmatrix}$ and B is the inverse of the matrix A

   (i) If B is the inverse of A, is A the inverse of B?

   (ii) By calculation or otherwise, write down $(A^{-1})^{-1}$, i.e. the inverse of the inverse of A.

   (iii) If A' denotes the transpose of the matrix A, calculate $(A^{-1})'$ and $(A')^{-1}$. Are these equal?

4. Under a transformation associated with the matrix M, the point $P(x, y)$ maps onto the point $P'(x', y')$ such that $x' = x-2y$ and $y' = -2x+3y$. Write down the 2×2 matrix M, such that $\begin{pmatrix} x' \\ y' \end{pmatrix} = M\begin{pmatrix} x \\ y \end{pmatrix}$

   The matrix N maps $P'(x', y')$ back onto $P(x, y)$, find the matrix N.

5. If $J = \begin{pmatrix} a & 0 \\ 0 & a \end{pmatrix}$, with $a \neq 0$, find the value of a such that $J = J^{-1}$.

6. If $P = \begin{pmatrix} \cos\theta & -\sin\theta \\ \sin\theta & \cos\theta \end{pmatrix}$ show that the inverse of P, namely P^{-1}, can be written in the form

$$\begin{pmatrix} \cos(-\theta) & -\sin(-\theta) \\ \sin(-\theta) & \cos(-\theta) \end{pmatrix}$$

   If matrix P rotates the point $A(x, y)$ through an angle θ about the origin, what should P^{-1} do to the point $A(x, y)$?

7. $P = \begin{pmatrix} 0 & 1 \\ 1 & 0 \end{pmatrix}$ $Q = \begin{pmatrix} 1 & 0 \\ 0 & 1 \end{pmatrix}$

   $R = \begin{pmatrix} -1 & 0 \\ 0 & -1 \end{pmatrix}$ $S = \begin{pmatrix} 0 & -1 \\ -1 & 0 \end{pmatrix}$

   Complete the following table for matrix multiplication.

×	P	Q	R	S
P	Q	P	S	
Q	P			
R				
S				

   From your table,

   (i) Which matrix is the identity matrix for the operation of matrix multiplication?

   (ii) Check that each matrix is its own inverse.

   (iii) Are the matrices P, Q, R, S commutative?

8. If $P = \begin{pmatrix} 2 & 1 \\ 2 & 4 \end{pmatrix}$, and $Q = \begin{pmatrix} 1 & 2 \\ 2 & 0 \end{pmatrix}$, find the inverse of matrix P and solve the equation $PX = Q$, where X is a 2×2 matrix.

2.4

9. Which of the following pairs of matrices are inverses of one another?

   (i) $\begin{pmatrix} 3 & -1 \\ 2 & -1 \end{pmatrix}$ and $\begin{pmatrix} 1 & -1 \\ 2 & -3 \end{pmatrix}$

   (ii) $\begin{pmatrix} 7 & 4 \\ 16 & 9 \end{pmatrix}$ and $\begin{pmatrix} -9 & 4 \\ 16 & -7 \end{pmatrix}$

   (iii) $\begin{pmatrix} 5 & -7 \\ -2 & 3 \end{pmatrix}$ and $\begin{pmatrix} 3 & 7 \\ 2 & 5 \end{pmatrix}$

   (iv) $\begin{pmatrix} 2 & -3 \\ 1 & 5 \end{pmatrix}$ and $\begin{pmatrix} 5 & 3 \\ -1 & 2 \end{pmatrix}$

10. If $A = \begin{pmatrix} -1 & 3 \\ 2 & -5 \end{pmatrix}$ and $B = \begin{pmatrix} 5 & 3 \\ 2 & 1 \end{pmatrix}$ are such that $AB = I$, which of the following statements are true?

    (i) $A = B^{-1}$

    (ii) $(A^2) = (B^2)^{-1}$

    (iii) $(A') = (B')^{-1}$

11. $P = \begin{pmatrix} 0 & 1 \\ 1 & 0 \end{pmatrix}$ and $Q = \begin{pmatrix} a & b \\ c & d \end{pmatrix}$ with $ad - bc = 1$. Write down PQ, QP, Q^{-1} and verify that $(PQ)^{-1} = Q^{-1}P^{-1}$

12. If $A = \begin{pmatrix} 3 & 6 \\ -2 & -4 \end{pmatrix}$ is singular (i.e. has no inverse), show that AB where $B = \begin{pmatrix} a & b \\ c & d \end{pmatrix}$ and $ad - bc \neq 0$ is also singular.

13. If matrix A has no inverse and matrix B has no inverse, will the product AB have no inverse? Examine this by constructing two matrices A and B with no inverses and then examine the product AB.

USING MATRICES TO SOLVE SYSTEMS OF LINEAR EQUATIONS

Remember: If A^{-1} is the inverse of A, then

$$A^{-1}A = AA^{-1} = I \quad \text{and} \quad AI = IA = A$$

Example 1

Solve the equations $\left.\begin{aligned} 2x - 4y &= 5 \\ 3x + 8y &= 11 \end{aligned}\right\}$ using matrices.

Write these as

$$\begin{pmatrix} 2 & -4 \\ 3 & 8 \end{pmatrix}\begin{pmatrix} x \\ y \end{pmatrix} = \begin{pmatrix} 5 \\ 11 \end{pmatrix}$$

which can be written $AX = H$, where

$$A = \begin{pmatrix} 2 & -4 \\ 3 & 8 \end{pmatrix}, \quad X = \begin{pmatrix} x \\ y \end{pmatrix}, \quad H = \begin{pmatrix} 5 \\ 11 \end{pmatrix}$$

Now
$$\begin{aligned} A^{-1}AX &= A^{-1}H \\ \Rightarrow IX &= A^{-1}H \\ \Rightarrow X &= A^{-1}H \end{aligned}$$

But
$$A^{-1} = \frac{1}{(16+12)}\begin{pmatrix} 8 & 4 \\ -3 & 2 \end{pmatrix} = \frac{1}{28}\begin{pmatrix} 8 & 4 \\ -3 & 2 \end{pmatrix}$$

Hence
$$X = \frac{1}{28}\begin{pmatrix} 8 & 4 \\ -3 & 2 \end{pmatrix} H$$

i.e.
$$\begin{pmatrix} x \\ y \end{pmatrix} = \frac{1}{28}\begin{pmatrix} 8 & 4 \\ -3 & 2 \end{pmatrix}\begin{pmatrix} 5 \\ 11 \end{pmatrix} = \frac{1}{28}\begin{pmatrix} 84 \\ 7 \end{pmatrix}$$

i.e.
$$\begin{pmatrix} x \\ y \end{pmatrix} = \begin{pmatrix} \frac{84}{28} \\ \frac{7}{28} \end{pmatrix} = \begin{pmatrix} 3 \\ \frac{1}{4} \end{pmatrix}$$

Therefore $\left.\begin{matrix} x = 3 \\ y = \frac{1}{4} \end{matrix}\right\}$ and the solution set of the system is $\{(3, \frac{1}{4})\}$.

Example 2

Solve the system of equations

$$\left.\begin{aligned} 2x - \ y &= \ \ 2 \\ 6x - 3y &= -12 \end{aligned}\right\}$$

If we write these equations in matrix form we get:

$$\begin{pmatrix} 2 & -1 \\ 6 & -3 \end{pmatrix}\begin{pmatrix} x \\ y \end{pmatrix} = \begin{pmatrix} 2 \\ -12 \end{pmatrix}$$

and here $ad - bc = -6 + 6 = 0$ and hence the matrix $\begin{pmatrix} 2 & -1 \\ 6 & -3 \end{pmatrix}$ has no inverse. Therefore the equations have no solution.

Since an equation of first degree in x and y represents a straight line, the two given equations represent two parallel straight lines which therefore do not intersect. Hence the two equations have no solution (figure 2).

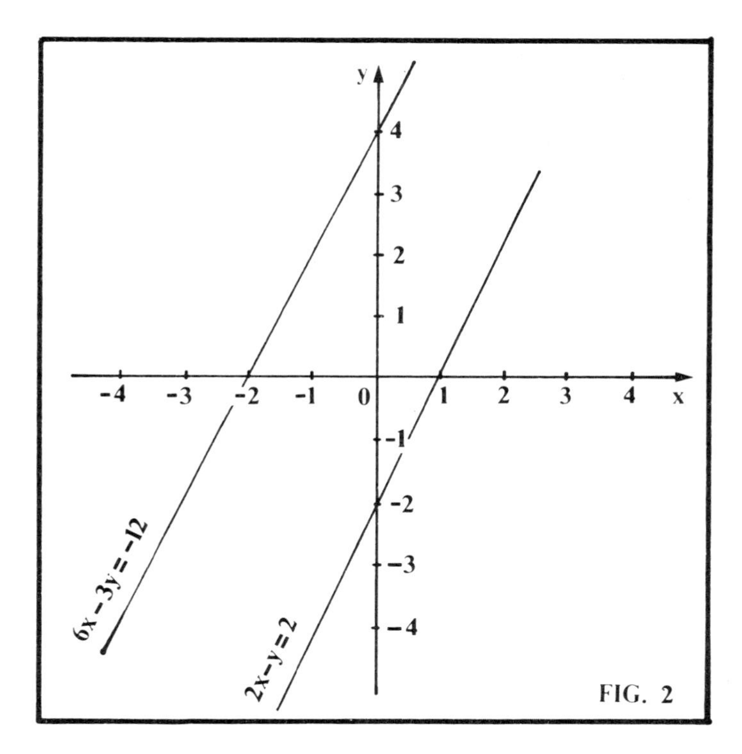

FIG. 2

ASSIGNMENT 2.4

1. Solve the following matrix equations to find the 2×2 matrix X, by multiplying by an appropriate inverse.

   (i) $\begin{pmatrix} 1 & 1 \\ 1 & 2 \end{pmatrix} X = \begin{pmatrix} 2 & 1 \\ 0 & -1 \end{pmatrix}$

   (ii) $\begin{pmatrix} 2 & -4 \\ 1 & -3 \end{pmatrix} X = \begin{pmatrix} 2 & 1 \\ 3 & -4 \end{pmatrix}$

   (iii) $X \begin{pmatrix} 1 & 1 \\ 1 & 2 \end{pmatrix} = \begin{pmatrix} 2 & 1 \\ 0 & -1 \end{pmatrix}$

2. If A, X and B are 2×2 matrices and $AX = B$, then $X = A^{-1}B$ or $X = BA^{-1}$. Which is the correct result? Give a reason for your answer.

3. $AX + B = C$, where A, X, B, and C are all 2×2 matrices. Which of the following statements are correct?

   (i) $AX = C - B$

   (ii) $X + A^{-1}B = A^{-1}C$

   (iii) $X = A^{-1}(C - B)$

   (iv) $X = (C - B)A^{-1}$

   (v) $A^{-1}AX + A^{-1}B = CA^{-1}$

4. Solve, where possible, the following equations for the matrix X.

(i) $\begin{pmatrix} 2 & -1 \\ 0 & 1 \end{pmatrix} X = \begin{pmatrix} 3 & 1 \\ 4 & -2 \end{pmatrix}$

(ii) $X \begin{pmatrix} 2 & 1 \\ 3 & -4 \end{pmatrix} = \begin{pmatrix} 5 & 0 \\ 1 & 2 \end{pmatrix}$

5. If $A = \begin{pmatrix} 2 & 1 \\ 3 & 4 \end{pmatrix}$, $B = \begin{pmatrix} 2 \\ -1 \end{pmatrix}$ and X is the column matrix $\begin{pmatrix} x \\ y \end{pmatrix}$ such that $AX = B$, write down a set of two simultaneous equations involving x and y.

6. Given that $\begin{pmatrix} 3 & x \\ y & 3 \end{pmatrix}\begin{pmatrix} 3 & 2 \\ 4 & 3 \end{pmatrix} = \begin{pmatrix} 1 & 0 \\ 0 & 1 \end{pmatrix}$, find a matrix equal to the matrix $\begin{pmatrix} 3 & x \\ y & 3 \end{pmatrix}$ and hence find the values of x and y.

7. Given that $\begin{pmatrix} 8 & 5 \\ 3 & 2 \end{pmatrix}\begin{pmatrix} 2 & a \\ b & 8 \end{pmatrix} = \begin{pmatrix} 1 & 0 \\ 0 & 1 \end{pmatrix}$, by multiplying each side of this equation by the inverse of an appropriate matrix, find a and b.

8. Use matrix methods to solve the following systems of equations.

(i) $3x + y = 5$, $x + y = 3$

(ii) $2x + 3y = 13$, $3x - 2y = 0$

(iii) $3x + 4y = 1$, $5x - 2y = -7$

(iv) $5x - y = 9$, $2x - y = -3$

(v) $7x + 3y = 6$, $9x - 5y = -10$

(vi) $6x + 5y = 11$, $5x - 2y = 3$

(vii) $2x - y - 4 = 0$, $3x + 2y + 4 = 0$

(viii) $ax + by = p$, $cx + dy = q$, if $ad - bc \neq 0$.

9. Express in matrix form the following equations,

(i) $$\begin{aligned} x + y + z &= 8 \\ 2x + y + z &= 7 \\ x + y - z &= 6 \end{aligned}$$

(ii) $$\begin{aligned} x + 3y - 2z &= 6 \\ 4x + y - z &= 7 \\ 8x - 2y + z &= 4 \end{aligned}$$

10. If $A = \begin{pmatrix} 1 & -1 & 1 \\ 1 & 1 & 2 \\ 2 & -1 & 3 \end{pmatrix}$ and $B = \begin{pmatrix} 5 & 2 & -3 \\ 1 & 1 & -1 \\ -3 & -1 & 2 \end{pmatrix}$

show that $AB = I$, where I is the 3×3 unit or identity matrix, and use this result to solve the equations

$$\left.\begin{aligned} x - y + z &= 1 \\ x + y + 2z &= 0 \\ 2x - y + 3z &= 2 \end{aligned}\right\}$$

11. If $P = \begin{pmatrix} 3 & -9 & 3 \\ -1 & 1 & 0 \\ -2 & 5 & -3 \end{pmatrix}$ and $Q = \begin{pmatrix} 1 & 4 & 1 \\ 1 & 1 & 1 \\ 1 & -1 & 2 \end{pmatrix}$

show that $PQ = -3I$, where I is the 3×3 unit or identity matrix, and use this result to solve the equations

$$\begin{aligned} x + 4y + z &= 6 \\ x + y + z &= 3 \\ x - y + 2z &= 2 \end{aligned}$$

ASSIGNMENT 2.5

Objective items testing Sections 2.1–2.4

Instructions for answering these items are given on page 22.

1. If Q is the matrix $\begin{pmatrix} 3 & 4 \\ -2 & -3 \end{pmatrix}$, the inverse of Q is

   A. $\begin{pmatrix} -3 & -2 \\ 4 & 3 \end{pmatrix}$

   B. $\begin{pmatrix} 3 & -2 \\ 4 & -3 \end{pmatrix}$

   C. $\begin{pmatrix} 3 & -4 \\ 2 & -3 \end{pmatrix}$

   D. $\begin{pmatrix} 3 & 4 \\ -2 & -3 \end{pmatrix}$

   E. non-existent

2. If $PQ = R$ where $Q = \begin{pmatrix} 1 & -1 \\ 2 & 3 \end{pmatrix}$ and $R = \begin{pmatrix} 1 & -1 \\ 0 & 5 \end{pmatrix}$, then P is

   A. $\begin{pmatrix} 1 & 2 \\ 0 & 1 \end{pmatrix}$

   B. $\begin{pmatrix} 1 & -1 \\ 0 & 5 \end{pmatrix}$

   C. $\begin{pmatrix} 1 & 0 \\ -2 & 1 \end{pmatrix}$

   D. $\begin{pmatrix} 1 & 0 \\ 2 & -1 \end{pmatrix}$

   E. $\begin{pmatrix} 1 & -5 \\ 0 & 1 \end{pmatrix}$

3. If $S = \begin{pmatrix} \cos\alpha & -\sin\alpha \\ \sin\alpha & \cos\alpha \end{pmatrix}$ and $T = \begin{pmatrix} \cos\alpha & \sin\alpha \\ -\sin\alpha & \cos\alpha \end{pmatrix}$ which one of the following statements is false?

   A. $S = T^{-1}$

   B. $S + T = T + S$

   C. $ST = TS$

   D. $S - T = 2\sin\alpha \begin{pmatrix} 0 & -1 \\ 1 & 0 \end{pmatrix}$

   E. $S' = T'$

4. The point $P(x, y)$ is mapped onto the point $P'(x', y')$ by by the operation of the matrix $\begin{pmatrix} 2 & 3 \\ 2 & 4 \end{pmatrix}$ such that

$$\begin{pmatrix} 2 & 3 \\ 2 & 4 \end{pmatrix}\begin{pmatrix} x \\ y \end{pmatrix} = \begin{pmatrix} x' \\ y' \end{pmatrix}$$

   The matrix of the mapping which maps $P'(x', y')$ back onto the point $P(x, y)$ is

   A. $\begin{pmatrix} 4 & -3 \\ -2 & 2 \end{pmatrix}$

   B. $\begin{pmatrix} 2 & -\frac{3}{2} \\ -1 & 1 \end{pmatrix}$

   C. $\begin{pmatrix} 2 & -3 \\ -2 & 4 \end{pmatrix}$

   D. $\begin{pmatrix} 2 & -3 \\ -1 & 1 \end{pmatrix}$

   E. $\begin{pmatrix} 2 & -\frac{3}{2} \\ -\frac{1}{2} & \frac{1}{2} \end{pmatrix}$

5. If $P = \begin{pmatrix} 1 & 0 \\ -1 & 2 \end{pmatrix}$, then $P^2\begin{pmatrix} x \\ y \end{pmatrix} = \begin{pmatrix} -1 \\ -1 \end{pmatrix}$ is satisfied by

A. $x = -1, y = -1$
B. $x = 1, y = 1$
C. $x = 0, y = -1$
D. $x = 2, y = -1$
E. $x = -2, y = -1$

6. If P is a square matrix of order 2, which of the equivalences in the following chain of statements is false?

$$P^2 = 7P + I$$

A. $\Leftrightarrow \quad P^{-1}P^2 = P^{-1}7P + P^{-1}I$
B. $\Leftrightarrow (P^{-1}P)P = 7P^{-1}P + P^{-1}I$
C. $\Leftrightarrow \quad IP = 7I + P^{-1}I$
D. $\Leftrightarrow \quad P = 7 + P^{-1}$
E. $\Leftrightarrow \quad P^{-1} = P - 7$

7. If $L = \begin{pmatrix} 0 & 1 \\ 2 & 3 \end{pmatrix}$ and $M = \begin{pmatrix} -1 & 1 \\ 1 & 1 \end{pmatrix}$, then $(LM)^{-1}$ is

A. $\begin{pmatrix} -1 & 1 \\ 1 & 1 \end{pmatrix}$

B. $\begin{pmatrix} -\frac{1}{4} & \frac{1}{4} \\ \frac{5}{4} & \frac{1}{4} \end{pmatrix}$

C. $\begin{pmatrix} 1 & -1 \\ 1 & -5 \end{pmatrix}$

D. $\begin{pmatrix} \frac{5}{4} & \frac{1}{4} \\ \frac{1}{4} & -\frac{1}{4} \end{pmatrix}$

E. $\begin{pmatrix} \frac{5}{4} & -\frac{1}{4} \\ -\frac{1}{4} & \frac{1}{4} \end{pmatrix}$

8. The system of equations

$$\begin{aligned} ax + y &= 1 \\ x - by &= 4 \end{aligned}$$

will have a solution if

A. $a - b \neq 0$
B. $a + b = 1$
C. $ab \neq 0$
D. $ab \neq -1$
E. some other relation exists between a and b.

9. Given the system of equations

$$\left.\begin{aligned} 6x - 3y &= -2 \\ 2x - y &= 4 \end{aligned}\right\}$$

(1) the equations have solution set $\{(1, -2)\}$
(2) the matrix $\begin{pmatrix} 6 & -3 \\ 2 & -1 \end{pmatrix}$ has inverse $\begin{pmatrix} -1 & 3 \\ -2 & 6 \end{pmatrix}$
(3) the equations represent two parallel straight lines.

10. Matrix P has inverse P^{-1}, matrix Q has inverse Q^{-1}
(1) Matrix PQ has inverse $(PQ)^{-1}$
(2) $(P+Q)^{-1} = P^{-1} + Q^{-1}$
(3) $(PQ)^{-1} = P^{-1}Q^{-1}$.

11. (1) Matrix A is of order 4×3, matrix B is of order 3×2.
(2) Matrix AB is of order 4×2.

12. (1) Matrix $P = \begin{pmatrix} 3 & 6 \\ 2 & p^2 \end{pmatrix}$ has an inverse.
(2) $p \neq 2$ or -2.

UNIT 3: MAPPING AND FUNCTIONS

REVISION OF RELATIONS AND FUNCTIONS

3.1

Consider the relation "is greater than" from the set $A = \{2, 4, 8\}$ to the set $B = \{3, 6, 7, 11\}$. This is illustrated by the arrow diagram in figure 3.

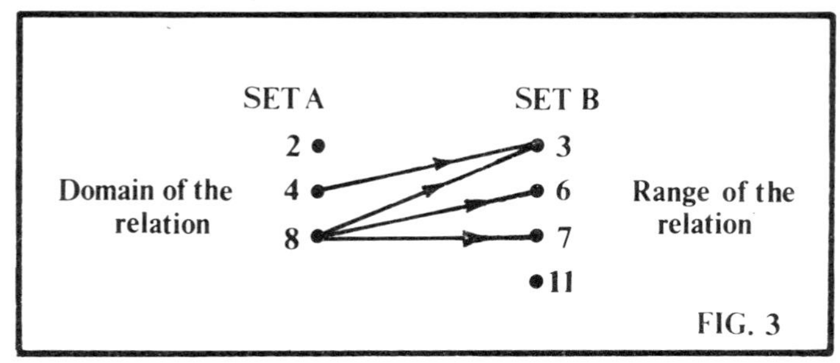

FIG. 3

The solution set is the set of ordered pairs

$$\{(4, 3), (8, 3), (8, 6), (8, 7)\}$$

In set builder notation the solution set becomes

$$\{(x, y): x > y, x \in A, y \in B\}.$$

If the relation "is greater than" is denoted by R, note the following points about relations:

(i) Not all the elements of A are related by R to the elements of B.

(ii) The elements of B need not all have elements of A related to them by R.

(iii) The **domain** of the relation R is the sub-set of A which has elements related to the elements of B, i.e. the sub-set $\{4, 8\}$.

Note that the domain of the relation R is the set of elements of A which appear as first coordinates of the set of ordered pairs in the solution set.

(iv) The **range** of the relation R is the sub-set of elements of B which have elements of A related to them by R, i.e. the sub-set $\{3, 6, 7\}$.

Note that the range of R is the set of elements of B which appear as second coordinates of the set of ordered pairs in the solution set.

FUNCTIONAL RELATIONS OR FUNCTIONS

If A is the set $\{-2, -1, 0, 1, 2\}$, B the set $\{0, 1, 2, 3, 4, 5\}$ and f the relation given by "a has as square b", where $a \in A$, $b \in B$, then this is illustrated by the arrow diagram in figure 4.

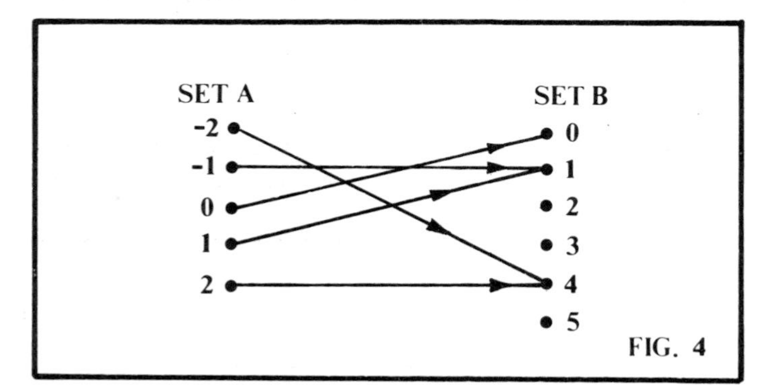

FIG. 4

Compare this with the previous diagram. *Note* that only one arrow leaves each element of set A, and that the domain of f is the set A.

A relation f from A to B is called a **function** (or **mapping**) from A into B, if,

(i) the domain of f is the entire set A.

(ii) each element of A is related by f to exactly one element of B.

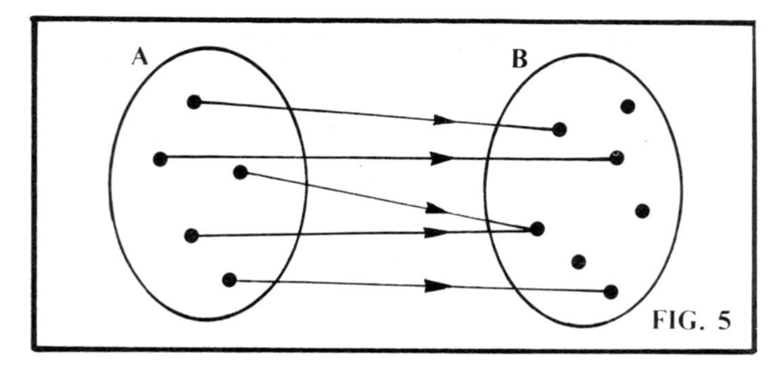

FIG. 5

Figure 5 shows a function or mapping from A into B.

Note that all the elements of A are used up and each element of A is related by f to exactly one element of B.

In the previous example we often write $f(-2) = 4$, $f(1) = 1$, etc. which means that 4 is the image of -2 under f etc., and we say that ***f* maps *A* into *B*** (f: $A \rightarrow B$).

Example 1

$f: x \rightarrow x^2 + 1$, $x \in A$, can be written $f(x) = x^2 + 1$, where $x^2 + 1$ is the image of x under f. This formula (or rule) can be used to find the image of each element of the domain.

If A is the set $\{-3, -1, 0, 2, 3\}$ find the range of f.

Let the range of f be the set B, then

Range is set $B = \{1, 2, 5, 10\}$.

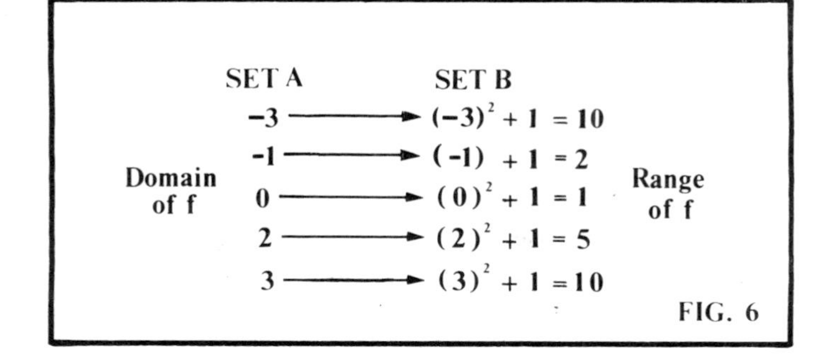

FIG. 6

THREE IMPORTANT MAPPINGS

(i) *One to one mapping*

Let f be a function from the set $A = \{-2, -1, 0, 1, 2\}$ to B, the set of integers, defined by the formula $f(x) = 2x$. Then using an arrow diagram:

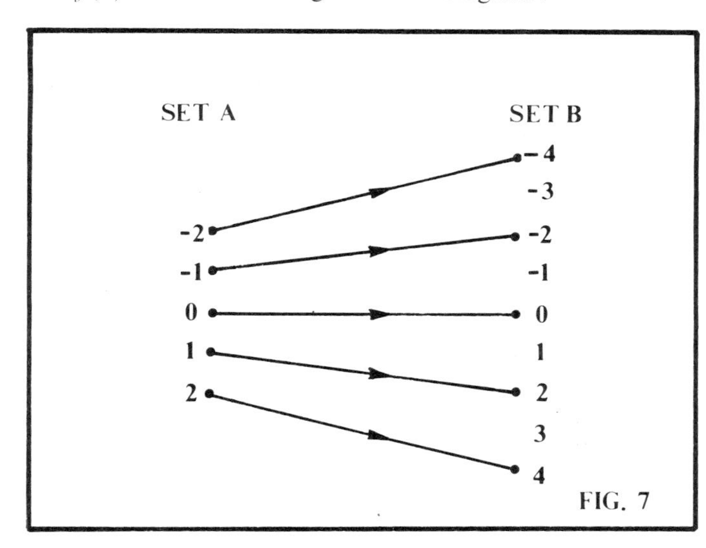

FIG. 7

In this case

(i) each element of A is mapped to exactly one element of B.

(ii) no two elements of A have the same image in B.

This mapping is called a **one to one** or **1–1 mapping.**

(ii) *One to one correspondence*

In (i) some of the elements of B have no elements of A related to them. If in a 1–1 mapping **every** element of B is the image of an element of A, then the mapping or function is called a **1–1 correspondence.**

Example: If $f: x \to 2x$, where $x \in N$, and $2x \in$ set of even positive integers,

then

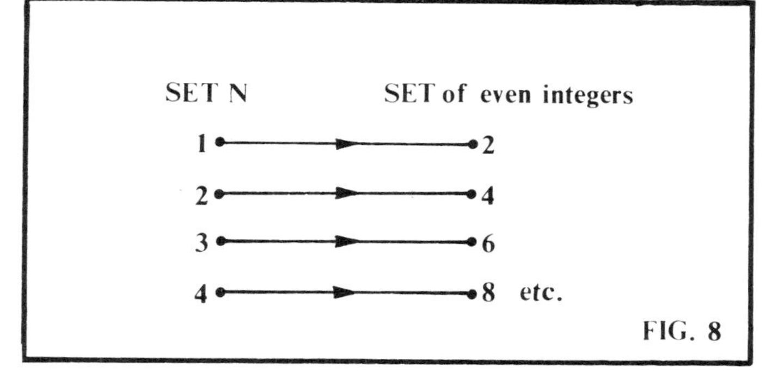

FIG. 8

This is a 1–1 correspondence.

(iii) *The identity function*

Let I be a function on the set of real numbers defined by the formula $I(x) = x$, then I maps each real number onto itself.

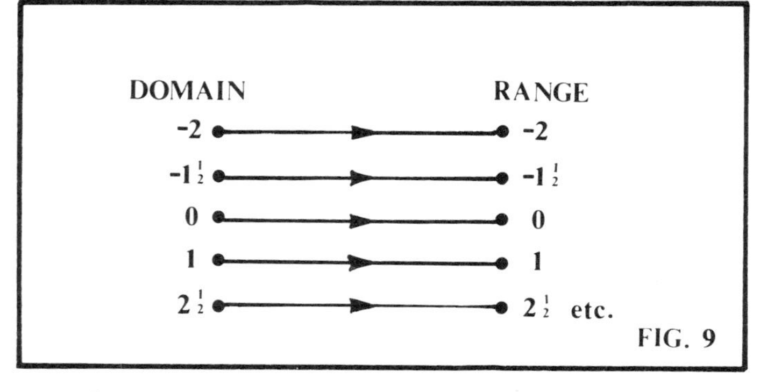

FIG. 9

I is called the **identity function** and maps each member of R onto itself. $I: x \to x$ is a 1–1 correspondence.

ASSIGNMENT 3.1

1. The relation "is greater than" is defined on the set $S = \{1, 2, 3, 4, 5\}$ [this means S or a sub-set of S is the domain of the relation and S or a sub-set of S is the range of the relation].
   Show this relation,
   (i) by an arrow diagram.
   (ii) as a set of ordered pairs.
   (iii) by points on a Cartesian diagram.
   (iv) in set-builder notation.

2. The relation "x is double y" is defined on the set $S = \{1, 2, 3, 4, 5, 6, 7, 8, 9, 10\}$. Show this as an arrow diagram. Write down the domain D and the range R of the relation.
   Write down the solution set,
   (i) as a set of ordered pairs.
   (ii) in set-builder notation.

3. A relation is defined on the set $S = \{1, 2, 3, 4\}$ by the set of ordered pairs $\{(1, 2), (1, 3), (1, 4), (2, 3), (2, 4), (3, 4)\}$.
   Examine the truth of each of the following statements.
   (i) the domain of the relation is $\{1, 2, 3\}$
   (ii) the relation may be defined as
   $$\{(x, y): x < y, x \in S, y \in S\}$$
   (iii) the range of the relation is $\{2, 3, 4\}$
   (iv) the relation is a 1–1 mapping on S.

4. Except for (iii), the following relations are defined on R, the set of real numbers.
   (i) $f: x \to 3 - x$ (ii) $f: x \to x^2$
   (iii) $f: x \to \pm\sqrt{(25 - x^2)}$, where $-5 \leqq x \leqq 5$
   (iv) $f: x \to 2x$ (v) $f: x \to x$

   For each relation write down five ordered pairs which are members of the solution set. Plot each set of ordered pairs on a cartesian diagram and sketch the graph of each relation.
   State whether each relation is also a 1–1 mapping, 1–1 correspondence or simply a mapping.

5. Which of the following functions are 1–1 mappings, 1–1 correspondences or simply mappings.
   (i) $g: x \to x - 1$, from the set Z to the set Z.
   (ii) $f: x \to x$, from set R to the set R.
   (iii) $R: x \to x^2$, from $A = \{-1, 0, 1, 2\}$ to the set W.
   (iv) $k: y \to \sin y$, from R into R.
   (v) $g: x \to |x|$, from $A = \{-2, -1, 0, 1, 2\}$ to the set of integers, where $|x|$ denotes the numerical value of x.

6. For the mapping $g: x \to x^2 + 1$ on the set of real numbers, write down $g(0)$, $g(-1)$, $g(3)$ and $g(-2)$. Sketch the graph of g.

7. In the following, the image of the mapping is specified by a different formula for different subsets of its domain. The domain of each is stated.
   (i) $f: x \to x$, when $x \geqq 0$
   and $f: x \to -x$, when $x < 0$, with domain R.
   Complete the following ordered pairs (1,), (−1,), (0,), (−2,), (2,). Sketch the graph of the function.
   (ii) $g(x) = \begin{cases} x + 1, & \text{when } x \geqq 0 \\ 2x - 1, & \text{when } x < 0, \end{cases}$
   with domain $\{x: x \in R, -4 \leqq x \leqq 4\}$
   Write down the ordered pairs with first member −4, −3, −2, −1, 0, 1, 2, 3, 4. Sketch the graph of the function.

8. Let $O = \{1, 3, 5, 7, \ldots\}$ and $N = \{1, 2, 3, 4, 5, \ldots\}$.

Write down examples of mappings from O to N such that,

(i) the mapping is a 1–1 mapping.

(ii) the mapping is neither a 1–1 mapping nor a 1–1 correspondence.

9. Let function $f: x \to 3$ with domain $\{x: x \in R\}$. Sketch the graph of the function. What is the range of f?

10. Let $f(x) = x^2 - 4$ define a function f whose domain is $\{x: -4 \leqq x \leqq 4, x \in R\}$. Write down the range of f. Sketch the graph of f and hence or otherwise find the minimum value of the function.

    Deduce the graph of $g(x) = 4 - x^2$ and the corresponding maximum value of this function defined by $g(x)$.

    What is the range of the function g?

11. Use an arrow diagram to show the following mappings.

    (a) The function with domain N and range R and the rule "f maps each member of N to its square."

    (b) The function g with domain and range the set of real numbers which maps any real number to twice that number minus 5.

12. For each of the following funtions with given domain find the image set or range of the function.

    (a) $f: x \to +\sqrt{x}$, $\{x: x \in R^+\}$

    (b) $g: x \to |x|$, $\{x: x \in R\}$

    (c) $k: x \to \frac{1}{x}$, $\{x: x \in R, x \neq 0\}$

13. For the function $f: x \to |x| + 4$ with domain R, what is the image of

    (i) 1 (ii) -2 (iii) $\{x: -1 \leqq x \leqq 1, x \in R\}$

    (iv) Z

3.2 COMPOSITION OF FUNCTIONS

If A, B and C are three sets and f and g two functions such that $f: A \to B$ and $g: B \to C$, then A is mapped onto C, via B, by some new function. This new function is called **the composite function** of g and f and is denoted by $g \circ f$.

Again If $a \in A$, $b \in B$, $c \in C$ such that $f(a) = b$ and $g(b) = c$ then $g[f(a)] = c$. *f is performed first, then g*, and we have $(g \circ f)(a) = g[f(a)] = c$.

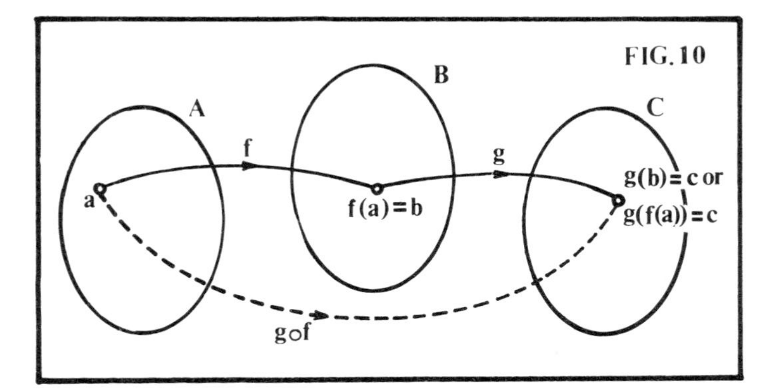

Thus we have the following definition: if f and g are functions with domains A and B respectively, then $g \circ f$ is a function defined by the formula $(g \circ f)(x) = g(f(x))$ whose domain is $\{x: x \in A, f(x) \in B\}$. This is shown in figure 10.

Note that if $f(x) \notin B$, i.e. if the image set of f is not a subset of the domain of g, then $g \circ f$ cannot be formed.

Example 1

If $f(x) = 2x$ and $g(x) = x - 7$, $x \in Z$, find $(g \circ f)(x)$ and $(g \circ f)(3)$.

$$(g \circ f)(x) = g[f(x)] = g(2x) = 2x - 7.$$

$$(g \circ f)(3) = g[f(3)] = g(6) = 6 - 7 = -1.$$

Note: $(f \circ g)(x) = f[g(x)] = f(x-7) = 2(x-7) = 2x - 14$.

Properties of Composition of Functions

It can be shown that the "composite function"

(i) is associative, i.e. $(g \circ f) \circ h = g \circ (f \circ h)$.

(ii) is not commutative, i.e. $g \circ f \neq f \circ g$ (in general).

(iii) has the identity property, i.e. $I \circ f = f \circ I = f$, where I is the identity function defined by $I(x) = x$.

ASSIGNMENT 3.2

1. f and g are two functions with domain the set of integers defined by $f(x) = x+1$ and $g(x) = x^2$.

   Show that $(f \circ g)(x) = x^2+1$ and
   $$(g \circ f)(x) = x^2+2x+1.$$
   Find $(f \circ g)(2)$ by first finding $g(2)$ and check your answer by using the formula for $(f \circ g)(x)$.

   Repeat this for $(g \circ f)(-2)$ by first finding $f(-2)$ and check by using the formula for $(g \circ f)(x)$.

2. The functions f: $X \to Y$ and g: $Y \to Z$ are defined in figure 11 below.

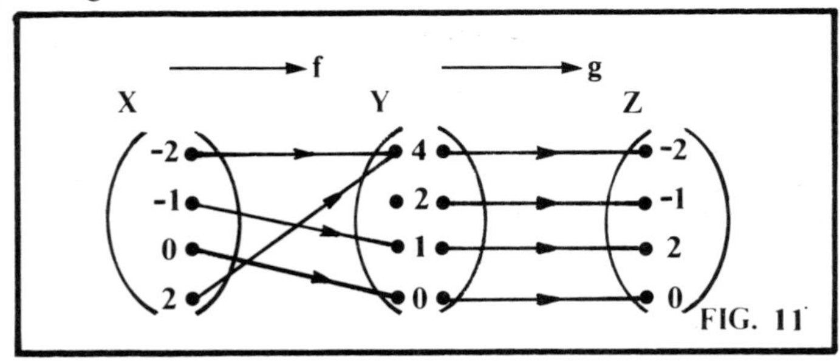

FIG. 11

   (i) Write down $f(-1)$ and $g(2)$.
   (ii) Find $(g \circ f)(-2)$ and $(g \circ f)(2)$.
   (iii) Find $(f \circ g)(1)$.
   (iv) State the domain and range of $(g \circ f)$.

3. The following functions have domain R unless otherwise stated. In each case find a formula and state the domain for $g \circ f$.

   (i) $f: x \to 2-x$, and $g: x \to x^2$
   (ii) $f: x \to x^2$ and $g: x \to x-1$
   (iii) $f(x) = x+1$ and $g(x) = x^2+x-1$
   (iv) $f(x) = \sqrt{x}$ and $g(x) = 3x^2+1$, where the domain of f is R^+
   (v) $f(x) = x^2+1$ and $g(x) = \sqrt{x}$, where $x \in R^+$

   What would be the range of $(g \circ f)(x)$ in parts (i), (ii) and (v).

4. The functions h and k are defined on the set of real numbers R.

   (i) $h: x \to x^2-1$, $k: x \to \dfrac{1}{x+1}$, $x \neq -1$
   (ii) $h: x \to -2x$, $k: x \to x^3$
   (iii) $h: x \to 2x+1$, $k: x \to 4$

   In each case find a formula for $(h \circ k)(x)$ and $(k \circ h)(x)$, if they exist.

5. The functions f and g are defined by $f: x \to 2-x$ and $g: x \to x-2$, with domains the set of real numbers. Show that $(g \circ f)(x) - (f \circ g)(x) = -4$.

6. X denotes the operation of reflection in the x-axis. Y denotes the operation of reflection in the y-axis.

   Find the images of the points $(3, 1)$ and $(-2, 4)$ under the mappings $Y \circ X$ and $X \circ Y$.

   Show that the image of (a, b) is the same under both mappings. What conclusion can you make about the mappings $Y \circ X$ and $X \circ Y$?

7. Functions f and g with domains the set of real numbers are defined by,
   $$f(x) = x^2 \text{ and } g(x) = \begin{cases} 2x+1, & \text{when } x \geqq 0 \\ x, & \text{when } x < 0 \end{cases}$$

Calculate (i) $(f \circ g)(3)$ (ii) $(f \circ g)(-2)$
Find a formula for $(g \circ f)(x)$.

8. Functions h and k with domains the set of real numbers are defined by,

$$h(x) = x^2 \text{ and } k(x) = \begin{cases} x-1, & \text{when } x \geqq 0 \\ -x, & \text{when } x < 0 \end{cases}$$

Calculate (i) $(k \circ h)(-3)$ (ii) $(k \circ h)(7)$, $(h \circ k)(7)$.
Find a formula for $(k \circ h)(x)$.

9. $f: x \to x^3$, $g: x \to x+1$ and $h: x \to 2x$ are mappings on the set of real numbers.
Find a formula for the image of y under $h \circ g \circ f$.
Calculate $(f \circ g \circ h)(-5)$.

10. $f: x \to x^2$, $g: x \to x$ and $h: x \to x+2$ are mappings on the set of real numbers.
Show that $(f \circ g \circ h)k - (h \circ g \circ f)k = 2(2k+1)$.

11. $f: x \to \dfrac{1}{x-1}$ with domain $\{x : x \in R, x \neq 1\}$

and $g: x \to \dfrac{1}{x}$ with domain $\{x : x \in R, x \neq 0\}$

Find a formula for $(g \circ f)(x)$ and state the domain and range of $g \circ f$.
Why can we not find a formula for $(f \circ g)(x)$?

12. I, f and g are functions on the set of real numbers defined by $I: x \to x$, $f: x \to -x$ and $g: x \to \dfrac{1}{x}$, $x \neq 0$.
(a) Write down the composite mappings defined by
(i) $(I \circ I)(x)$ (ii) $(I \circ f)(x)$ (iii) $(f \circ I)(x)$
(iv) $(I \circ g)(x)$ (v) $(g \circ I)(x)$
(b) Calculate $(I \circ I)(-3)$, $(I \circ f)(-3)$, $(f \circ I)(-3)$.
(c) Which of the following statements are true?
(i) $(I \circ f \circ g)(x) = (f \circ g)(x)$
(ii) $(I \circ I)(x) = I(x)$
(iii) $(g \circ g)(x) = I(x)$, $x \neq 0$
(iv) $(f \circ f)(x) = I(x)$
(d) If we write $(I \circ I)(x) = I^2(x)$ and $(f \circ f)(x) = f^2(x)$, etc. Write down the set of values of $n \in N$ for which $f^n(x) = I(x)$.
Does $g^5(x) = I(x)$, $x \neq 0$?

13. f and g are two functions defined by $f: x \to x^2$ with domain R and $g: x \to (x+1)^2$ with domain R. Find $f \circ g$ and $g \circ f$ and state the domain of each.

14. If $f: x \to (x+1)$, $x \in R$ and $g: x \to \dfrac{1}{x}$, $x \neq 0$, $x \in R$, which of $f \circ g$ and $g \circ f$ cannot be found and why?

15. f and g are the functions $f: x \to (x+1)^2$, $x \in R$ and $g: x \to \dfrac{1}{x}$, $x \in R$, $x \neq 0$. Find $f \circ g(x)$ and say why $g \circ f(x)$ does not exist.

3.3 INVERSE FUNCTIONS

Given a function $f: A \to B$ there exists a function $g: B \to A$, which reverses the mapping f, if f is a one-to-one correspondence.

Note:

(i) In this case f and g are called **inverse functions**.
(ii) g is called the inverse of f and is denoted by f^{-1} (f inverse).
(iii) f is the inverse of g.
(iv) g is also a one-to-one correspondence.
(v) The range of f is the domain of g, and the domain of f is the range of g.

(vi) $(f \circ f^{-1})(x) = I(x)$ and $(f^{-1} \circ f)(x) = I(x)$.

(vii) Given a function $f: A \to B$, it is not always possible to find a function g from B to A which reverses the mapping f.

Example 1

$f: x \to 2x$, $x \in R$, has inverse $f^{-1}: x \to \frac{x}{2}$, since it is evident from figure 12 that to reverse the effect of f we must divide $f(x)$ by 2.

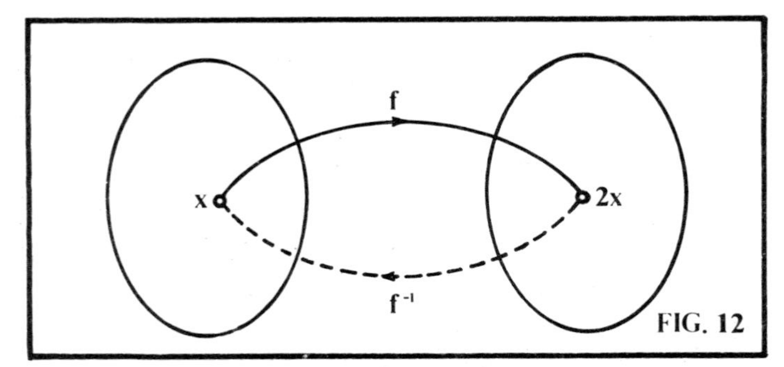

FIG. 12

Note that $(f^{-1} \circ f)(x) = f^{-1}[f(x)] = f^{-1}(2x) = x$

and $(f \circ f^{-1})(x) = f[f^{-1}(x)] = f\left(\frac{x}{2}\right) = x$

Compare inverses with: In the morning we *pull on our socks, put on our shoes* and *tie the laces*. To undo this we *untie the laces, take off our shoes*, then *pull off our socks*.

Hence the inverse of *"pull on, put on, tie"* is *"untie, take off, pull off"*, in that order.

Example 2

If the function $f: R \to R$ is defined by $f(x) = 2x + 3$, find f^{-1}.

[i.e. $f: x \to 2x + 3$, where $x \in R$]

$f(x) = 2x + 3$ means "we first double x and then add 3."

$\therefore$ Inverse is "we first subtract 3 from x and then half the result."

$\therefore$ $f^{-1}(x) = \frac{1}{2}(x - 3)$

Hence if $f(x) = 2x + 3$, then $f^{-1}(x) = \frac{1}{2}(x - 3)$.

Finding the Inverse Function

Note: The inverse operation of *adding* is *subtracting*, the inverse operation of *multiplying* is *dividing* and the inverse operation of *inverting* a fraction is *inverting* the fraction.

Example 3

To x add 3 giving $x + 3$; inverse is "from x subtract 3 giving $x - 3$."

Multiply x by 4 giving $4x$; inverse is "divide x by 4 giving $\frac{x}{4}$."

Invert x giving $\frac{1}{x}$; inverse is "invert x giving $\frac{1}{x}$."

Example 4

If $f: x \to \frac{x}{x - 3}$, where $x \in R$, $x \neq 3$, find f^{-1}

METHOD I

Let $$f: x \to \frac{x}{x - 3} = 1 + \frac{3}{x - 3}$$

To obtain the image of x under f we have to perform the following operations:

"from x subtract 3, invert the result, multiply by 3 and add 1."

To find f^{-1}, we have to perform the inverse operations, namely, "from x subtract 1, divide by 3, invert the result and then add 3."

Hence:

$$x \to x-1 \to \frac{x-1}{3} \to \frac{3}{x-1} \to \frac{3}{x-1} + 3 = \frac{3+3x-3}{x-1} = \frac{3x}{x-1}$$

hence $\quad f^{-1}: x \to \frac{3x}{x-1}, x \in R, x \neq 1$

METHOD II

$$f: x \to \frac{x}{x-3}, x \neq 3$$

Let $\quad \frac{x}{x-3} = a$ and solve for x

$$\Leftrightarrow \quad x = xa - 3a$$

$$\Leftrightarrow 3a = xa - x$$

$$= x(a-1)$$

$$\Rightarrow \quad x = \frac{3a}{a-1}, a \neq 1$$

hence $\quad f^{-1}: a \to \frac{3a}{a-1}$

$$\therefore \ f^{-1}: x \to \frac{3x}{x-1}, x \neq 1$$

Now check that $(f \circ f^{-1})(x) = x$ and $(f^{-1} \circ f)(x) = x$.

ASSIGNMENT 3.3

All the functions in this assignment are on the set of real numbers, unless otherwise stated.

1. Find the inverse of the functions.

   (i) $f: x \to 2x$ (ii) $f: x \to 3x+1$

   (iii) $g: x \to 2(x+2)$ (iv) $g: y \to -y$

   (v) $h: t \to -2t$ (vi) $h: s \to 1-2s$

2. $h: x \to 3x-2$ and $k: x \to \frac{x+2}{3}$

   By finding the composite functions $(h \circ k)$ and $(k \circ h)$, show that $h^{-1} = k$ and $k^{-1} = h$.

3. Functions $f: A \to B$ and $g: B \to C$ are defined by the arrow diagram in figure 13.

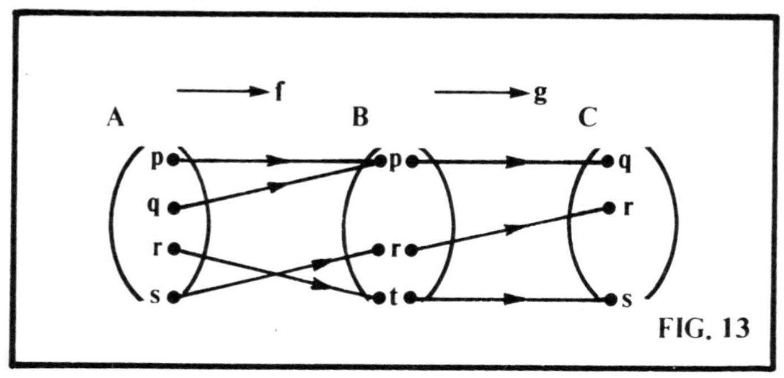

FIG. 13

   Write down, if they exist,

   (i) The image of p in B under f.

   (ii) The image of t in A under f^{-1}.

   (iii) The image of s in B under g^{-1}.

   Write down, if they exist, the values of

   (iv) $(f \circ g)(r)$

   (v) $(g \circ f)(r)$

   (vi) $(g \circ g^{-1})(s)$

   (vii) $(f^{-1} \circ g^{-1})(s)$

   (viii) $(g^{-1} \circ f^{-1})(r)$

4. Find the inverse of the functions defined by the following formulae, if the inverse exists. State the domain and range of the inverse function.

   [*Note:* If f^{-1} is the inverse of f then the domain of f

is the range of f^{-1} and the range of f is the domain of f^{-1}.]

(i) $f(x) = x-1$ (ii) $f(x) = 1-x$

(iii) $f(x) = x$ (iv) $f(x) = x^2$

(v) $f(x) = \frac{1}{x}$, $x \neq 0$ (vi) $f(x) = 1-3x$

(vii) $f(x) = \frac{1}{2}x-3$ (viii) $f(x) = \frac{1}{1-x}$, $x \neq 1$

(ix) $f(x) = x^3-1$

5. Functions g and h are defined as

$$g: x \to 2x+3,\ x \in R \text{ and } h: x \to \frac{1}{1-x},\ x \in R,\ x \neq 1$$

Show that $(g \circ h)(x) = \dfrac{5-3x}{1-x}$, $x \neq 1$

Find formulae for $g^{-1}(x)$ and $h^{-1}(x)$ and state values of x (if any) for which g^{-1} and h^{-1} are not defined.

6. $f: x \to x^2$, $g: x \to x+2$, $h: x \to 2x$ are mappings. Find a formula for the image of y under
   (i) $h \circ g \circ f$
   (ii) $h^{-1} \circ g^{-1}$.
   Calculate $(h \circ g \circ f)(-3)$ and $(h^{-1} \circ g^{-1})(4)$.

7. Find the inverses f^{-1} and g^{-1} of the functions $f: x = 2x+1$ and $g: x \to x-2$.
   Find a formula for $g \circ f$, and show that

$$(g \circ f)^{-1} = f^{-1} \circ g^{-1}.$$

8. Functions f and g are defined by $f: x \to \dfrac{1}{1-x^3}$, $x \neq 1$ and $g: x \to x-1$. Find the inverse functions f^{-1} and g^{-1}. For what values of x does f^{-1} or g^{-1} not exist?

   Calculate the image of $\frac{1}{9}$ under the inverse mapping $g^{-1} \circ f^{-1}$.

9. Which of the following functions have inverses? If the inverse function exists, find it and state its domain and range.
   (i) $f: x \to x^2$
   (ii) $g: x \to 2x+1$
   (iii) $h: x \to \dfrac{x+1}{x-1}$, $x \neq 1$
   (iv) $k: x \to |x|$

10. Show that

$$\text{if } f: x \to \frac{x+1}{x-1},\ x \neq 1 \text{ then } f^{-1}: x \to \frac{x+1}{x-1},\ x \neq 1.$$

   Check this by showing that $f \circ f^{-1}(x) = f^{-1} \circ f(x) = x$.

11. Find the inverse of the function $f: x \to \dfrac{x-1}{x+1}$, $x \neq -1$.

   For what value of x is $f^{-1}(x)$ not defined?

ASSIGNMENT 3.4

Objective items testing Sections 3.1–3.3

Instructions for answering these items are given on page 22.

1. The functions f and g are defined by $f: x \to 2-x$ and $g: x \to x-2$, $x \in R$. Which one of the following statements is **false**?

   A. $g = -f$
   B. $g = f^{-1}$
   C. $g \circ f(x) - f \circ g(x) = -4$
   D. $f^{-1} = f$
   E. $f(2) = g(2)$

2. If $f: x \to 5x$ and $g: x \to x-3$, $x \in R$, then $f \circ g: x \to$

A. $5x-15$

B. $5x-3$

C. $5x+3$

D. $3-5x$

E. $3x-5$

3. If $f: x \to x+1$ and $g: x \to (x-1)^2$ with domain R, then the function $h: x \to (x+1)^2$ is given by

A. $h = g \circ g \circ f$

B. $h = g \circ f \circ g$

C. $h = f \circ f \circ g$

D. $h = g \circ f \circ f$

E. $h = f \circ g \circ f$

4. If f is the function defined by $f: x \to 3-x$, $x \in R$, then the inverse function f^{-1} is

A. $\dfrac{1}{3-x}$

B. $-\dfrac{1}{3-x}$

C. $3x-1$

D. $3-x$

E. $3+x$

5. If $f: x \to 1-x^2$, $x \in R$ and $g: x \to \dfrac{1}{x}$, $x \in R$, $x \neq 0$, then the solution set of the equation $(f \circ g)(x) = \frac{3}{4}$ is

A. $\{0\}$

B. $\{2, -2\}$

C. $\{\frac{1}{2}, -\frac{1}{2}\}$

D. $\varnothing$

E. $\left\{\dfrac{\sqrt{3}}{2}, -\dfrac{\sqrt{3}}{2}\right\}$

6. Functions f and g with domains the set of real numbers are defined by $f: x \to x^2$ and

$$\begin{cases} g: x \to 2x+1, & \text{if } x \geqq 0 \\ g: x \to x, & \text{if } x < 0 \end{cases}$$

Find a formula for $(g \circ f)(x)$.

A. x^2

B. $(2x+1)^2$

C. $2x^2+1$

D. $2x^2+1$ for $x \geqq 0$ and x^2 for $x < 0$

E. No formula exists.

7. For which values of $x \in R$ is the function

$$f: x \to \frac{1}{\sqrt{(4-x^2)}}$$

defined?

A. All x except $x = 2$ and $x = -2$

B. $x > 2$ only

C. $x < -2$ only

D. $x > 2$ and $x < -2$ only

E. $-2 < x < 2$ only

8. If $f: x \to \dfrac{x+2}{2}$ and $g: x \to 2x-1$, $x \in R$, then $(g \circ f)^{-1}: x \to$

A. $x+1$

B. $x-1$

C. $1-x$

D. $\frac{1}{2}(2x+1)$

E. $\frac{1}{2}(2x-1)$

9. Consider the graphs in figure 14.

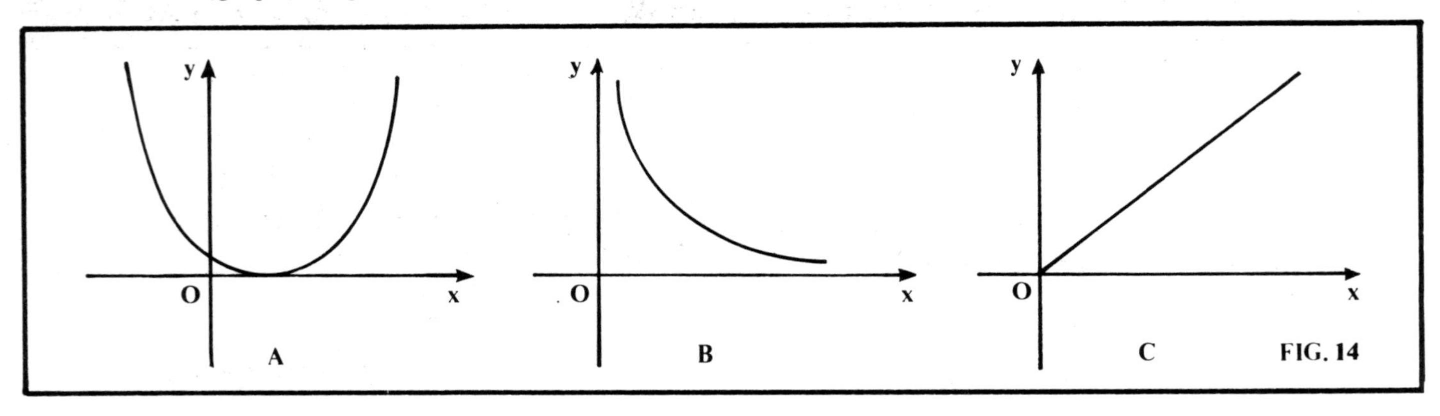

FIG. 14

(1) (a), (b), (c) all illustrate functions which have inverses.

(2) Only (b) illustrates a function which has an inverse.

(3) Only (a) illustrates a function which has no inverse.

10. A function f is defined by

$$f(x) = \frac{3}{x^2 - 4}, (x \in R, x \neq 2 \text{ or } -2).$$

(1) The function will have an inverse.

(2) $x = -1$ has image -1 under f.

(3) $x = 2$ and $x = -2$ are not members of the domain of f.

11. (1) $f(x) = x^n - 1$, n a positive integer, $x \in R$.

(2) $f^{-1}(x) = \sqrt[n]{(x+1)}$.

12. (1) f, g and h are one to one correspondences such that $h = f \circ g$.

(2) $h^{-1} = f^{-1} \circ g^{-1}$.

UNIT 4: POLYNOMIALS AND APPLICATIONS 

POLYNOMIALS: EVALUATION OF A POLYNOMIAL

4.1

A **polynomial in x** is an expression involving the algebraic sum of powers of x. The **degree** of the polynomial is the index of the highest power of x in the polynomial.

Example 1

(i) x^2+6x-2 is a polynomial of degree 2.

(ii) $2x^5-6x^3+11$ is a polynomial of degree 5.

Note that in a Polynomial

(a) All the powers of the indices $\in W$.

(b) The number in front of each power of x is called the coefficient of that term.

In (i) 6 is the coefficient of the term in x
-2 is called the constant term.

In (ii) 2 is the coefficient of the term in x^5
-6 is the coefficient of the term in x^3, etc.

Two polynomials are equal $\Leftrightarrow$ they have the same degree and their corresponding coefficients are equal.

Example 2

$$2x^3+5x^2-3x+7 = 2x^3+ax^2-3x+b$$
$$\Leftrightarrow a = 5 \text{ and } b = 7$$

An expression of the form $(x-1)^2(2+x)-5$ is also a polynomial since it can be written as x^3-3x-3.

Example 3

For what values of p and q are the polynomials $(x+3)(px^2+qx+2)$ and x^3-7x+6 equal?

If the polynomials are equal then,

$$\begin{aligned} x^3-7x+6 &= (x+3)(px^2+qx+2) \\ &= x(px^2+qx+2)+3(px^2+qx+2) \\ &= px^3+qx^2+2x+3px^2+3qx+6 \\ &= px^3+(q+3p)x^2+(2+3q)x+6 \end{aligned}$$

This requires,

$$p = 1 \quad \ldots\ldots\ldots\ldots\ldots\ldots (1)$$
$$q+3p = 0 \quad \ldots\ldots\ldots\ldots\ldots\ldots (2)$$
$$2+3q = -7 \ldots\ldots\ldots\ldots\ldots\ldots (3)$$

From (1) and (2), $p = 1$ and $q = -3$ and these values of p and q satisfy equation (3).

Hence the polynomials are equal if $p = 1$ and $q = -3$.

EVALUATING A POLYNOMIAL

We often denote a polynomial in x by $f(x)$.

Example 4

If $\quad f(x) = x^2-x+5$

then $f(2) = 2^2-2+5 = 7$ is the value of the polynomial when $x = 2$.

Example 5

Calculate the value of $3x^3-8x^2+5x-6$ when $x = 8$.

Let $\quad f(x) = 3x^3-8x^2+5x-6$

Hence $\quad f(8) = 3(8)^3-8(8)^2+5(8)-6$

$= 1058$

An easier way to find $f(8)$ is shown below

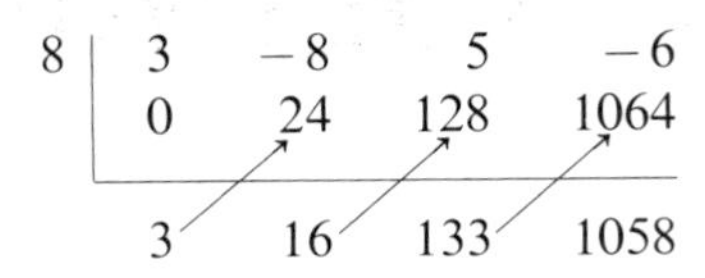

8	3	−8	5	−6
	0	24	128	1064
	3	16	133	1058

$\therefore\ f(8) = 1058$

Note: First row contains the coefficients of every power of x in descending order. If a particular power is missing, i.e. if its coefficient is zero, then a zero must be inserted in the corresponding position in the first row. Each arrow indicates multiplication by 8 followed by an addition.

ASSIGNMENT 4.1

1. State the degree of each of the following polynomials:
   (i) x^3-3x^2+2 (ii) $4-6x+2x^2-x^3$
   (iii) x^4+5 (iv) $(x-2)^2$
   (v) $(x-2)(x^2+4x)+9$

2. In each of the polynomials give the coefficient of the terms stated
   (i) x^3-2x+4—state the coefficient of x.
   (ii) $(2x+3)^2$—state the coefficient of x^2.
   (iii) $(x+2)(2x-1)$—state the constant term.
   (iv) $3+2x-x^3$—state the coefficients of x^2 and x^3.

3. For what values of a, b, c, p, q are the following polynomials equal
   (i) $3x+2 = px+2$
   (ii) $x^3-2x^2+x-3 = x^3+px^2+x-q$

(iii) $ax^2+bx+c=(x-1)^2$

(iv) $(x+2)(x^2+bx+c)=x^3+8$

4. Find $f(2)$ if $f(x)=x^3-2x^2+x-1$
5. Find $f(-1)$ if $f(y)=y^4-y^3-y^2+y$
6. Find $f(\frac{1}{2})$ if $f(x)=2x^3-4x^2+6x-1$
7. Find $f(7)$ if $f(u)=u^3-8u^2+9u+4$

Calculate:

8. $g(-3)$ when $g(x)=x^4-3x^3+8x^2-1$
9. $F(-2)$ when $F(z)=z^5+4z^4-4z^3+5$
10. $f(2k)$ when $f(x)=x^3-3x^2+8x-4$
11. If $f(x)=2x^3+8x^2-ax+4$ and $f(-3)=7$, calculate the value of a.
12. If $f(x)=x^3+px^2+3x-q$ and $f(-1)=1$ and $f(2)=7$, find p and q.
13. The polynomial $f(x)$ can be written as
$$f(x)=(x-2)(3x+1)+12$$
find $f(2)$, $f(-\frac{1}{2})$ and $f(0)$.
14. Show that for the polynomial
$$f(x)=2x^3-x^2-13x+4, \quad f(-2)=f(-\tfrac{1}{2})=f(3).$$
15. If $f(x)=2x^2+6x-5$, show that $f(2)=f(-5)$.
Find the value of a for which $f(a+2)=f(a-2)$ and show that for this value of a, $f(a+k)=f(a-k)$ for all values of k.
16. If $f(a)=2a^2-3a+1$, solve the equation $f(2a)=f(3a)$.
17. If $g(k)=k^2-k+1$, show that $g(1-k)=g(k)$.

DIVISION OF A POLYNOMIAL

Example 1

Divide $4x^2-3x+2$ by $x-2$. We can set this out as follows:

$$\begin{array}{r|l} & 4x\ +5 \\ \hline x-2 & 4x^2-3x+2 \\ & 4x^2-8x \\ \hline & \qquad 5x+2 \\ & \qquad 5x-10 \\ \hline & \qquad\quad 12 \end{array}$$

We call $x-2$ the divisor,
$4x+5$ the quotient
and 12 the remainder.

The division ends when a remainder is reached which is of lower degree than the divisor. In the above example division ends since 12 is of lower degree than $x-2$.

Note that since we are only dealing with divisors of the form $x-h$, where h may be positive or negative the remainder will always be a constant, i.e. will not contain x.

Example 2

If $f(x)=4x^2-3x+2$

then $f(x)=(x-2)(4x+5)+12$, so that if a function defined by $f(x)$ is divided by $x-2$, then $f(x)=(x-2)Q(x)+R$, where the quotient $Q(x)$ is another polynomial.

DIVISION OF ax^3+bx^2+cx+d by $x-h$

Denote $f(x)=ax^3+bx^2+cx+d$, then our result must be of the form $f(x)=(x-h)Q(x)+R$. We have to find $Q(x)$ and R.

$$
\begin{array}{r|l}
 & ax^2+(ah+b)x \ +(ah^2+bh+c) \\
\hline
x-h & ax^3+\ bx^2+\ cx+\quad d \\
 & ax^3-ahx^2 \\
 & (ah+b)x^2+\ cx \\
 & (ah+b)x^2-(ah^2+bh)x \\
 & \qquad (ah^2+bh+c)x+d \\
 & \qquad (ah^2+bh+c)x-(ah^3+bh^2+ch) \\
 & \qquad\qquad ah^3+bh^2+ch+d
\end{array}
$$

Hence, quotient $Q(x) = ax^2+(ah+b)x+(ah^2+bh+c)$ and remainder $R = ah^3+bh^2+ch+d$

Note the coefficients in the quotient, and that the remainder is $f(h)$.

A quicker method to find $f(h)$ is the method used at the end of 4.1.

$$
\begin{array}{c|cccc}
h & a & b & c & d \\
 & 0 & ah & ah^2+bh & ah^3+bh^2+ch \\
\hline
 & a & ah+b & ah^2+bh+c & ah^3+bh^2+ch+d=f(h)
\end{array}
$$

The quotient is $ax^2+(ah+b)x+(ah^2+bh+c)$ and the remainder is $f(h) = ah^3+bh^2+ch+d$ when the polynomial $f(x) = ax^3+bx^2+cx+d$ is divided by $x-h$. This method of calculation is known as **synthetic division.**

This last piece of work is called **The Remainder Theorem** which states that, "If a polynomial $f(x) = ax^3+bx^2+cx+d$ is divided by $(x-h)$ the final remainder is

$$f(h) = ah^3+bh^2+ch+d.$$

Remainders

(i) Dividing $f(x)$ by $x-h$, the remainder is $f(h)$.

(ii) Dividing $f(x)$ by $x+h$, the remainder is $f(-h)$, for $x+h = x-(-h)$.

(iii) Dividing $f(x)$ by $2x-1$, the remainder is $f(\frac{1}{2})$.

(iv) Dividing $f(x)$ by $3x+2$, the remainder is $f(-\frac{2}{3})$.

(v) Dividing $f(x)$ by $ax\pm b$, the remainder is $f\left(\mp\frac{b}{a}\right)$.

Example 3

Find the remainder when $2x^3+x^2+3$ is divided by $x+2$.

Let $f(x) = 2x^3+x^2+3$.

Put $x = -2$, then

$$f(-2) = 2(-2)^3+(-2)^2+3 = -16+4+3 = -9.$$

or

$$
\begin{array}{c|cccc}
-2 & 2 & 1 & 0 & 3 \\
 & 0 & -4 & 6 & -12 \\
\hline
 & 2 & -3 & 6 & -9
\end{array}
= f(-2). \text{ Remainder} = -9.
$$

ASSIGNMENT 4.2

Find the quotient and remainder on dividing,

1. $8x-7$ by $x+1$
2. u^2-2u+4 by $u-1$
3. $2x^2+4x-3$ by $x+2$
4. $4x^2-8x+5$ by $x-2$
5. $6y^2+y+3$ by $2y+1$
6. x^3-2x^2+3x+1 by $x+2$
7. $4x^3+4x^2-4x-2$ by $2x-1$

If on dividing $f(x)$ by $(x-h)$ the quotient is $Q(x)$ and the remainder is R, by finding $Q(x)$ and R write the following polynomials in the form

$$f(x) = (x-h)Q(x)+R.$$

8. $f(x) = x^3-2x^2+6x-1$ on division by $x-2$.
9. $f(x) = 2x^3-x^2+2x+3$ on division by $x+3$.
10. $f(x) = x^3+kx^2-k^2x+k^3$ on division by $x-k$.
11. $g(t) = t^3-3t-2$ on division by $t+2$.
12. The polynomial x^3-px^2+qx-2 has remainder 4 on

division by $x+1$ and remainder 16 on division by $x-2$, find p and q.

13. The polynomial $f(x) = 2x^3-8x^2+ax-3$ has the same remainder on division by $2x-1$ and by $x-2$. Find a.

14. Divide $x^4+x^3-5x^2-x-7$ by $x+3$ using long division. Check your answer for the quotient and remainder by using synthetic division.

15. Without actual division show that,
    (i) $8x^9-9x^4+1$ is divisible by $x-1$
    (ii) x^3+3x^2-2x-8 is divisible by $x+2$

16. Find the remainder when $3x^3-4x^2+2x+6$ is divided by $x-1$, $x+2$ and $2x+1$ respectively.

17. Find the quotient and remainder on dividing,
    (i) $u^4-5u^2+6u-15$ by $u+3$
    (ii) $2y^4-y^3-2y^2+y-12$ by $y-2$

18. Find the value of k for which the polynomial $2x^4+3x^3-4x^2+kx+2$ is divisible by $x+2$ and show that when k has this value, the polynomial is also divisible by $2x-1$.

THE FACTOR THEOREM AND ROOTS OF A POLYNOMIAL

4.3

(a) If when $f(x)$ is divided by $(x-h)$, the remainder is zero, then $(x-h)$ is a factor of $f(x)$ and the converse is true, i.e. $(x-h)$ is a factor of $f(x) \Leftrightarrow f(h) = 0$.

(b) If $(x-h)$ is a factor of $f(x)$, i.e. if $f(h) = 0$, then $x = h$ is a root of the equation $f(x) = 0$.

Example 1

Show that $x-2$ is a factor of x^3-x^2-5x+6 and factorise.

$$\begin{array}{r|rrrr} 2 & 1 & -1 & -5 & 6 \\ & 0 & 2 & 2 & -6 \\ \hline & 1 & 1 & -3 & 0 = f(2) \end{array}$$

Remainder on division by $x-2$ is zero, $\Rightarrow x-2$ is a factor of $f(x)$.

Hence $x^3-x^2-5x+6 = (x-2)(x^2+x-3)$.

Example 2

Find the factors of x^3+2x^2-5x-6.

Let $f(x) = x^3+2x^2-5x-6$.

Start by trying ± 1, ± 2, ± 3, ± 6, i.e. factors of 6.

Try $x = -1$, $f(-1) = (-1)^3+2(-1)^2-5(-1)-6$
$$= -1+2+5-6$$
$$= 0$$

or Try $x = -1$,

$$\begin{array}{r|rrrr} -1 & 1 & 2 & -5 & -6 \\ & 0 & -1 & -1 & 6 \\ \hline & 1 & 1 & -6 & 0 = f(-1) \end{array}$$

Since $f(-1) = 0$, $x+1$ is a factor and

$$x^3+2x^2-5x-6 = (x+1)(x^2+x-6)$$
$$= (x+1)(x+3)(x-2)$$

Example 3

Solve the equation $2x^3-7x^2+7x-2 = 0$.

Let $f(x) = 2x^3-7x^2+7x-2$, then $f(x) = 0$.

Try $x = 1$, $f(1) = 2-7+7-2 = 0$.

$\Rightarrow x-1$ is a factor of $f(x)$.

Hence

$$2x^3-7x^2+7x-2 = (x-1)(2x^2-5x+2)$$
$$= (x-1)(x-2)(2x-1), \text{ and}$$

$$2x^3-7x^2+7x-2=0 \Leftrightarrow (x-1)(x-2)(2x-1)=0$$
$$\Rightarrow x=1, \text{ or } x=2 \text{ or } x=\tfrac{1}{2}$$

Thus solution set is $\{\frac{1}{2}, 1, 2\}$.

ASSIGNMENT 4.3

Reminders

(a) If $x-h$ is a factor of $f(x)$ then $f(h)=0$.

(b) If $f(h)=0$ then $(x-h)$ is a factor of $f(x)$.

(c) If $(x-h)$ is a factor of $f(x)$ then $x=h$ is a root of the equation $f(x)=0$.

(d) If $x=h$ is a root of $f(x)=0$, then $(x-h)$ is a factor of $f(x)$.

1. Show that $(x+2)$ is a factor of x^2-x-6.
2. Show that $(x-3)$ and $(x+4)$ are factors of $x^3+3x^2-10x-24$.
3. Which, if any, of $(x+1)$, $(x+2)$, $(x-1)$, $(x-4)$ are factors of $x^3-x^2-10x-8$?
4. Show $3x-2$ is a factor of $3x^2+10x-8$.
5. Show that $(2z-1)$ and $(2z+1)$ are factors of $4z^3-12z^2-z+3$.
6. Show that $u-3$ is a factor of $u^2+2u-15$ and write down the other factor.
7. For the polynomial $f(y)=2y^4+9y^3-4y^2-36y-16$, $f(2)=f(-4)=f(-\frac{1}{2})=0$, write down 3 factors of $f(y)$ and hence or otherwise show that $(y+2)$ is also a factor.
8. Form a polynomial $f(x)$ of degree 3 in x, such that $f(1)=f(-2)=f(3)=0$.
9. If $f(x)=a(b-c)x^2+b(c-a)x+c(a-b)$, which of $(x-1)$, $(x+1)$ are factors of $f(x)$?
10. For what value of p is $(x+1)$ a factor of x^3-2x^2+8x+p?
11. For what values of p and q are $(x-2)$ and $(x+2)$ factors of x^3-px^2+qx-1?

In questions 12–16, find all the factors of the polynomials.

12. $x^2-8x+15$.
13. $x^3-9x^2+23x-15$, given $x-5$ is a factor.
14. y^3-3y^2-y+3.
15. z^3+2z^2-5z-6.
16. $4t^3-8t^2-t+2$.
17. Show that $x-2$ and $x-3$ are factors of $x^4-4x^3-x^2+16x-12$ and find the other factors.
18. Factorise fully x^4-5x^2+4.
19. Factorise $f(x)=x^3+6x^2+11x+6$ and hence find the solution set of the equation $f(x)=0$.
20. Show that $x=2$ is a solution of the equation $2x^3+3x^2-11x-6=0$ and hence find the solution set of the equation.

In questions 21–23 solve the equations by first factorising the polynomial.

21. $x^3-7x-6=0$.
22. $x^3+2x^2-5x-6=0$.
23. $3x^3-4x^2-5x+2=0$.

24. For what value of k has the polynomial $f(x) = 2x^3 - 11x^2 + 17x + k$ zero remainder on division by $x - 2$?
Solve the equation $2x^3 - 11x^2 + 17x + k = 0$ for this value of k.

25. Show that $x = a$ is a solution of the equation $x^3 - 6ax^2 + 11a^2x - 6a^3 = 0$ and hence find the solution set of the equation.

26. For what value of p is $y + 5$ a factor of $2y^3 + py^2 - 2y + 15$?

27. For what values of k is $x + 1$ a factor of $f(x) = 9x^4 - 25x^2 - 10kx - k^2$? For what values of k is $x - 2$ a factor of $f(x)$?
Show that there is one value of k for which $(x+1)(x-2)$ is a factor of $f(x)$.

28. Verify that $2 - \sqrt{5}$ is a root of the equation $x^2 - 4x - 1 = 0$.

29. If $x^4 + ax^2 + b$ and $3x^2 + a$ have a common factor, prove that $9b = 2a^2$.

ASSIGNMENT 4.4

Objective items testing Sections 4.1–4.3

Instructions for answering these items are given on page 22.

1. The value of $5 - x - 2x^2 + x^3$ when x is -1 is

   A. -2
   B. 1
   C. 3
   D. 5
   E. 7

2. If $x^3 - 2x^2 + 3x + 1 = (x-2)(x^2+3) + 2p$ then the value of p is

   A. -6
   B. $-\frac{7}{2}$
   C. $\frac{5}{2}$
   D. $\frac{7}{2}$
   E. 7

3. Which of the following are both **not** factors of $2x^3 + x^2 - 5x + 2$?

   A. $(x+2)$ and $(x-2)$
   B. $(2x+1)$ and $(x-1)$
   C. $(x-2)$ and $(x+1)$
   D. $(2x+1)$ and $(2x-1)$
   E. $(x-1)$ and $(x+1)$

4. If $f(x) = (x+2)(x-2)$, then the equation $f(2x) = 0$ has solution set

   A. $\{4, -4\}$
   B. $\{2, -2\}$
   C. $\{1, -1\}$
   D. $\{0, 0\}$
   E. some other solution set.

5. The set of factors of $x^3 - 12x - 16$ contains which of the following

   I. $(x-2)$ II. $(x+2)$ III. $(x-4)$ IV. $(x+4)$

   A. I and III only
   B. I, II and III
   C. I and IV only
   D. II and III only
   E. I, II and IV.

6. The polynomial $f(x) = 5x^3 - 13x + 12$ when divided by $x-4$ can be written in the form $f(x) = (x-4)Q(x) + R$. Find the value of R.

   A. -360
   B. -256
   C. 280
   D. 384
   E. It cannot be found.

7. When the polynomial $3x^3 - 7x^2 - 15$ is divided by $x+3$ the remainder is

   A. 21
   B. 3
   C. -33
   D. -105
   E. -159

8. If $x = 2$ is a solution of the equation $2x^3 - x^2 - 5x - 2 = 0$ then the solution set of the equation is

   A. $\{2, -2, 1\}$
   B. $\{2\}$
   C. $\{2, 2, 2\}$
   D. $\{2, -1, \frac{1}{2}\}$
   E. $\{2, -1, -\frac{1}{2}\}$

9. $f(x^2) = x^4 - 3x^2 - 4$ is a polynomial in x^2
   (1) $f(4) = 0$.
   (2) $f(x^2)$ has linear factors $(x-2)$ and $(x+2)$.
   (3) a quadratic factor is x^2+1.

10. The function $f(x) = x^3 + x^2 - 8x + 4$ when divided by $(x-1)$ can be written as $f(x) = (x-1)Q(x) + R$ where $Q(x)$ is a polynomial in x and R is the remainder.
    (1) $R = -2$
    (2) $Q(x) = x^2 + 2x - 6$.
    (3) $f(1) = -2$.

11. (1) For the polynomial $f(x) = x^3 - 4x^2 + 4x - 2$, $f(1) = -1$ and $f(3) = 1$.
    (2) Equation $f(x) = 0$ has a solution $x = 2$.

12. (1) $x-2$ is a factor of the polynomial $f(x)$.
    (2) $x = 2$ is a root of the equation $f(x) = 0$.

4.4 APPROXIMATE SOLUTION OF EQUATIONS

This topic may be omitted until the relevant work on the calculus has been completed.

Newton's approximation to an irrational root of an equation

Let $x = a$ be an approximation to an irrational root of the equation $f(x) = 0$, then if $x = a_1$ is a better approximation we have that,

$$a_1 = a - \frac{f(a)}{f'(a)}$$

Similarly a better approximation than a_1 would be

$$a_2 = a_1 - \frac{f(a_1)}{f'(a_1)}$$

Example 1

Show that the equation $4x^3-6x^2-17x+9=0$ has a root between 2 and 3 and find a closer approximation to the root.

Let $f(x) = 4x^3-6x^2-17x+9$

$$\begin{aligned} f(2) &= 4(2)^3-6(2)^2-17(2)+9 \\ &= 32-24-34+9 \\ &= -26+9 \\ &= -17 \end{aligned}$$

$$\begin{aligned} f(3) &= 4(3)^3-6(3)^2-17(3)+9 \\ &= 108-54-51+9 \\ &= 117-105 \\ &= 12 \end{aligned}$$

The roots of an equation $f(x)=0$ are the values of x at the points where the graph of $f(x)$ crosses the x-axis. If $x=a$ is a root of $f(x)=0$ then the graph crosses the x-axis at the point where $x=a$ and *therefore* $f(a)=0$.

Hence as x passes through the value a, the graph passes, from below the x-axis to above the x-axis or from above the x-axis to below the x-axis.

Hence the value of $f(x)$ changes from negative to positive or the value of $f(x)$ changes from positive to negative, except at points where the graph of $f(x)$ touches the x-axis. We are only considering points where the graph actually crosses the x-axis.

Hence, in this case, there is a root between $x=2$ and $x=3$ and, from the figure 15, the root appears to be nearer $x=3$.

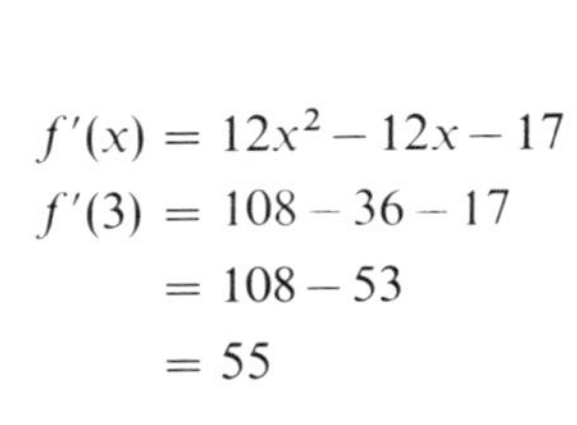

$$\begin{aligned} f'(x) &= 12x^2-12x-17 \\ f'(3) &= 108-36-17 \\ &= 108-53 \\ &= 55 \end{aligned}$$

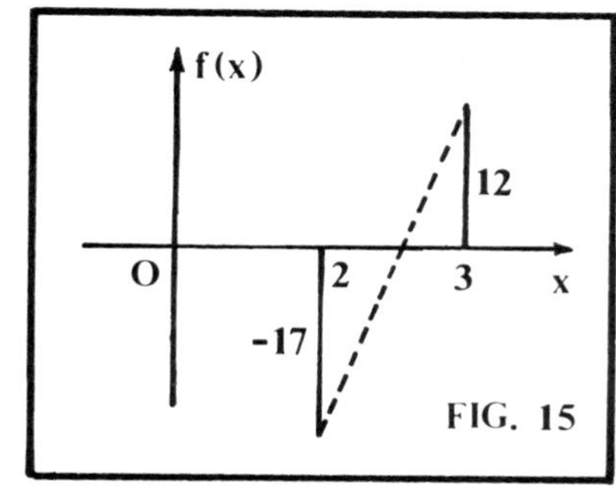

FIG. 15

Hence, if $x=a_1$ is a better approximation than $x=3$, we have

$$\begin{aligned} a_1 &= a - \frac{f(a)}{f'(a)} = 3 - \frac{f(3)}{f'(3)} = 3 - \tfrac{12}{55} \\ &= 3-0{\cdot}218 \\ &= 2{\cdot}782. \end{aligned}$$

Two difficulties may arise. These are shown in the following two examples.

Example 2

Show that the equation $x^3+10x^2-1=0$ has a root between $x=0$ and $x=1$ and find a closer approximation to this root.

Let
$$\begin{aligned} f(x) &= x^3+10x^2-1 \\ f(0) &= 0+0-1 = -1 \text{ and} \\ f(1) &= 1+10-1 = 10. \end{aligned}$$

$\Rightarrow$ There is a root between $x=0$ and $x=1$ and the root appears to be nearer $x=0$.

$$\begin{aligned} f'(x) &= 3x^2+20x \\ f'(0) &= 0 \end{aligned}$$

Hence $a_1 = a - \dfrac{f(a)}{f'(a)} = 0 - \dfrac{-1}{0}$ which has no value.

We must therefore try to narrow the interval within which the root lies.

Now half the interval between $x = 0$ and $x = 1$ is $x = \frac{1}{2}$, so it might be helpful to find $f(\frac{1}{2})$.

$$f(\tfrac{1}{2}) = \tfrac{1}{8} + \tfrac{10}{4} - 1 = \tfrac{21}{8} - 1 = \tfrac{13}{8}$$

$\Rightarrow$ the root does not lie between $x = \frac{1}{2}$ and $x = 1$.

Try $x = \frac{1}{4}$

$$f(\tfrac{1}{4}) = \tfrac{1}{64} + \tfrac{10}{16} - 1 = \tfrac{41}{64} - 1 = -\tfrac{23}{64}$$

$\Rightarrow$ the root lies between $x = \frac{1}{4}$ and $x = 1$.

In fact the root lies between $x = \frac{1}{4}$ and $x = \frac{1}{2}$ and appears nearer $x = \frac{1}{4}$.

Now $f'(\frac{1}{4}) = \frac{3}{16} + \frac{20}{4} = \frac{83}{16}$.

Hence a better approximation is

$$\begin{aligned} a_1 &= \tfrac{1}{4} - \frac{-\frac{23}{64}}{\frac{83}{16}} = \tfrac{1}{4} + \tfrac{23}{64} \times \tfrac{16}{83} \\ &= \tfrac{1}{4} + \tfrac{23}{332} \\ &= 0{\cdot}25 + 0{\cdot}069 \\ &= 0{\cdot}319. \end{aligned}$$

Example 3

Show that the equation $x^3 + x^2 - 1 = 0$ has a root between $\frac{1}{2}$ and 1 and evaluate the root correct to two decimal places.

Let
$$\begin{aligned} f(x) &= x^3 + x^2 - 1 \\ f(\tfrac{1}{2}) &= \tfrac{1}{8} + \tfrac{1}{4} - 1 = -\tfrac{5}{8} \text{ and} \\ f(1) &= 1 + 1 - 1 = 1 \end{aligned}$$

$\Rightarrow$ There is a root between $x = \frac{1}{2}$ and $x = 1$ and the root appears to be nearer $x = \frac{1}{2}$.

$$\begin{aligned} f'(x) &= 3x^2 + 2x \\ f'(\tfrac{1}{2}) &= \tfrac{3}{4} + 1 = \tfrac{7}{4} \end{aligned}$$

Hence a better approximation is

$$\begin{aligned} x &= \tfrac{1}{2} - \frac{-\frac{5}{8}}{\frac{7}{4}} \\ &= \tfrac{1}{2} + \tfrac{5}{8} \times \tfrac{4}{7} \\ &= \tfrac{1}{2} + \tfrac{5}{14} \\ &= 0{\cdot}5 + 0{\cdot}257 \\ &= 0{\cdot}857 \end{aligned}$$

But the value $x = 0{\cdot}857$ is nearer $x = 1$ than $x = \frac{1}{2}$ and contradicts our first assumption that it was nearer $x = \frac{1}{2}$.

We proceed therefore as we did in example 2, i.e. halve the interval between $x = \frac{1}{2}$ and $x = 1$, i.e. $x = \frac{3}{4}$ and find $f(\frac{3}{4})$.

$$\begin{aligned} f(\tfrac{3}{4}) &= \tfrac{27}{64} + \tfrac{9}{16} - 1 \\ &= \tfrac{27 + 36 - 64}{64} \\ &= -\tfrac{1}{64} \end{aligned}$$

Thus we have $f(\frac{3}{4}) = -\frac{1}{64}$ and $f(1) = 1$, so that the graph of $f(x)$ cuts the x-axis between $x = \frac{3}{4}$ and $x = 1$. i.e. the root is actually nearer $x = 1$ than $x = \frac{1}{2}$. In fact it is nearer $x = \frac{3}{4}$. Also

$$f'(\tfrac{3}{4}) = \tfrac{27}{16} + \tfrac{6}{4} = \tfrac{51}{16}$$

and a better approximation is

$$\begin{aligned} x &= \tfrac{3}{4} - \frac{-\frac{1}{64}}{\frac{51}{16}} \\ &= \tfrac{3}{4} + \tfrac{16}{51 \times 64} \\ &= \tfrac{3}{4} + \tfrac{1}{204} \\ &= 0{\cdot}75 + 0{\cdot}004 \\ &= 0{\cdot}754 \\ &= 0{\cdot}75 \text{ to two decimal places.} \end{aligned}$$

ASSIGNMENT 4.5

1. Show that the equation $x^3-x^2-5x+2=0$ has a root $x=-2$ and that the other two roots lie between 0 and 1, and 2 and 3.

2. Show that $x=-1$ is the only real root of the equation $x^3-x^2+3x+5=0$.

3. The equation $2x^3+7x^2-5x-4=0$ has two roots $x=1$ and $x=-\frac{1}{2}$. Find the third root.

4. Show that the equation $10x^3+x+6=0$ has a root between 0 and -1 and find the root correct to two decimal places.

5. A root of the equation $x^3-4x-1=0$ lies between $x=2$ and $x=3$. Show that it is nearer $x=2$ and find a better approximation to the root than $x=2$.

6. Find correct to two decimal places a positive root of the equation $x^3-6x^2+2=0$.

7. If $f(x)=x^3-2x^2-x+1$, show that $f(0)=1$ and $f(1)=-1$. Does a root of the equation $f(x)=0$ lie between 0 and $\frac{1}{2}$ or between $\frac{1}{2}$ and 1? Hence find the value of this root correct to two decimal places.

8. Find to two decimal places the root of the equation $x^3+3x-7=0$ which is near $x=1{\cdot}5$.

9. Verify that the equation $x^3+2x-5=0$ has a root between 1 and 1·5 and find it correct to two decimal places.

10. Show that one root of the equation $x^4-4x-7=0$ lies between -1 and -2 and another lies between 1 and 2 and find each root correct to two decimal places.

11. Show that the equation $x^3-3x-1=0$ has a root between 1 and 2 and find the root correct to two decimal places.
    If $f(x)=x^3-3x-1$ show that the equation $f(-x)=0$ has a root between -1 and -2 and find the root correct to two decimal places.

12. Show that the equation $x^3-4x+1=0$ has a root between 0 and 1. If $x=2y$ is substituted in this equation show that the resulting equation,
    (i) is $8y^3-8y+1=0$.
    (ii) has a root between 0 and $\frac{1}{2}$ and hence find this root correct to two decimal places.

ASSIGNMENT 4.6

Supplementary Examples

1. In a certain series the sum of the first n terms is given by the formula $S_n=\frac{1}{3}n(n+1)(n+2)$. Calculate S_1, S_2 and S_3 and hence find the first three terms of the series. Find an expression for u_n the nth term of this series.

2. u_r is the rth term of a sequence satisfying the recurrence relation $u_{r+1}=2u_r-3$. If $u_1=1$, find u_2, u_3, u_4 and hence or otherwise deduce that u_r may be written $3-2^r$.

3. The nth term of a geometric sequence is 3^{n-3}, what is
    (a) the first term?
    (b) the second term?
    (c) the value of p if the sum to p terms is $\frac{364}{9}$?

4. The first three terms of a geometric sequence are $2+3k$, $3k-2$ and $2-k$ respectively. Show that this is true for two values of k.

Find the sum to n terms of the resulting sequences and deduce that one sequence has a sum to infinity. Explain why the other sequence does not have a sum to infinity.

5. The second term of a geometric sequence exceeds the first term by $\frac{8}{9}$. If the first term is twice the common ratio show that there are two possible sequences but only one has a sum to infinity, and find this sum to infinity.

6. If $A = \begin{pmatrix} -2 & -5 \\ 1 & 3 \end{pmatrix}$ and $B = \begin{pmatrix} 2 & -1 \\ 0 & -2 \end{pmatrix}$ find in simplest form.

   (i) A', A^{-1} and $3A$ (ii) B', B^{-1} and $-B$

   (iii) $A+B$ and $A-B$.

   Show that

   (v) $(AB)' = B'A'$ (vi) $(AB)^{-1} = B^{-1}A^{-1}$.

7. Show that the matrix $P = \begin{pmatrix} p+2 & p \\ q & q+2 \end{pmatrix}$ has no inverse if $p+q = -2$.

   If $2Q = QP$, where Q is the row matrix (p, q) show that $p^2+q^2 = 0$.

8. The matrix $R = \begin{pmatrix} \cos\theta & -\sin\theta \\ \sin\theta & \cos\theta \end{pmatrix}$ gives the mapping

   $P(x, y) \rightarrow P'(x', y')$ under a rotation of θ about the origin. Write down the matrix T which gives the mapping under a rotation of 45° about the origin. Show that

   $T^2 = \begin{pmatrix} 0 & -1 \\ 1 & 0 \end{pmatrix}$ and state what mapping this gives.

   Hence or otherwise find the least number $n \in N$ such that $T^n = I$.

9. The transformation represented by the matrix $M = \begin{pmatrix} 4 & -2 \\ 3 & -1 \end{pmatrix}$ maps a point (a, b) into the point $(-2, 4)$. Write this mapping as an equation in matrix form. By using M^{-1} the inverse of M find the point (a, b).

10. $f: x \rightarrow 2x+3$ and $g: x \rightarrow x^2$ are functions with domain R.

    (i) Find f^{-1}, g^{-1}, $f \circ g$ and $g \circ f$ if they exist.

    (ii) Find $f^{-1}(3)$, $(f \circ g)(0)$ and $(g \circ f)(-2)$.

11. The functions g and f with domain R the set of real numbers are defined by $g: x \rightarrow x^2+1$ and $f: x \rightarrow x-1$. Find in its simplest form $(g \circ f)(k)-(f \circ g)(k)$. Solve the equation $(g \circ f)(x)+(f \circ g)(x) = 6$.

12. $f: x \rightarrow x-2$ with domain R and $g: x \rightarrow \dfrac{x}{x-1}$ with domain $\{x: x \in R, x \neq 1\}$.

    (i) Obtain a formula for $(f \circ g)(x)$ and $(g \circ f)(x)$ if they exist or say why if they do not exist.

    (ii) By evaluating $(g \circ g)(x)$ or otherwise, prove that g is its own inverse.

13. The functions $h: R \rightarrow R$ and $k: R \rightarrow R$ are defined by $h(x) = 2x+1$ and $k(x) = x^2-x-1$.

    (i) Find $(h \circ k)(x)$ and $(k \circ h)(x)$ and hence find replacements for x such that $(h \circ k)(x) = (k \circ h)(x)$.

    (ii) Find $(h \circ h)(x)$ and $(h \circ h \circ h)(x)$ and hence if $(h \circ h)(x)$ and $(h \circ h \circ h)(x)$ are written $h^2(x)$ and $h^3(x)$ prove that $h^n(x) = 2^n x+2^n-1$.

14. A function f is defined by $f(x) = x^3+px+q$. Where p and q are constants. When $f(x)$ is divided by $x-2$ the remainder is 2 and when divided by $x+2$ the remainder is -14. Find p and q and hence solve the equation $f(-3x) = -4$.

UNIT 5: THE QUADRATIC EQUATION AND QUADRATIC FUNCTION

THE GENERAL QUADRATIC EQUATION: NATURE OF THE ROOTS

5.1

$ax^2+bx+c=0\ldots(1)$ is the general quadratic equation.

$$
\begin{aligned}
& ax^2+bx+c=0 \text{ is true} \\
\Leftrightarrow \quad & x^2+\frac{b}{a}x+\frac{c}{a}=0 \\
\Leftrightarrow \quad & x^2+\frac{b}{a}x=-\frac{c}{a} \\
\Leftrightarrow \quad & x^2+\frac{b}{a}x+\left(\frac{b}{2a}\right)^2=\frac{b^2}{4a^2}-\frac{c}{a} \\
\Leftrightarrow \quad & \left(x+\frac{b}{2a}\right)^2=\frac{b^2-4ac}{4a^2} \\
\Leftrightarrow \quad & x+\frac{b}{2a}=\frac{\pm\sqrt{(b^2-4ac)}}{2a} \\
\Leftrightarrow \quad & x=-\frac{b}{2a}\pm\frac{\sqrt{(b^2-4ac)}}{2a}
\end{aligned}
$$

Hence roots are $x=\dfrac{-b\pm\sqrt{(b^2-4ac)}}{2a}$.

This method of finding the formula for the roots of a quadratic equation is called "completing the square".

NATURE OF THE ROOTS OF A RATIONAL QUADRATIC EQUATION

A quadratic equation has two roots given by,

$$x = \frac{-b+\sqrt{(b^2-4ac)}}{2a} \quad \text{and} \quad x = \frac{-b-\sqrt{(b^2-4ac)}}{2a}$$

Consider the term under the root sign, namely b^2-4ac which is called the **Discriminant** of ax^2+bx+c

(i) If $b^2-4ac > 0$, the roots of equation (1) are **real** and **unequal** (they may be rational or irrational).

Example 1

(a) $x^2-4x+3=0$

$$b^2-4ac = 16-12 = 4$$

$$\therefore x = \frac{4\pm\sqrt{4}}{2}$$

Roots are **real, unequal,** and **rational,** since the square root of 4 can be found exactly.

(b) $2x^2-3x-4=0$

$$b^2-4ac = 9+32 = 41$$

$$\therefore x = \frac{3\pm\sqrt{41}}{4}$$

Roots are **real, unequal,** and **irrational,** since the square root of 41 cannot be found exactly.

Hence, if the discriminant is positive *and* the square of a rational number the roots are real, unequal and rational, but if positive and *not* the square of a rational number they are real, unequal and irrational.

(ii) If $b^2-4ac=0$, the roots are **real, equal** and **rational.**

Example 2

$2x^2+12x+18=0$

$b^2-4ac = 144-144 = 0$

$$\therefore x = \frac{-12\pm 0}{4} = -3$$

Two equal roots $x=-3$.

(iii) If $b^2-4ac<0$, there are **no real** roots.

Example 3

$x^2-2x+5=0$

$b^2-4ac = 4-20 = -16$

$$\therefore x = \frac{2\pm\sqrt{-16}}{2}$$

$\therefore$ No real roots, since there is no real square root of -16.

Example 4

State the nature of the roots of the equation

$$x^2-3x-7=0.$$

We have to examine b^2-4ac, where $a=1$, $b=-3$, $c=-7$

$$\begin{aligned} b^2-4ac &= 9-4\times1\times(-7) \\ &= 9+28 \\ &= 37 \\ &> 0 \end{aligned}$$

$\therefore$ the roots are real and unequal.

Since 37 is not the square of a rational number the roots are real, unequal and irrational.

Example 5

If the equation $3x^2-8x+p=0$ has no real roots find the value of p.

For no real roots,

$$b^2-4ac<0$$

$$\Rightarrow 64 - 12p < 0$$
$$\Rightarrow \quad 64 < 12p$$
$$\Rightarrow \quad 12p > 64$$
$$\Rightarrow \quad p > \tfrac{64}{12} = \tfrac{16}{3}$$

Hence, if p is any number greater than $\frac{16}{3}$, the equation $3x^2 - 8x + p = 0$ will have no real roots.

Example 6

Find k given that the quadratic equation

$$kx^2 + (k-6)x + 2 = 0$$

has equal roots.

For equal roots $b^2 - 4ac = 0$, where $a = k$, $b = k-6$, $c = 2$.

$$\Rightarrow \quad (k-6)^2 - 4.k.2 = 0$$
$$\Rightarrow k^2 - 12k + 36 - 8k = 0$$
$$\Rightarrow \quad k^2 - 20k + 36 = 0$$
$$\Rightarrow \quad (k-2)(k-18) = 0$$
$$\Rightarrow \quad k = 2 \text{ or } 18$$

Example 7

Show that the expression $x^2 - 4x + 3$ has no rational factors but that the expression $x^2 - 3x - 10$ has, and find the rational factors of the second expression.

For $x^2 - 4x + 3$, $b^2 - 4ac = 16 - 6 = 10$.

But 10 is not the square of a rational number and hence the expression has no rational factors.

For $x^2 - 3x - 10$, $b^2 - 4ac = 9 + 40 = 49$, and the expression has rational factors since 49 is the square of ± 7, both of which are rational numbers.

Also since $x^2 - 3x - 10 = (x+2)(x-5)$, the rational factors are $(x+2)$ and $(x-5)$.

Example 8

x is a real number and the equation $k = \dfrac{1}{x^2 - x + 1}$ has real and distinct (i.e. unequal) roots. Within what interval must the value of k lie?

$$k = \frac{1}{x^2 - x + 1}$$
$$\Leftrightarrow \quad k(x^2 - x + 1) = 1$$
$$\Leftrightarrow kx^2 - kx + (k-1) = 0$$

For this equation to have real and distinct roots,

$b^2 - 4ac > 0$, where $a = k$, $b = -k$ and $c = k-1$

$$\Rightarrow k^2 - 4k(k-1) > 0$$
$$\Rightarrow k^2 - 4k^2 + 4k > 0$$
$$\Rightarrow \quad -3k^2 + 4k > 0$$
$$\Rightarrow \quad 3k^2 - 4k < 0$$
$$\Rightarrow \quad k(3k-4) < 0$$
$$\Rightarrow \quad 0 < k < \tfrac{4}{3}$$

Hence, for real and distinct roots $0 < k < \frac{4}{3}$.

Example 9

Find the equations of the lines through the point $(0, 5)$ which are tangents to the circle $x^2 + y^2 = 9$.

Any line through the point $(0, 5)$ with gradient m has equation $y = mx + 5$.

This line cuts or touches the circle $x^2 + y^2 = 9$ at points whose x-coordinates are given by the equation

$$x^2 + (mx+5)^2 = 9$$

i.e. $$x^2 + m^2x^2 + 10mx + 25 = 9$$

i.e. $$(1+m^2)x^2 + 10mx + 16 = 0$$

The line will be a tangent if and only if this equation has equal roots in x.

i.e.
$$\begin{aligned}
&\Leftrightarrow & b^2-4ac &= 0\\
&\Leftrightarrow & 100m^2-4(1+m^2).16 &= 0\\
&\Leftrightarrow & 100m^2-64-64m^2 &= 0\\
&\Leftrightarrow & 36m^2-64 &= 0\\
&\Leftrightarrow & 9m^2-16 &= 0\\
&\Leftrightarrow & m^2 &= \tfrac{16}{9}\\
&\Leftrightarrow & m &= \pm\tfrac{4}{3}
\end{aligned}$$

Hence the two tangents have equation $y=\frac{4}{3}x+5$ and $y=-\frac{4}{3}x+5$.

i.e. $3y-4x=15$ and $3y+4x=15$

ASSIGNMENT 5.1

1. Solve the following equations by completing the square where x is a variable on the set of real numbers. (Answers may be left in surd form).

   (i) $x^2+4x-9=0$ (ii) $2x^2-3x-2=0$

   (iii) $4x^2-12x+9=0$ (iv) $2x^2-3x-1=0$

2. Use the quadratic formula to solve the following equations where x is a variable on the set of real numbers. (Answers may be left in surd form).

   (i) $x^2-3x-5=0$ (ii) $x^2-7x+11=0$

   (iii) $2x^2-9x+7=0$ (iv) $2x^2-x+4=0$

   (v) $9x^2-12x+4=0$ (vi) $x^2+4x-7=0$

   (vii) $4x^2-20x+25=0$ (viii) $x^2+3x-28=0$

3. By examining the discriminant b^2-4ac state the nature of the roots of each of the following equations.

   (i) $x^2+5x+3=0$ (ii) $2x^2-x+1=0$

   (iii) $3x^2+x-4=0$ (iv) $x^2-3x+7=0$

   (v) $9x^2-12x+4=0$ (vi) $2x^2-4x-9=0$

   (vii) $x(x-3)=2$ (viii) $y^2-7y+10=0$

   (ix) $4x^2-12x+9=0$ (x) $t^2+2t-15=0$

   (xi) $2x^2+4x+3=0$ (xii) $9y^2+5y+2=0$

4. Each of the equations $x^2-8x+p=0$, $3x^2-8x+q=0$, $x^2-rx+25=0$ and $sx^2+10x+5=0$ has equal roots. Find the values of p. q, r and s.

5. Find k if $x^2+(k-1)x-(2k+1)=0$ has equal roots.

6. The equation $x+\dfrac{1}{x}=p$ has real roots. Find the values of p.

7. Find the greatest value of p for which the equation $x^2-6x+p=0$ has real roots.

8. For the equation $kx^2+x(5k+1)+4k=0$, show that $b^2-4ac=(9k+1)(k+1)$. For what values of k will the equation have real roots in x?

9. For what values of k will the equation $x^2-kx+(k+3)=0$ have,

   (i) equal roots? (ii) real roots?

10. Show that the equation $x^3-3x^2+5x-3=0$ has only one real root.

11. The equation $px^2+qx+r=0$ has no real roots. Write down the relation between p, q and r for this to be true.

    If $q^2=4pr$ what would be the nature of the roots of the equation $px^2+qx+r=0$?

12. The equation $4x^2+kx+4=0$ has real roots. Within what range must k lie?

13. $n = \dfrac{x^2 - 4x + 10}{5 - 2x}$ where $n, x \in R, x \neq \dfrac{5}{2}$.

Write this equation as a quadratic equation in x and show that if this quadratic equation has real roots in x then n cannot lie between -3 and 2.

14. For what range of values of p will the equation $p = \dfrac{x^2 - x - 1}{2 - x}$, $x \neq 2$, when written as a quadratic in x, have no real roots in x?

15. Which of the following expressions have rational factors? If rational factors exist, find them.

(i) $2x^2 + 2x - 1$ (ii) $2x^2 - x - 1$
(iii) $x^2 - x - 12$ (iv) $3x^2 + 4x + 1$
(v) $x^2 + 3x - 2$ (vi) $2x^2 + 5x - 1$

16. Show, by obtaining a quadratic equation in y which has equal roots, that the line $x = 3y + 10$ is a tangent to the circle $x^2 + y^2 = 10$.

17. For what values of m will the line $y = mx + 4$ be a tangent to the circle with equation $x^2 + y^2 = 8$.

18. The line $y = kx$ intersects the circle

$$x^2 + y^2 - 6x - 2y + 1 = 0$$

in two distinct points. Find the range of values of k. For what value of k would the line be a tangent to the circle?

19. Find the equations of the tangents from the point $(0, 3)$ to the curve with equation $x^2 + 2y^2 = 12$.

20. Form the equation whose roots are

(i) $5, -2$ (ii) $-7, -11$ (iii) $a + b, a - b$

21. The equation $x^2 + 2(1 + p)x + p^2 = 0$ has equal roots in x. What is the value of p?

22. Show that for $p \neq 0$, the roots of the equation $p(x^2 - 1) = (q - r)x$ are always real.

23. Which of the following expressions have rational factors and in the case of rational factors find them.

(i) $x^2 + 8x + 11$ (ii) $3x^2 + 4x - 5$
(iii) $3x^2 - 10x - 8$ (iv) $4x^2 - 12x + 9$
(v) $2x^2 + 5x + 2$

24. For what values of p has the equation $x^2 - 2px = 12 - 7p$ equal roots?

THE GRAPH OF $f(x) = ax^2 + bx + c$ 5.2

1. If $a > 0$ then the graph has a Minimum Turning Point. If $a < 0$ then the graph has a Maximum Turning Point.

2. Find the point where the graph crosses the y-axis by putting $x = 0$.

3. Find the points (if any) where the graph crosses the x-axis by putting $f(x) = 0$ and solving the equation $ax^2 + bx + c = 0$.
   If $b^2 - 4ac > 0$ the graph cuts the x-axis in two distinct points.
   If $b^2 - 4ac = 0$ the graph touches the x-axis.
   If $b^2 - 4ac < 0$ the graph does not cross the x-axis.

4. If we want to be more accurate we find the turning point, by calculus or by completing the square.

Example 1

Sketch the graph of $f(x) = x^2 + 5x + 7$.

1. $a = 1 > 0$, $\therefore$ a minimum turning point.

2. When $x = 0$, $f(x) = 7$, $\therefore$ point $(0, 7)$ lies on the graph.

3. When $f(x) = 0$, $x^2+5x+7 = 0$ has

$$b^2-4ac = 25-28 = -3$$

$\therefore$ No real roots and the graph does not cut the x-axis.

4. Turning Point:

$$\begin{aligned} f(x) &= x^2+5x+7 \\ &= x^2+5x+(\tfrac{5}{2})^2+7-\tfrac{25}{4} \\ &= (x+\tfrac{5}{2})^2+\tfrac{3}{4} \end{aligned}$$

Hence the minimum value of the function f defined by $f(x) = x^2+5x+7$ is $\frac{3}{4}$ and occurs when $x = -\frac{5}{2}$.

i.e. Minimum Turning Point is $(-\frac{5}{2}, \frac{3}{4})$. It also follows that the line $x = -\frac{5}{2}$ is an axis of symmetry.

Hence figure 16:

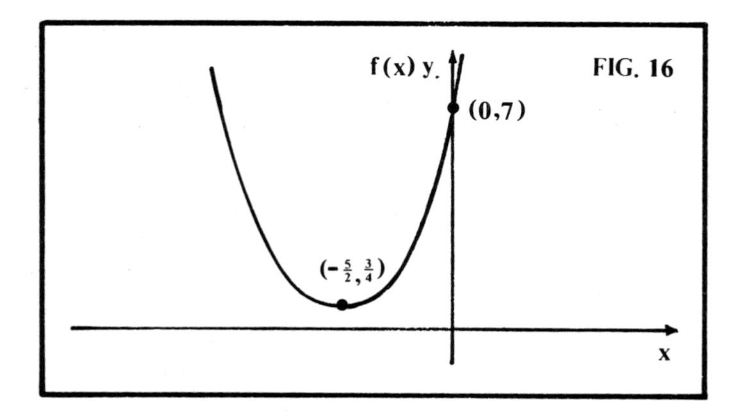

Example 2

Sketch the graph of the function f defined by

$$f(x) = 2+x-x^2$$

1. $a = -1 < 0$, $\therefore$ graph has a maximum turning point.
2. When $x = 0, f(x) = 2$, $\therefore$ the point $(0, 2)$ lies on the graph.
3. When $f(x) = 0$,

$$\begin{aligned} & 2+x-x^2 = 0 \\ \Leftrightarrow \quad & x^2-x-2 = 0 \quad \text{and } b^2-4ac = 1+8 = 9 \end{aligned}$$

$\Rightarrow$ roots are real, unequal and rational

$$\begin{aligned} \text{and } & x^2-x-2 = 0 \\ \Leftrightarrow & (x-2)(x+1) = 0 \\ \Rightarrow & \quad x = 2 \text{ or } -1 \end{aligned}$$

Hence the points $(2, 0)$ and $(-1, 0)$ lie on the graph.

4. Turning point

$$\begin{aligned} f(x) &= 2+x-x^2 \\ &= 2-(x^2-x) \\ &= 2-(x^2-x+\tfrac{1}{4})+\tfrac{1}{4} \\ &= 2\tfrac{1}{4}-(x-\tfrac{1}{2})^2 \end{aligned}$$

Hence the maximum value is $2\frac{1}{4}$ and occurs when $x = \frac{1}{2}$.

i.e. the maximum turning point is $(\frac{1}{2}, 2\frac{1}{4})$.

It follows that the line $x = \frac{1}{2}$ is an axis of symmetry. Hence figure 17:

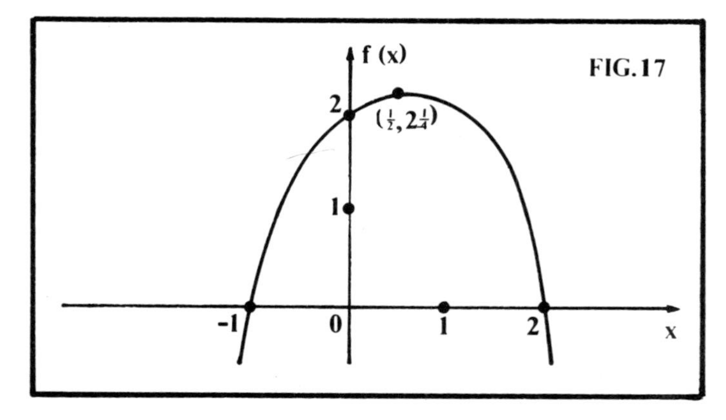

ASSIGNMENT 5.2

1. Write each of the following functions defined by $f(x)$, $g(x)$ or $h(x)$ in the form $f(x) = k(x+a)^2+b$ or $f(x) = b-k(x+a)^2$ where $a, b, k \in R$.

   (i) $f(x) = x^2+2x+3$ (ii) $f(x) = x^2+4x-5$

   (iii) $g(x) = 2-x-x^2$ (iv) $g(x) = 1+2x-x^2$

   (v) $h(x) = 2x^2-4x+5$ (vi) $h(x) = 2x^2-3x+1$

2. For each of the following functions defined by $f(x)$, $g(x)$ or $h(x)$ find,

   (a) the maximum or minimum value of each function.

   (b) the value of x for which the maximum or minimum value occurs.

   (c) the maximum or minimum turning point of each function.

   Sketch the graph of each function and write down an axis of symmetry for each graph.

   (i) $f(x) = x^2-x-2$ (ii) $g(x) = 2x^2+5x-3$

   (iii) $h(x) = 4x-x^2+5$ (iv) $g(x) = x^2-3x+4$

   (v) $f(x) = 3x-4-x^2$ (vi) $h(x) = x(x-2)$

   (vii) $f(x) = -4-x^2$ (viii) $h(x) = x^2-4x-4$

3. Each graph of figure 18 corresponds to one of the functions defined by $f(x)$ below. Find which function corresponds to each graph.

   A: $f(x) = x^2-x+2$

   B: $f(x) = 6x-x^2-9$

   C: $f(x) = -(x^2+4x+5)$

   D: $f(x) = x^2+2x+1$

   E: $f(x) = 5+3x-2x^2$

   F: $f(x) = x^2-7x+6$

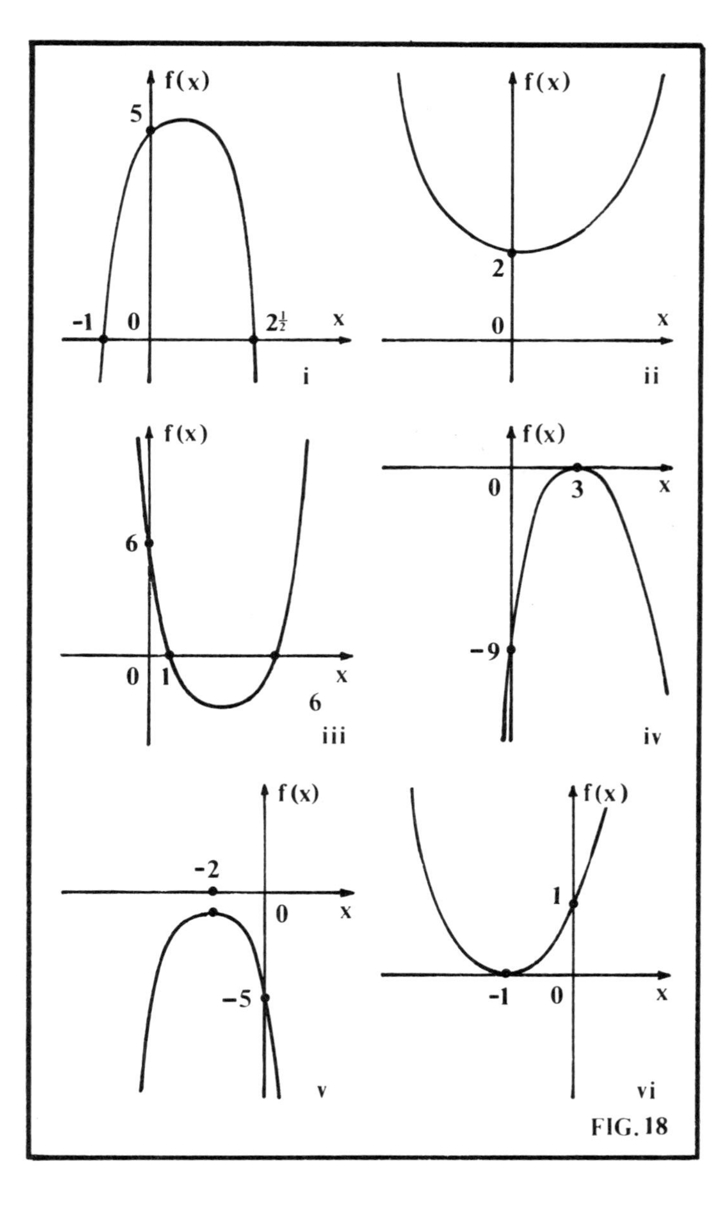

FIG. 18

4. Sketch the graph of the function defined by

$$f(x) = x^2 - 3x - 4.$$

On the same diagram draw the graph of the function defined by $g(x) = 4 + 3x - x^2$. Note that the graph of g is a reflection of the graph of f in the x-axis.

5. Sketch the graph of the function f defined by $f(x) = x^2 - 5x + 4$ and hence deduce the graph of g defined by $g(x) = -x^2 + 5x - 4$.

5.3 THE SUM AND PRODUCT OF THE ROOTS OF A QUADRATIC EQUATION

If $x = \alpha$ is a root of the equation $ax^2 + bx + c = 0$ then $(x - \alpha)$ is a factor of $ax^2 + bx + c$.

Similarly if $x = \beta$ is a root then $(x - \beta)$ is a factor of $ax^2 + bx + c$.

$\therefore$ $(x - \alpha)$ and $(x - \beta)$ are two factors of $ax^2 + bx + c$ and the general quadratic equation with these roots is of the form $a(x - \alpha)(x - \beta) = 0$.

Hence

$$a(x - \alpha)(x - \beta) = ax^2 + bx + c$$
$$a[x^2 - x(\alpha + \beta) + \alpha\beta] = ax^2 + bx + c$$

Comparing coefficients gives

$$-a(\alpha + \beta) = b \Rightarrow \alpha + \beta = -\frac{b}{a}$$

and $$a\alpha\beta = c \Rightarrow \quad \alpha\beta = \frac{c}{a}$$

Hence if α and β are the roots of $ax^2 + bx + c = 0$,

then $\alpha + \beta$ (i.e. the sum of the roots) $= -\frac{b}{a}$

and $\alpha\beta$ (i.e. the product of the roots) $= \frac{c}{a}$

Alternative method

The equation $ax^2 + bx + c = 0$ has roots given by,

$$x = \frac{-b \pm \sqrt{(b^2 - 4ac)}}{2a}$$

Let the two roots of the equation be $x = \alpha$ and $x = \beta$. Then

$$\alpha = \frac{-b + \sqrt{(b^2 - 4ac)}}{2a} \quad \text{and} \quad \beta = \frac{-b - \sqrt{(b^2 - 4ac)}}{2a}$$

$\alpha + \beta$ (i.e. the sum of the roots)

$$= \frac{-b + \sqrt{(b^2 - 4ac)}}{2a} + \frac{-b - \sqrt{(b^2 - 4ac)}}{2a}$$
$$= \frac{-b + \sqrt{(b^2 - 4ac)} - b - \sqrt{(b^2 - 4ac)}}{2a}$$
$$= -\frac{2b}{2a}$$
$$= -\frac{b}{a}$$

$\alpha\beta$ (i.e. the product of the roots)

$$= \left[\frac{-b}{2a} + \frac{\sqrt{(b^2 - 4ac)}}{2a}\right]\left[\frac{-b}{2a} - \frac{\sqrt{(b^2 - 4ac)}}{2a}\right]$$
$$= \left[\left(-\frac{b}{2a}\right)^2 - \frac{b^2 - 4ac}{4a^2}\right]$$
$$= \frac{b^2}{4a^2} - \frac{b^2 - 4ac}{4a^2}$$
$$= \frac{4ac}{4a^2}$$
$$= \frac{c}{a}$$

Example 1

If α and β are the roots of the equation $2x^2-3x-4=0$, find the value of $\frac{1}{\alpha}+\frac{1}{\beta}$.

α and β are the roots of $2x^2-3x-4=0$

$$\Rightarrow \qquad \alpha+\beta = -\frac{b}{a} = -\left(\frac{-3}{2}\right) = \frac{3}{2}$$

and
$$\alpha\beta = \frac{c}{a} = \frac{-4}{2} = -2$$

So
$$\frac{1}{\alpha}+\frac{1}{\beta} = \frac{\beta+\alpha}{\alpha\beta} = \frac{\frac{3}{2}}{-2} = -\frac{3}{4}$$

Example 2

Find the quadratic equation whose roots are $\frac{1}{2}$ and $-\frac{3}{4}$.

Let $f(x)=0$ be the equation

$x=\frac{1}{2}$ and $x=-\frac{3}{4}$ are roots of $f(x)=0$

$\therefore$ $(x-\frac{1}{2})$ and $(x+\frac{3}{4})$ are factors of $f(x)$.

Hence the equation is

$$(x-\tfrac{1}{2})(x+\tfrac{3}{4})=0$$
$$\Leftrightarrow \quad x^2-\tfrac{1}{2}x+\tfrac{3}{4}x-\tfrac{3}{8}=0$$
$$\Leftrightarrow \quad 8x^2-4x+6x-3=0$$
$$\Leftrightarrow \quad 8x^2+2x-3=0$$

Or let the roots be $\alpha=\frac{1}{2}$ and $\beta=-\frac{3}{4}$

Then
$$\alpha+\beta = \tfrac{1}{2}-\tfrac{3}{4} = -\tfrac{1}{4}$$

and
$$\alpha\beta = \tfrac{1}{2}(-\tfrac{3}{4}) = -\tfrac{3}{8}$$

i.e.
$$\alpha+\beta = \frac{-b}{a} = \frac{-2}{8} \quad \text{and} \quad \alpha\beta = \frac{c}{a} = \frac{-3}{8}$$

Hence the new equation is

$$x^2-\left(\frac{-b}{a}\right)x+\left(\frac{c}{a}\right)=0$$
$$\Rightarrow x^2-\left(\frac{-2}{8}\right)x+\left(\frac{-3}{8}\right)=0$$
$$\Leftrightarrow \qquad 8x^2+2x-3=0$$

Example 3

If α and β are the roots of the equation $3x^2+x-2=0$ find the equation whose roots are α^2 and β^2.

α and β are the roots of $3x^2+x-2=0$

$$\Rightarrow \alpha+\beta = \frac{-b}{a} = \frac{-1}{3} \quad \text{and} \quad \alpha\beta = \frac{c}{a} = \frac{-2}{3}$$

The new equation has to have roots $x=\alpha^2$ and $x=\beta^2$.

Sum of the roots of the new equation

$$= \alpha^2+\beta^2 = \alpha^2+2\alpha\beta+\beta^2-2\alpha\beta$$
$$= (\alpha+\beta)^2-2\alpha\beta$$
$$= \tfrac{1}{9}+\tfrac{4}{3} = \tfrac{13}{9}$$

Product of the roots of the new equation $=\alpha^2\beta^2=(\alpha\beta)^2=\frac{4}{9}$

Hence sum of the roots $=\dfrac{13}{9}=\dfrac{-b}{a}$

and product of the roots $=\dfrac{4}{9}=\dfrac{c}{a}$

Hence the new equation is

$$x^2-(\tfrac{13}{9})x+(\tfrac{4}{9})=0$$
$$\Leftrightarrow 9x^2-13x+4=0$$

ASSIGNMENT 5.3

1. If α and β are the roots of the given equations, write down
   (a) the sum of the roots of each equation (i.e. $\alpha+\beta$) and
   (b) the product of the roots of each equation (i.e. $\alpha\beta$)
   (i) $x^2-3x+4=0$ (ii) $2x^2+4x-5=0$
   (iii) $px^2+qx+r=0$ (iv) $ax^2+(b+1)x=1$
   (v) $(x-a)(x-b)=0$ (vi) $x-\frac{2}{x}=3$

2. The sum of the roots of the equation $ax^2+bx+2=0$ is -2 and the product of the roots is 12. Find the values of a and b.

3. Write down the equation in simplest form whose roots are
   (i) 2 and 3 (ii) 3 and -1 (iii) $\frac{1}{2}$ and -2
   (iv) $-\frac{1}{2}$ and $-\frac{1}{3}$ (v) p and q

4. α and β are the roots of the equation $3x^2-4x+9=0$. Find the values of
   (i) $\alpha+\beta$ and $\alpha\beta$ (ii) $\frac{1}{\alpha}+\frac{1}{\beta}$ (iii) $\alpha^2+\beta^2+2\alpha\beta$
   (iv) $\alpha^2+\beta^2-2\alpha\beta$

5. α and β are the roots of the equation $2x^2-4x+1=0$. Find the values of
   (i) $\alpha+\beta$ and $\alpha\beta$ (ii) $\alpha^2+\alpha\beta+\beta^2$ (iii) $\alpha^2-\alpha\beta+\beta^2$

6. If α and β are the roots of $x^2-px+q=0$, find the values in terms of p and q of
   (i) $\alpha+\beta$ and $\alpha\beta$ (ii) $\frac{1}{\alpha}+\frac{1}{\beta}$ (iii) $\alpha^2+\beta^2$
   (iv) $\alpha^2+\alpha\beta+\beta^2$ (v) $\left(\alpha+\frac{1}{\beta}\right)\left(\beta+\frac{1}{\alpha}\right)$
   (vi) $(\alpha-1)(\beta-1)$ (vii) $\frac{\alpha}{\beta}+\frac{\beta}{\alpha}$

7. Find the values of a if one root of the equation $x^2+ax+12=0$ is three times the other.

8. Find the value of p if one root of the equation $3x^2-4x+p=0$ is three times the other.

9. One root of the equation $px^2-qx+r=0$ is four times the other. Find a relation between p, q and r.

10. If α and β are the roots of the equation $2x^2+3x+6=0$, find the equation whose roots are,
    (i) $\frac{1}{\alpha}$ and $\frac{1}{\beta}$ (ii) α^2 and β^2.

11. α and β are the roots of the equation $2x^2-6x-1=0$. Find the equation whose roots are $\alpha-1$ and $\beta-1$.

12. If $p=\alpha+2\beta$ and $q=\beta+2\alpha$, prove that
    $$p+q=3(\alpha+\beta) \quad \text{and} \quad pq=2(\alpha+\beta)^2+\alpha\beta.$$
    If α and β are the roots of the equation $2x^2+x-3=0$ write down the values of $p+q$ and pq.
    Form the equation whose roots are $\alpha+2\beta$ and $\beta+2\alpha$.

13. If $p=\frac{\alpha}{\beta}$ and $q=\frac{\beta}{\alpha}$, prove that
    $$p+q=\frac{\alpha^2+\beta^2}{\alpha\beta}=\frac{(\alpha+\beta)^2-2\alpha\beta}{\alpha\beta} \quad \text{and} \quad pq=1$$
    Write down the value of $p+q$ if α and β are the roots of the equation $2x^2-3x+2=0$, and hence form the

equation whose roots are $\frac{\alpha}{\beta}$ and $\frac{\beta}{\alpha}$.

14. Without solving the equation $3x^2-2x-4=0$, find the sum, the product, and the sum of the squares, of the roots.

15. Find the value of p if one root of the equation $2x^2-21x+p=0$ is six times the other root.

16. Given that one root of the equation $2x^2-kx+45=0$ is ten times the other. Find the value of k.

17. The roots of the equation $x^2+kx+1=0$ are in the ratio $4:1$. Find the value of k.

18. The sum of the roots of a quadratic equation is 3 and the sum of the squares of the roots is 19. Find the equation.

IRRATIONAL NUMBERS OF THE FORM $p \pm \sqrt{q}$

5.4

The roots of the equation $ax^2+bx+c=0$ are given by

$$x = \frac{-b \pm \sqrt{(b^2-4ac)}}{2a} = \frac{-b}{2a} \pm \frac{\sqrt{(b^2-4ac)}}{2a}$$

This can be written as $x = \frac{-b}{2a} \pm \sqrt{\left(\frac{b^2-4ac}{4a^2}\right)}$

If we let $\frac{-b}{2a} = p$ and $\frac{b^2-4ac}{4a^2} = q$ then we get the roots in the form $x = p \pm \sqrt{q}$.

Hence if $p+\sqrt{q}$ is a root of a quadratic equation, then $p-\sqrt{q}$ is the other root, i.e. if one root is irrational, then the other is also irrational.

Two irrational numbers of the form $p+\sqrt{q}$ and $p-\sqrt{q}$ are called **conjugate irrationals.**

Note that their product $(p+\sqrt{q})(p-\sqrt{q}) = p^2-q$ is a rational number since p and q are rational.

Examples of conjugate irrationals are

$$\begin{array}{rcl} 2+\sqrt{3} & \text{and} & 2-\sqrt{3} \\ \sqrt{5} & \text{and} & -\sqrt{5} \\ \sqrt{2}-1 & \text{and} & -\sqrt{2}-1 \end{array}$$

Example 1

Simplify $\frac{6}{5-\sqrt{2}}$ by multiplying numerator and denominator by the conjugate irrational.

$$\frac{6}{5-\sqrt{2}} = \frac{6(5+\sqrt{2})}{(5-\sqrt{2})(5+\sqrt{2})} = \frac{6(5+\sqrt{2})}{25-2}$$
$$= \frac{6(5+\sqrt{2})}{23}$$

Example 2

Simplify $(3\sqrt{2}-4)^2$

$$(3\sqrt{2}-4)^2 = (3\sqrt{2})^2 - 2.4.3\sqrt{2}+16$$
$$= 18-24\sqrt{2}+16$$
$$= 34-24\sqrt{2}$$

Example 3

Evaluate $\frac{a^2}{a+1}$ when $a = 1+\sqrt{3}$

$$\frac{a^2}{a+1} = \frac{(1+\sqrt{3})^2}{2+\sqrt{3}} = \frac{(1+\sqrt{3})^2}{(2+\sqrt{3})} = \frac{4+2\sqrt{3}}{2+\sqrt{3}} = \frac{2(2+\sqrt{3})}{(2+\sqrt{3})} = 2$$

Example 4

Show that 1, $\frac{1}{\sqrt{2}+1}$ and $3-2\sqrt{2}$ are the first three terms of a geometric sequence.

If we let $a = 1$, $b = \frac{1}{\sqrt{2}+1}$ and $c = 3-2\sqrt{2}$ then we require to show that $\frac{b}{a} = \frac{c}{b}$ or $b^2 = ac$

$$b = \frac{1}{\sqrt{2}+1} = \frac{1}{\sqrt{2}+1} \cdot \frac{\sqrt{2}-1}{\sqrt{2}-1} = \frac{\sqrt{2}-1}{1} = \sqrt{2}-1$$
$$b^2 = (\sqrt{2}-1)^2 = 3-2\sqrt{2}$$
$$ac = 1.(3-2\sqrt{2}) = 3-2\sqrt{2}$$

Hence $b^2 = ac$ and therefore a, b and c are the first three terms of a geometric sequence.

ASSIGNMENT 5.4

1. Write down the conjugate irrational of each of the following irrational numbers.
   $\sqrt{2}$, $-2\sqrt{3}$, $3+\sqrt{2}$, $-3+\sqrt{2}$, $\sqrt{3}+1$, $-\sqrt{3}+1$, $a+\sqrt{b}$, $\sqrt{b}-1$, $3-2\sqrt{3}$, $\frac{1-4\sqrt{2}}{2}$, $\frac{4+2\sqrt{2}}{5}$.

2. A quadratic equation has a root $x = \sqrt{3}$. Write down the other root of this equation and hence form the quadratic equation.

3. A quadratic equation has a root $1-\sqrt{2}$. Write down the other root of the equation and form the equation.

4. Simplify the following expressions.
   (i) $(3-\sqrt{5})^2$ (ii) $(3+2\sqrt{2})^2$ (iii) $(2\sqrt{5}-\sqrt{3})^2$
   (iv) $(\sqrt{6}-\sqrt{3})(\sqrt{12}-\sqrt{2})$
   (v) $(\sqrt{3}+\sqrt{2})(3\sqrt{3}-2\sqrt{2})$
   (vi) $(\sqrt{7}-\sqrt{5})(\sqrt{7}+\sqrt{5})$
   (vii) $(2\sqrt{3}+\sqrt{5})(2\sqrt{3}-\sqrt{5})$

5. Find the value of k if the equation $x^2-4x+k = 0$ has a root $x = 2-\sqrt{3}$.

6. The equation $ax^2+bx+c = 0$ has a root $3-\sqrt{2}$. Write down the other root and find the values of a, b, and c.

7. Simplify each of the following by multiplying numerator and denominator by the conjugate of the denominator.
   (i) $\frac{1}{\sqrt{2}-1}$ (ii) $\frac{1}{\sqrt{5}+1}$ (iii) $\frac{1}{2\sqrt{3}-3}$
   (iv) $\frac{7}{4+\sqrt{3}}$ (v) $\frac{5}{2\sqrt{5}-10}$

8. Simplify the following by expressing with rational denominators.
   (i) $\frac{2-\sqrt{3}}{2+\sqrt{3}}$ (ii) $\frac{\sqrt{2}}{\sqrt{6}-\sqrt{2}}$ (iii) $\frac{\sqrt{7}-\sqrt{5}}{\sqrt{7}+\sqrt{5}}$
   (iv) $\frac{1}{\sqrt{3}+1}+\frac{1}{\sqrt{3}-1}$.

9. If $a = 3-\sqrt{5}$ find the value of $a+\frac{4}{a}$ and $a^2+\frac{16}{a^2}$.

10. Simplify
$$\frac{\sqrt{10}}{5\sqrt{2}-2\sqrt{5}}+\frac{\sqrt{5}}{2+\sqrt{10}}$$

11. Simplify
$$\frac{2\sqrt{3}+\sqrt{2}}{4\sqrt{3}-3\sqrt{2}}-\frac{2\sqrt{3}-\sqrt{2}}{4\sqrt{3}+3\sqrt{2}}$$

12. Three successive terms of a sequence are $\sqrt{3}+\sqrt{2}$, $2+\sqrt{6}$ and $\frac{2}{\sqrt{3}-\sqrt{2}}$. Show that the sequence is a geometric one.
    Find the common ratio of the sequence.

13. The two smaller sides of a right-angled triangle have lengths $6\sqrt{3}+3$ units and $6-3\sqrt{3}$ units. Calculate the exact length of the hypotenuse of the triangle.

14. Write down the equations whose roots are,

 (i) $3+\sqrt{5}, 3-\sqrt{5}$ (ii) $-2+\sqrt{3}, -2-\sqrt{3}$

15. Show that the numbers $\sqrt{3}$, $3+\sqrt{6}$ and $6\sqrt{2}+5\sqrt{3}$ form a geometric sequence.

16. Simplify

$$\frac{\sqrt{6}}{2\sqrt{3}-3\sqrt{2}}+\frac{\sqrt{2}}{3+\sqrt{6}}$$

17. Express $\frac{4}{5+\sqrt{5}}$ and $\frac{5-\sqrt{5}}{3-\sqrt{5}}$ with rational denominators and find b such that $\frac{4}{5+\sqrt{5}}, b, \frac{5-\sqrt{5}}{3-\sqrt{5}}$ form a geometric sequence.

ASSIGNMENT 5.5

Objective items testing Sections 5.1–5.4

Instructions for answering these items are given on page 22.

1. The equation $2x^2+(k-1)x+8=0$ has equal roots in x. Find the value(s) of k.

 A. 1
 B. 0 or 1
 C. 4 or -4
 D. 3 or -3
 E. Some other value or values.

2. For $x \in R$, the minimum value of x^2-9x is

 A. $\frac{-81}{4}$
 B. $\frac{-9}{2}$
 C. 0
 D. 3
 E. 9

3. The condition that the graph of the function $f(x) = ax^2+bx+c$ cuts the x-axis in 2 distinct points is

 A. $b = 0$
 B. $b^2-4ac = 0$
 C. $-\frac{b}{a} = \frac{c}{a}$
 D. $b^2-4ac > 0$
 E. $b^2-4ac < 0$

4. The equation $3x^2+px+3=0$ has real roots. All the values of p satisfy

 A. $-6 \leqq p \leqq 6$
 B. $0 \leqq p \leqq 3$
 C. $-3 \leqq p \leqq 0$
 D. $p \geqq 6$ or $p \leqq -6$
 E. $p \geqq 3$ or $p \leqq -3$

5. The equation $f(x) = 0$ has roots α and β, such that $\alpha+\beta = \frac{2}{3}$ and $\alpha\beta = \frac{1}{2}$. The equation $f(x) = 0$ is

 A. $3x^2-2x+1=0$
 B. $6x^2-4x+3=0$
 C. $6x^2+4x+3=0$
 D. $3x^2+2x-1=0$
 E. $2x^2-2x+1=0$

6. The set of values of $p \in R$ for which the equation in x $(1+p)x^2-2(1+3p)x+3 = 0$ has real roots in x is,

A. $p \geqq \frac{1}{3}$ or $p \leqq -\frac{2}{3}$
B. $-\frac{2}{3} \leqq p \leqq \frac{1}{3}$
C. $-\frac{2}{3} \leqq p \leqq 0$
D. $p = \frac{1}{3}$ or $p = -\frac{2}{3}$
E. $p \geqq 1$ or $p \leqq -2$

7. If the equation $x^2-5x+3 = 0$ has roots α and β then the equation whose roots are 2α and 2β is,

A. $4x^2-10x+3 = 0$
B. $4x^2-10x+6 = 0$
C. $4x^2+10x+12 = 0$
D. $x^2-10x+12 = 0$
E. $2x^2-10x+12 = 0$

8. In the equation $x^2+px+q = 0$, p and q are variables on $\{1, 2, 3, 4, 5\}$. How many such equations can be formed if the roots are *not* real?

A. 2
B. 5
C. 12
D. 13
E. More than 13.

9. If the graph of the function f, defined by $f(x) = 3x^2-12x+8$ cuts the x-axis at the points with coordinates $(p, 0)$ and $(q, 0)$, then

(1) $p+q = 4$
(2) $pq = \frac{8}{3}$
(3) $p > 0$ and $q > 0$ or $p < 0$ and $q > 0$

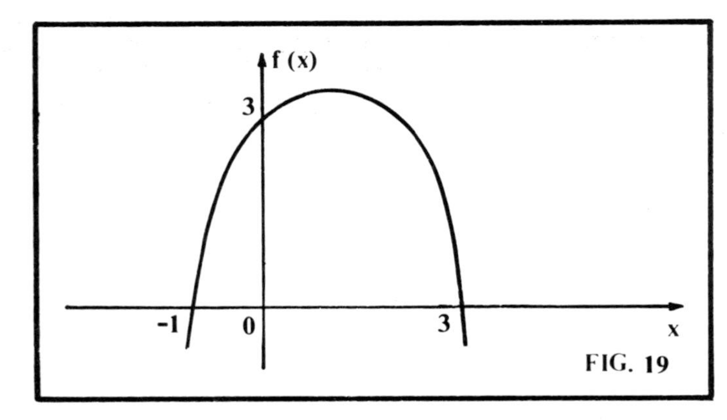

FIG. 19

10. The graph of a function f is shown in figure 19.
(1) $f(0) = 3$.
(2) $f(x) = 3+2x-x^2$.
(3) $\alpha+\beta \doteq 2$ if α and β are the roots of the equation $f(x) = 0$.

11. (1) $\dfrac{1}{\sqrt{2}+1}$ and $\dfrac{\sqrt{3}}{\sqrt{2}-1}$ are two irrational fractions.
(2) $\dfrac{1+\sqrt{3}}{(\sqrt{2}+1)(\sqrt{2}-1)}$ is the sum of the two fractions.

12. (1) Polynomial $f(x)$ of degree 2 is such that $f(-\frac{1}{2}) = 0$.
(2) $(2x+1)$ is a factor of $f(x)$.

UNIT 6: SYSTEMS OF EQUATIONS IN TWO OR THREE VARIABLES

INTRODUCTION

6.1

We have dealt with and solved several different kinds of equations, for example:

(i) simple equations involving only one variable, x, such as

$$2(x-1)+3x-2=9.$$

(ii) an equation of the type $2x^2+3x-1=0$, which is of degree two in one variable.

(iii) linear equations in two variables x and y such as

$$x+2y=3 \quad \text{and} \quad 3x-y=8.$$

It may be necessary in mathematics, especially in applied mathematics, to deal with more difficult equations of two variables or with equations of more than two variables. These equations usually arise as a result of trying to solve some practical or physical problem.

We deal now with three different systems of equations.

Systems of three simultaneous linear equations in 3 variables.

Example 1

Find the solution set of the system of equations,

$$4x-y-3z=-9 \quad (1)$$

$$x+2y-z=3 \quad (2)$$

$$5x-y-2z=-16 \quad (3)$$

Method is to eliminate one of the variables from two pairs of (1), (2) and (3), and then solve the resulting system of two equations in two variables.

$$\begin{array}{lrl}
\text{From equation (1)} \times 2 & 8x-2y-6z = & -18 \\
\text{From equation (2)} & x+2y-\ z = & 3 \\
\hline
\text{Add} & 9x \qquad -7z = & -15 \qquad (4) \\
\text{From equation (1)} & 4x-y-3z = & -9 \\
\text{From equation (3)} & 5x-y-2z = & -16 \\
\hline
\text{Subtract} & -x \qquad -z = & 7 \\
& \Rightarrow x+z \qquad = & -7 \qquad (5)
\end{array}$$

$$\begin{array}{lrl}
\text{From equation (4)} & 9x-7z = & -15 \\
\text{From equation (5)} \times 7 & 7x+7z = & -49 \\
\hline
\text{Add } 16x = & 16x \qquad = & -64 \\
& \Rightarrow x = & -4
\end{array}$$

Substitute $x = -4$ in equation (5)

$$x+z = -7$$

$$-4+z = -7$$

$$\Rightarrow z = -3$$

Substitute $x = -4$ and $z = -3$ in equation (1)

$$4x-y-3z = -9$$

$$-16-y+9 = -9$$

$$\Rightarrow \quad -y = -9-9+16 = -2$$

$$\Rightarrow \quad y = 2$$

Hence the solution set is $\{(-4, 2, -3)\}$.

Geometrical interpretation

A linear equation in two variables represents a line in 2-dimensions. Therefore two such equations represent two lines which will intersect in a point or two lines which will be parallel.

Example 2

$$\left.\begin{array}{r} 3x+4y = 7 \\ 4x-3y = 1 \end{array}\right\} \text{ intersect at } (1, 1)$$

but

$$\left.\begin{array}{r} 3x+4y = 7 \\ 3x+4y = 1 \end{array}\right\}$$

are parallel, since both lines have gradient $-\frac{3}{4}$.

A linear equation in three variables represents a plane in 3-dimensions.

Three planes can:

(i) intersect in a point, hence giving one unique solution.

(ii) intersect in a line, hence giving an infinite set of solutions.

(iii) be parallel, hence giving no solutions.

(iv) be such that each pair of planes intersect in a line, thus forming three faces of a prism.

ASSIGNMENT 6.1

Solve the systems of equations in questions 1–8, where the variables are on the set of real numbers.

1. $\begin{aligned} x+2y+3z &= 2 \\ 2x-y+2z &= -2 \\ x+y-2z &= 5 \end{aligned}$

2. $\begin{aligned} 3x-4y-4z &= -2 \\ 2x+y-2z &= 1 \\ x-y-z &= -1 \end{aligned}$

3. $\begin{aligned} 3x+y-2z &= 4 \\ x-3y-3z &= -4 \\ 2x+2y+z &= 2 \end{aligned}$

4. $\begin{aligned} 4x-3y+2z &= 11 \\ 3x-2y-3z &= 0 \\ x+y-2z &= 3 \end{aligned}$

5. $\begin{aligned} x-2y+2z &= 2 \\ 3x-4y+2z &= 3 \\ x-y+3z &= 0 \end{aligned}$

6. $\begin{aligned} a+4b-2c &= 12 \\ 3a+2b-4c &= 9 \\ 2a-6b+3c &= -18 \end{aligned}$

7. $\begin{aligned} 3x+5y-2z &= 1 \\ 3x+2y &= -3 \\ 4y+3z &= -6 \end{aligned}$

8. $\begin{aligned} 2a-3b-4c &= -12 \\ -3a+4b-2c &= 11 \\ 4a-2b+3c &= -21 \end{aligned}$

9. The points $(1, 1)$, $(-2, -14)$ and $(2, 2)$ lie on the parabola with equation $y = ax^2+bx+c$. Find the values of a, b and c and write down the equation of the parabola.

10. The circle with equation $x^2+y^2+2gx+2fy+c = 0$ passes through the points $(2, 0)$, $(3, -2)$, $(4, 1)$, find the

values of g, f and c and write down the equation of the circle.

11. Three points $(2, -1, 3)$, $(3, -2, -\frac{1}{2})$ and $(-2, 1, -4)$ lie on the plane with equation $ax+by+cz = d$. Find a, b and c in terms of d and write down the equation of the plane.

12. Find the solution set of the system of equations

$$\begin{aligned} 2x-2y+\ z &= 3 \\ 3x-4y+2z &= 8 \\ 4x-6y-3z &= -23 \end{aligned}$$

13. Find the solution set of the system of equations

$$\begin{aligned} x+\ y+\ z &= 4 \\ 2x+3y+4z &= 11 \\ 3x+2y-5z &= -3 \end{aligned}$$

14. The system of equations

$$\begin{aligned} ax+by-cz &= 6 \\ 4x-ay+\ z &= 3 \\ 2x+cy-bz &= 2 \end{aligned}$$

has solution set $\{1, -2, -3\}$. Find the value of a, b and c.

SYSTEMS OF EQUATIONS IN TWO VARIABLES, ONE LINEAR AND ONE QUADRATIC

6.2

Example 1

Find the solution set of the system of equations,

$$2y^2+3xy-2x^2 = 12 \quad (1)$$

$$3x+\ 2y = 4 \quad (2)$$

The method is to substitute for x or y, from equation (2), into equation (1) to obtain an equation in one variable.

From equation (2)

$$\begin{aligned} 3x+2y &= 4 \\ \Rightarrow \quad 2y &= 4-3x \\ \Rightarrow \quad y &= \tfrac{1}{2}(4-3x) \end{aligned}$$

Substitute this expression for y in equation (1)

$$\begin{aligned} 2y^2+3xy-2x^2 &= 12 \\ \Leftrightarrow \quad 2.\tfrac{1}{4}(4-3x)^2+3x.\tfrac{1}{2}(4-3x)-2x^2 &= 12 \\ \Leftrightarrow \quad \tfrac{1}{2}(4-3x)^2+\tfrac{3}{2}x(4-3x)-2x^2 &= 12 \\ \Leftrightarrow \quad (4-3x)^2+3x(4-3x)-4x^2 &= 24 \\ \Leftrightarrow \quad 16-24x+9x^2+12x-9x^2-4x^2-24 &= 0 \\ \Leftrightarrow \quad -8-12x-4x^2 &= 0 \\ \Leftrightarrow \quad x^2+3x+2 &= 0 \\ \Leftrightarrow \quad (x+1)(x+2) &= 0 \\ \Leftrightarrow \quad x = -1 \text{ or } x &= -2 \end{aligned}$$

When $x = -1$, $y = \frac{1}{2}(4-3x) = \frac{1}{2}(4+3) = \frac{7}{2}$.
When $x = -2$, $y = \frac{1}{2}(4-3x) = \frac{1}{2}(4+6) = 5$.
Hence solution set is $\{(-1, \frac{7}{2}), (-2, 5)\}$.

Geometrical interpretation

A second degree equation represents a curve and a first degree equation represents a line.

Note

(i) A line intersects a curve in two points, hence two solutions.

or

(ii) A line touches a curve in two coincident points, hence one solution (i.e. two equal solutions).

or

(iii) A line does not touch or intersect a curve, hence no solutions.

Example 2

Find whether the line with equation $2x+y-7=0$ intersects or touches the circle with equation

$$x^2+y^2+4x-2y-15=0.$$

$$2x+y-7=0 \quad (1)$$

$$x^2+y^2+4x-2y-15=0 \quad (2)$$

From equation (1)

$$2x+y-7=0$$

$$\Rightarrow \quad y=7-2x$$

Substitute this value of y in equation (2).

$$x^2+y^2+4x-2y-15=0$$

$$\Leftrightarrow \quad x^2+(7-2x)^2+4x-2(7-2x)-15=0$$

$$\Leftrightarrow x^2+49-28x+4x^2+4x-14+4x-15=0$$

$$\Leftrightarrow \quad 5x^2-20x+20=0$$

$$\Leftrightarrow \quad x^2-4x+4=0$$

$$\Leftrightarrow \quad (x-2)(x-2)=0$$

$$\Leftrightarrow \quad x=2 \text{ twice}$$

Hence the line touches (i.e. is a tangent) to the circle. When $x=2$, $y=7-2x=7-4=3$.

Hence (2, 3) is the point of contact.

ASSIGNMENT 6.2

Find the solution set of each of the systems of equations in questions 1–6.

1. $y=x^2$
   $y=x+2$

2. $y=4x-x^2$
   $y=x-4$

3. $x+y=7$
   $xy=-8$

4. $y=2x+1$
   $y=3x^2-4x+4$

5. $x=2y-3$
   $x^2+y^2=2$

6. $x^2+y^2=2$
   $y-2x=1$

7. The line $3y-2x=4$ cuts the curve $y=2x^2$ at the points A and B. Find the coordinates of A and B. Illustrate your solutions by a sketch.

8. Find the coordinates of the points of intersection of the line $y=2x-3$ and the circle $x^2+y^2=18$. Sketch the graph of the line and circle showing the points of intersection.

Solve each of the systems of equations in questions 9–14.

9. $2x+y=5$
   $x^2-2xy+5=0$

10. $x^2-8xy+2y^2=-13$
    $x+3y=6$

11. $y-2x=1$
    $3y^2=4x^2-1$

12. $2x-y=5$
    $x^2+xy+2y^2=46$

13. $2x^2+xy+y^2=8$
    $2x+3y=2$

14. $y-x=6$
    $2y-\dfrac{1}{x+3}=5$

15. Find the coordinates of the points in which the line $3y-x=4$ cuts the circle $x^2+y^2+3x+1=0$.

16. By solving the following systems of equations find which lines intersect the given circle, touch the given circle (i.e. are tangents) or do not intersect or touch the given circle.

    State the coordinates (if any) of the points of intersection or the point of contact.

    (i) $2x-y-8=0$
    $x^2+y^2-4x-2y=0$

    (ii) $y+x-1=0$
    $x^2+y^2+2x-2y+1=0$

(iii) $$2y+x+7 = 0$$
$$x^2+y^2+2x-4y+1 = 0$$

(iv) $$2x+y = 0$$
$$x^2+y^2+4x+2y = 0$$

17. Find the solution set of the system of equations
$$3y^2+xy-3x^2 = y+3x$$
$$3y+2x = 2$$

18. Show that the solution set of the equations
$$x^2+y^2 = 20x-60 \text{ and } 3y-x = 10$$
has only one member. If the quadratic equation represents a circle and the linear equation a straight line, what does this mean?

19. By finding the points of intersection or the point of contact, say whether the line with equation $2x+y = 10$ is or is not a tangent to the circle with equation $x^2+y^2+20y+20 = 0$.

20. Find the solution set of the system of equations
$$5x^2-8xy+3y^2 = 60$$
$$x+3y = 6$$

21. Find the solution set of the system of equations
$$x+y = 6$$
$$2y+\frac{1}{x-3} = 5$$

22. Find the solution set of the system of equations
$$(3x+y)(x+2y) = 12$$
$$2x-y = -4$$

FREEDOM OR PARAMETRIC EQUATIONS

6.3

If the coordinates (x, y) of any point P on a curve be given in terms of another variable (usually t or θ), by the equations,
$$x = f(t),\ y = g(t) \quad \text{or} \quad x = f(\theta),\ y = g(\theta)$$
then both x and y depend for their value on the value we give to t or θ.

t or θ is called a **parameter** and the equations are called the **freedom equations** or **parametric equations** of the curve.

The equation in x and y found by eliminating t or θ from the freedom equations is called the **constraint** or **cartesian equation** of the curve.

Example 1

$x = 2t^2 \ldots 1(a)$ and $y = 4t \ldots 1(b)$, where t varies, are the freedom equations of a curve.

When $t = \frac{1}{2}$, $x = \frac{1}{2}$ and $y = 2$, $\therefore$ the point $(\frac{1}{2}, 2)$ lies on the curve.

When $t = -2$, $x = 8$ and $y = -8$, $\therefore$ the point $(8, -8)$ lies on the curve.

When $x = 2$, $2 = 2t^2 \Rightarrow t^2 = 1 \Rightarrow t = \pm 1$, and when $t = \pm 1$, $y = \pm 4$, $\therefore$ the points $(2, 4)$ and $(2, -4)$ lie on the curve.

To find the cartesian equation of the curve we eliminate t from $1(a)$ and $1(b)$.

From equation $1(b)$
$$y = 4t \quad \Rightarrow t = \frac{y}{4}$$

Substitute $t = \dfrac{y}{4}$ into equation $1(a)$.
$$x = 2t^2 = 2 \cdot \frac{y^2}{16} = \frac{y^2}{8}$$
$$\Rightarrow y^2 = 8x \quad \text{is the cartesian equation of the curve.}$$

Example 2

The freedom equations of a circle are $x = r\cos\theta \ldots 1(a)$ and $y = r\sin\theta \ldots 1(b)$, where r is a constant and θ varies.

When $\theta = 0$, $x = r$ and $y = 0$, $\therefore$ the point $(r, 0)$ lies on the curve.

When $\theta = \frac{\pi}{2}$, $x = 0$ and $y = r$, $\therefore$ the point $(0, r)$ lies on the curve.

To obtain the constraint or cartesian equation eliminate θ.

From equation 1(*a*) $\quad x = r\cos\theta \Rightarrow \cos\theta = \frac{x}{r}$.

From equation 1(*b*) $\quad y = r\sin\theta \Rightarrow \sin\theta = \frac{y}{r}$.

But

$$\cos^2\theta + \sin^2\theta = 1$$

$$\therefore \quad \frac{x^2}{r^2} + \frac{y^2}{r^2} = 1$$

$$\Rightarrow \quad x^2 + y^2 = r^2$$

which is the cartesian equation of a circle centre $(0, 0)$, radius r.

ASSIGNMENT 6.3

1. In each of the following freedom equations t or θ is a parameter. Find the constraint (i.e. cartesian) equation of the curves with freedom equations given by,

   (i) $x = 3t^2$, $y = 6t$

   (ii) $x = 3\cos\theta$, $y = 3\sin\theta$

   (iii) $x = 2 + t$, $y = 3 - 2t$

   (iv) $x = 2\cos\theta$, $y = 4\sin\theta$

   (v) $x = at^2$, $y = 2at$

   (vi) $x = ct$, $y = \frac{c}{t}$

   (vii) $x = 3 - 2\cos t$, $y = 1 + \sin t$

   (viii) $x = a\cos\theta$, $y = b\sin\theta$

   (ix) $x = 2t^2 - 1$, $y = 3t$

   (x) $x = 4\cos^2 t°$, $y = 1 - 2\sin t°$

2. With reference to rectangular axes OX and OY, the coordinates of a point on a curve are given by,

   $x = 4\cos t°$, $y = 1 - 2\sin^2 t°$, where t is a variable.

   (i) Find the cartesian coordinates of the points on the curve given by $t = 60$ and $t = 30$.

   (ii) If $x = 2\sqrt{2}$, find the values of the parameter t and the corresponding values of y $(0 \leqq t \leqq 360)$.

   (iii) If $y = \frac{1}{2}$, find the values of the parameter t and the corresponding values of x. (Leave your answers in surd form.)

   (iv) By eliminating t find the cartesian equation of the curve.

3. A curve is given by the equations $x = 4t$, $y = 2t^2 + 1$, where t is a parameter.

   Tabulate values of x and y for $t = 0, 0{\cdot}2, 0{\cdot}4, 0{\cdot}8, 1, 1{\cdot}2$.

   With the usual x and y axes sketch the part of the curve for $0 \leqq t \leqq 1{\cdot}2$.

   Show that the cartesian equation of the curve is $8(y - 1) = x^2$.

4. A curve has equation given by $x = 4p^2$, $y = 8p$, where p varies. Find the gradient of the chord joining the points where $p = -1$ and $p = 2$.

5. Find the length of the chord joining the points on the curve $x = 1 + \cos\theta$, $y = \sin\theta$ at which θ has the values 0 and $\frac{\pi}{2}$.

6. Find the cartesian equations of the curves defined by the following parametric equations.

   (i) $x = \cos 2\theta$, $y = \sin\theta$

   (ii) $x = 2 - t$, $y = t^3 + 4$

(iii) $x = 2\cos\theta$
$y = \cos 2\theta$

(iv) $x = \cos\theta + \sin\theta$
$y = \cos\theta - \sin\theta$

7. A curve is defined by the equations

$$x = 3 + 2\cos\theta \quad \text{and} \quad y = 2\sin^2\theta,$$

where θ varies in the range $0 \leqq \theta < 2\pi$.

(i) Find the coordinates of the points on the curve where $y = 0$.

(ii) Find the coordinates of the point on the curve where $\theta = \dfrac{\pi}{3}$.

(iii) Construct a table showing values of x and y for $\theta = 0, \dfrac{\pi}{6}, \dfrac{\pi}{4}, \dfrac{\pi}{3}, \dfrac{\pi}{2}, \dfrac{2\pi}{3}, \dfrac{3\pi}{4}, \dfrac{5\pi}{6}, \pi$, and sketch the curve for the interval $0 \leqq \theta \leqq \pi$.

(iv) Find in its simplest form the gradient of the line joining the points corresponding to $\theta = \alpha$ and $\theta = \beta$.

(v) Find the equation of the curve in cartesian coordinates.

8. By eliminating t from the equations $\sin\theta = \dfrac{2t}{1+t^2}$ and $\cos\theta = \dfrac{1-t^2}{1+t^2}$ show that $\sin^2\theta + \cos^2\theta = 1$.

For what values of t is $\sin\theta = \cos\theta$? Hence show that for these values of t, $\sin\theta = \cos\theta = \pm\dfrac{1}{\sqrt{2}}$.

9. If x and y are related by the equations

$$x\sin^2 t = 1 + \cos t$$
$$y = 1 - \cos t,$$

where t is a parameter, show that $y^2x - 2yx + 2 - y = 0$ and hence by using the distributive law that $xy = 1$.

10. If $\cos\theta + \sin\theta = a$ and $\cos^2\theta - \sin^2\theta = b$ show that $a^4 = 2a^2 - b^2$.

ASSIGNMENT 6.4

Objective items testing Sections 6.1–6.3

Instructions for answering these items are given on page 22.

1. In the equation

$$p(x-2y+2z-2) + q(2x+y-3z-6) + r(4x-2y+z-6) = 0$$

where x, y, z are real, respective values of p, q and r which make the coefficient of z equal to zero could be

A. $-3, 2, 0$
B. $0, -1, 3$
C. $0, 1, -3$
D. $1, 0, 2$
E. $3, 2, 0$

2. The straight line with equation $x - y = 2$ touches or intersects the curve with equation $x^2 - y^2 = 8$. Which one of the following statements is *false*?

A. $x + y = 4$.
B. $(3, 1)$ is a member of the solution set of the equations.
C. $(-3, -1)$ is a member of the solution set of the equations.
D. The line is not a tangent to the curve.
E. The solution set of the equations is $\{(3, 1)\}$.

3. In the solution set of the system of linear equations

$$\begin{aligned} p - 2q + r &= 4 \\ p + 2q + 3r &= -2 \\ 2p - q + r &= 8 \end{aligned}$$

where p, q, r are real, the value of r is

A. 0
B. -1

C. 1 D. $-\frac{1}{3}$ E. none of these.

4. Solving the system of equations

$$\begin{cases} x^2+y^2+2x-4=0 \\ x+2y=1 \end{cases}$$

by substituting for x, leads to which one of the following equations?

A. $5y^2+8y+1=0$
B. $5y^2-8y-1=0$
C. $5y^2-1=0$
D. $5y^2-8y+7=0$
E. $5y^2+8y-1=0$

5. The solution set of the system of simultaneous equations

$$\begin{aligned} x+y-z &= 1 \\ x-y+z &= 1 \\ -x+y+z &= 1 \end{aligned}$$ is

A. $\varnothing$ B. $\{(1, 0, -1)\}$
C. $\{(1, 0, 0)\}$ D. $\{(1, 1, 1)\}$
E. none of these.

6. $X = \{(x, y): y = 2x\}$ and $Y = \{(x, y): x^2+y^2 = 5\}$ where $x, y \in R$. Find $X \cap Y$.

A. $\varnothing$ B. $\{(1, 2)\}$
C. $\{(1, 2), (-1, -2)\}$ D. R
E. none of these.

7. If $x = \cos\theta$, $y = \cos 2\theta$ then for all θ

A. $y^2 = 1-2x^2$ B. $y^2 = 2x^2-1$
C. $y^2 = x^2-1$ D. $y = 1-2x^2$
E. $y = 2x^2-1$

8. The equation in x and y for the curve with parametric equations $x = 1-2t^2$, $y = \frac{2}{t}$ is

A. $y^2 = 8(1-x)$
B. $y^2(1-x) = 8$
C. $y^2(x+1) = 8$
D. $y^2 = 2(1-x)$
E. $y^2 = 4(1+x)$

9. L is the system of equations

$$\begin{aligned} x+2y-z &= 3 \\ 3x-y+2z &= -5 \\ 2x+y-3z &= 7 \end{aligned}$$

The set of values of x, y, z given by $(-1, 2, 0)$

(1) satisfies two of the equations only
(2) satisfies one equation only
(3) is the solution set of L.

10. If the straight line $x+y = 2$ meets the curve whose equation is $xy = 1$, then

(1) the line intersects the curve in two distinct points.
(2) the line touches the curve at one point.
(3) the line is a tangent to the curve.

11. (1) The point P has coordinates (x, y) such that $x = t+2$, $y = 3t^2$, $t \in R$.
(2) The equation of the locus of P is $y = 3(x-2)^2$.

12. (1) The line with equation $y = k$ cuts the circle with equation $x^2+y^2 = 36$ in two distinct points.
(2) $k = -5$.

UNIT 7: THE EXPONENTIAL AND LOGARITHMIC FUNCTIONS

REVISION OF INDICES

7.1

For all $m, n \in Q$

(i) $a^m \times a^n = a^{m+n}$.

(ii) $a^m \div a^n = a^{m-n}$.

(iii) $(a^m)^n = a^{mn}$.

(iv) $a^{-m} = \dfrac{1}{a^m}$

(v) $a^{p/q} = \sqrt[q]{(a^p)} = (\sqrt[q]{a})^p$, p an integer, q a positive integer.

(vi) $a^0 = 1$, $a \neq 0$.

Example 1

a. $(2a)^2 = 4a^2$

b. $(2a)^{-2} = \dfrac{1}{(2a)^2} = \dfrac{1}{4a^2}$

c. $2a^{-2} = 2 \times a^{-2} = \dfrac{2}{a^2}$

d. $\dfrac{1}{2a^{-2}} = \dfrac{a^2}{2}$

e. $\left(\dfrac{a}{b}\right)^{-3} = \left(\dfrac{b}{a}\right)^3$—inverting a fraction changes the sign of the index.

Example 2

If $p = 7q^{\frac{1}{3}}r^{-\frac{1}{2}}$, find the value of r, when $p = 7$ and $q = 64$.

$$\begin{aligned} p &= 7q^{\frac{1}{3}}r^{-\frac{1}{2}} \\ \Rightarrow \quad 7 &= 7(64)^{\frac{1}{3}}r^{-\frac{1}{2}} \\ \Rightarrow \quad 1 &= 64^{\frac{1}{3}}r^{-\frac{1}{2}} \\ \Rightarrow \quad \frac{1}{r^{-\frac{1}{2}}} &= 64^{\frac{1}{3}} \\ \Rightarrow \quad \frac{1}{r^{-\frac{1}{2}}} &= 4 \\ \Rightarrow \quad r^{\frac{1}{2}} &= 4 \\ \Rightarrow \quad r &= 4^2 \\ \Rightarrow \quad r &= 16 \end{aligned}$$

Example 3

Evaluate y, given that $(y^2-3)^{-\frac{1}{2}} = 4$.

$$\begin{aligned} (y^2-3)^{-\frac{1}{2}} = 4 \Rightarrow (y^2-3)^{(-\frac{1}{2})\cdot(-\frac{2}{1})} &= 4^{(-\frac{2}{1})} \\ \Rightarrow \quad (y^2-3)^1 &= 4^{-2} \\ \Rightarrow \quad y^2-3 &= \frac{1}{4^2} \\ \Rightarrow \quad y^2-3 &= \tfrac{1}{16} \\ \Rightarrow \quad y^2 &= \tfrac{1}{16}+3 \\ &= \tfrac{49}{16} \\ y &= \pm\tfrac{7}{4} \end{aligned}$$

ASSIGNMENT 7.1

Simplify the following expressions in questions 1–21 by expressing each with a positive index.

1. $a^3 \times a^9$
2. $a^{\frac{1}{2}} \times a^{\frac{1}{4}}$
3. $p^{\frac{3}{4}} \div p^{\frac{1}{4}}$
4. $q^5 \div q^{-3}$
5. $(x^{\frac{1}{4}})^8$
6. $(a^{\frac{1}{4}})^{\frac{1}{2}}$
7. $(a^0)^3$
8. $(2x)^{-1}$
9. $(4x)^{-3}$
10. $3x^{-2}$
11. $3x^2 \times 4x^{-\frac{1}{2}}$
12. $8x^{\frac{1}{4}} \times 3x^{-\frac{1}{4}}$
13. $6a^{\frac{2}{3}} \div 3a^{\frac{1}{2}}$
14. $(a^{\frac{2}{3}})^{-3}$
15. $3s^{-\frac{1}{2}} \times (3s)^{-1}$
16. $x^{\frac{1}{2}}(x^{\frac{1}{2}}+x^{\frac{3}{2}})$
17. $a^{\frac{1}{3}}(a^{-\frac{1}{3}}-a^{\frac{2}{3}})$
18. $(x^{\frac{1}{2}}-x^{-\frac{1}{2}})^2$
19. $\dfrac{3x^0}{(2x)^{-2}}$
20. $\left(\dfrac{a}{b}\right)^{-1} \times \left(\dfrac{b}{a}\right)^{-2}$
21. $\left(\dfrac{2a}{b}\right)^{-4}$
22. If $a = 16$, $b = 9$ and $c = 27$, evaluate the following.
    (i) $a^{\frac{1}{4}}$ (ii) $b^{-\frac{1}{2}}$ (iii) $c^{\frac{2}{3}}$ (iv) $a^{-\frac{3}{4}}$
    (v) $3c^{-\frac{1}{3}} \times 4b^{-\frac{1}{2}}$ (vi) $(a+b)^{\frac{1}{2}}$ and $(a^{\frac{1}{2}}+b^{\frac{1}{2}})$
    (vii) $(b^{-\frac{1}{2}}+c^{\frac{1}{3}})^{-1}$ (viii) $(a^{\frac{1}{4}}-c^{\frac{2}{3}})^{-2}$
23. If $a = 5b^{-2}c^{-\frac{1}{2}}$ find c when $a = 5$, and $b = -2$.
24. Find the value of y, when $(y^2-1)^{\frac{1}{3}} = 2$.
25. Simplify and express the following with positive indices
    (i) $\left(\dfrac{x^{-\frac{1}{2}}}{3y^2}\right)^{-2}$ (ii) $(4a^{-2} \div 9b^2)^{-\frac{1}{2}}$
    (iii) $\left(\dfrac{16a^2}{b^{-2}}\right)^{-\frac{1}{4}}$ (iv) $\left(\dfrac{8a^3}{27b^{-3}}\right)^{-\frac{2}{3}}$
    (v) $(a^m b^{-n})^3 \times (a^3b^2)^{-m}$ (vi) $(\sqrt{x^2y^3})^6$
26. Find the value of
    (i) $(\frac{27}{8})^{-\frac{1}{3}}$ (ii) $\dfrac{1}{25^{-\frac{1}{2}}}$ (iii) $(\frac{16}{81})^{\frac{3}{4}}$
    (iv) $32^{-\frac{2}{5}}$ (vi) $27^{\frac{4}{3}}$
27. Find the value of $(16^{\frac{3}{4}}+49^{-\frac{1}{2}})^{-1}$.
28. If $a = 64$ and $b = 81$ calculate the value of $(a^{\frac{2}{3}}+b^{\frac{1}{2}})^{-\frac{3}{2}}$.
29. If $a = 9$ and $b = 16$ find the value of
    (i) $(3ab^{\frac{3}{4}})^{\frac{1}{3}}$ (ii) $a^{\frac{1}{2}}b^{-\frac{1}{2}}$ (iii) $a^{-\frac{1}{2}}+b^{\frac{1}{4}}$ (iv) $(a+b)^{-\frac{1}{2}}$

30. Simplify $\left(\frac{16}{125}\right)^k \times \frac{2^k}{5^{-k}}$.

31. If $\frac{3^p \times 27^{p+q}}{3^{p+q}} = 3$, show that $3p+2q = 1$.

32. If $\frac{x^{p-2}}{x^{p+2}} = \frac{x^{p-3}}{y}$ find y in terms of x in its simplest form.

33. Find the value of $\frac{a^{\frac{2}{3}}b^{\frac{1}{3}} + a^{\frac{1}{3}}b^{\frac{2}{3}}}{a^{\frac{4}{3}}b^{-\frac{1}{3}} + a}$ when $a = 8b$.

34. If $a = 16b^{-\frac{4}{3}}c^{\frac{2}{3}}$ find b in terms of a and c. Calculate b when $a = c = 81$.

THE EXPONENTIAL FUNCTION
$f: x \to a^x (a > 0)$

7.2

If $a > 0$ and $a \neq 1$ this function maps each rational number x on to a power of a. The number x here is an index or exponent. The function is the exponential function with base a and domain Q. We may extend the domain to R, since from the graph we may assume a^x exists for x irrational.

Example 1

The function $f: x \to 2^x$ is the exponential function with base 2.

Draw the graph of the exponential function with base 2 and with domain $\{x: -2 \leqq x \leqq 3\}$.

First tabulate several values of x in the domain and the corresponding values of $f(x)$

x	-2	-1	0	1	2	3 — Domain of f
$f(x)$	$\frac{1}{4}$	$\frac{1}{2}$	1	2	4	8 — Range of f

The graph of the function $f: x \to 2^x$ with domain $\{x: -2 \leqq x \leqq 3\}$ is shown in figure 20.

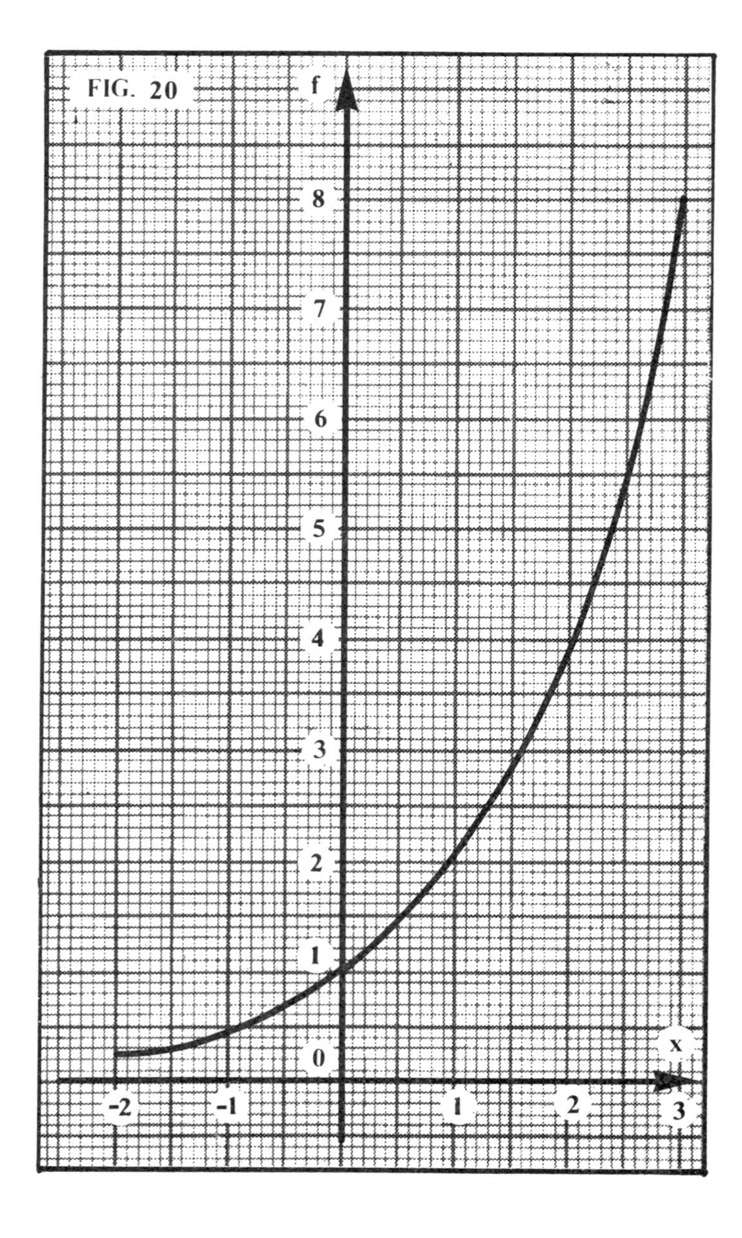

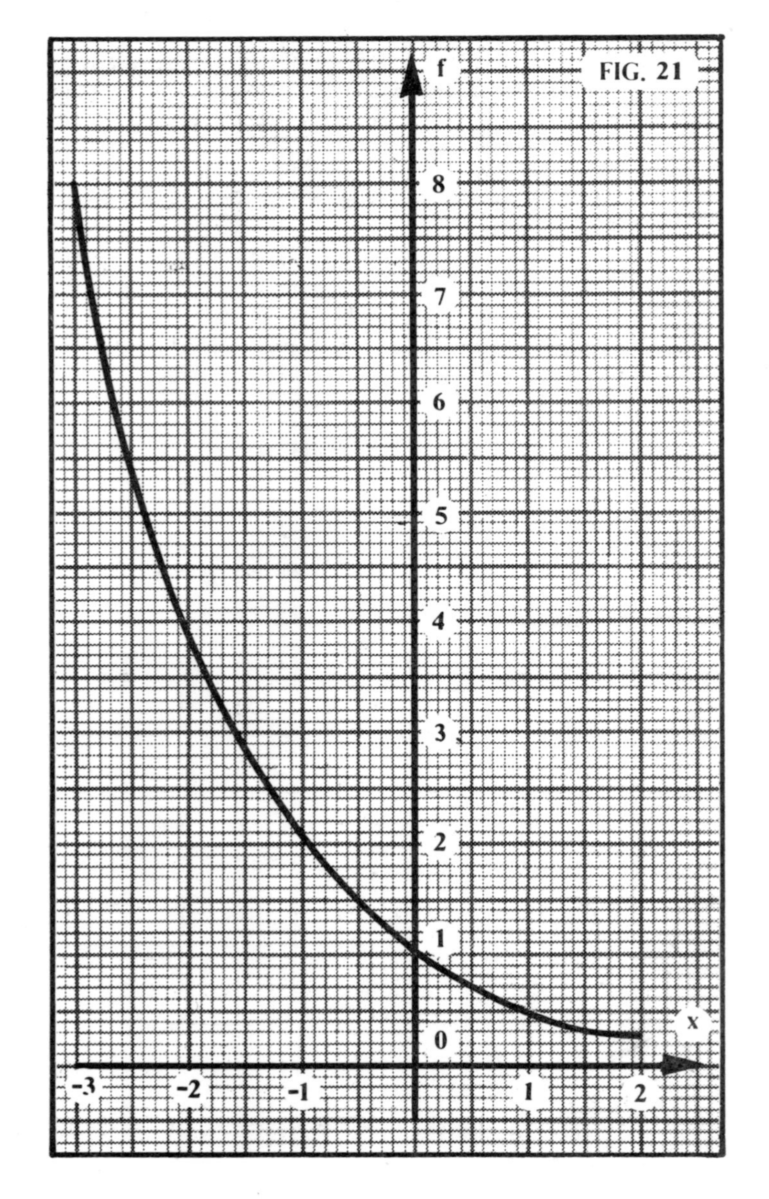

From the graph in example 1, note that as x increases from $x = -2$ to $x = 3$ the function defined by $f(x) = 2^x$ increases from $\frac{1}{4}$ to 8.

In general, if $a > 1$ then the function defined by $f(x) = a^x$ increases as x increases.

Example 2

The function $f: x \rightarrow (\frac{1}{2})^x$ is the exponential function with base $\frac{1}{2}$.

Draw the graph of the exponential function with base $\frac{1}{2}$ and with domain $\{x: -3 \leqq x \leqq 2\}$.

First tabulate several values of x in the domain and the corresponding values of $f(x)$.

x	-3	-2	-1	0	1	2	—Domain of f
$f(x)$	8	4	2	1	$\frac{1}{2}$	$\frac{1}{4}$	—Range of f

The graph of the function $f: x \rightarrow (\frac{1}{2})^x$ with domain $\{x: -3 \leqq x \leqq 2\}$ is shown in figure 21.

From the graph in example 2, note that as x increases from $x = -3$ to $x = 2$, the function defined by $f(x) = (\frac{1}{2})^x$ decreases from 8 to $\frac{1}{4}$.

In general if $0 < a < 1$ the function defined by the formula $f(x) = a^x$ decreases as x increases.

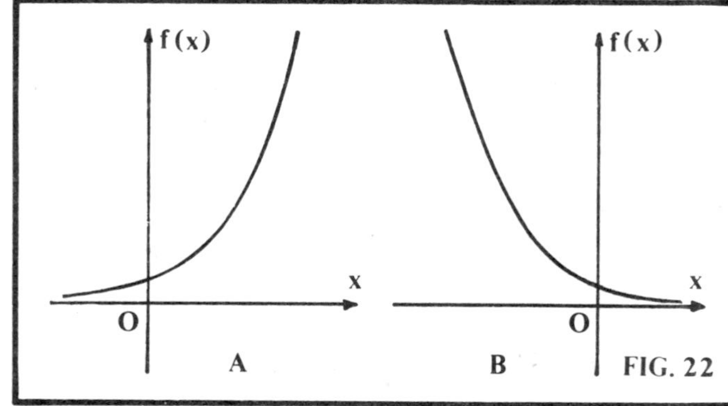

FIG. 22

Figures 22(a) and (b) show the shape of the graphs of $f: x \to a^x$ when $a > 1$ and when $0 < a < 1$.

Note: (i) The functions $f: x \to 2^x$ and $g: x \to (\frac{1}{2})^x$ have the same domain and range.

Hence in general the function $f: x \to a^x$, $a > 1$ and $g: x \to a^x$, $0 < a < 1$ have the same domain and range.

(ii) Reflection in the y-axis maps $P(x, y) \to P'(-x, y)$ i.e. maps $P[x, f(x)] \to P'(-x, f(-x)]$.

Hence reflection in the y-axis maps the function defined by the formula $f(x) = a^x$ onto the function defined by $g(x) = a^{-x} = \left(\dfrac{1}{a}\right)^x$.

Thus the graph of $g: x \to (\frac{1}{2})^x$ may be obtained from the graph of $f: x \to 2^x$ by reflection in the y-axis.

ASSIGNMENT 7.2

1. By choosing a suitable scale (which need not be the same on both axes) draw on the same diagram the graphs of the functions,

   (i) $f: x \to 2^x$

   (ii) $g: x \to 3^x$, both with domain $\{x: -2 \leqq x \leqq 3, x \in R\}$

   Estimate from your graph the value of $2^{\frac{1}{2}}$ and $3^{\frac{1}{2}}$.

2. Make a table of values for x and $f(x)$ for the mapping $f: x \to 10^x$ with domain $\{x: -1 \leqq x \leqq 1\}$. By choosing a suitable scale draw the graph of the function $f: x \to 10^x$ for this domain. From your graph,

   (a) Estimate the value of $10^{\frac{1}{4}}$ and $10^{-\frac{1}{4}}$.

   (b) If (i) $2 = 10^x$ write down the value of x.

   (ii) $10^x = 50$ write down the value of x.

   (c) Write 6 as a power of 10.

3. By writing $4^x = (2^2)^x = 2^{2x} = (2^x)^2$, make a table of x and $f(x)$ for the mapping $g: x \to 4^x$ with domain $\{x: -2 \leqq x \leqq 3\}$. Draw the graph of the mapping.

4. Name a common point of all the graphs in questions 1, 2 and 3. Show that the point $(0, 1)$ lies on the graph of

$$f: x \to a^x, a > 0, a \in R.$$

5. By choosing a suitable scale (which need not be the same on both axes) draw on the same diagram the graphs of the functions,

   (i) $f: x \to 3^x$

   (ii) $g: x \to (\frac{1}{3})^x$, both with domains

$$\{x: -3 \leqq x \leqq 3, x \in R\}$$

6. The function $f: x \to (\frac{1}{4})^x$ has domain

$$\{x: -2 \leqq x \leqq 2, x \in R\}.$$

Copy and complete the following table.

x	-2	$-\frac{3}{2}$	-1	$-\frac{1}{2}$	0	$\frac{1}{2}$	1	$\frac{3}{2}$	2
$(\frac{1}{4})^x$	16		4			$\frac{1}{2}$			

By choosing a suitable scale draw the graph of the function $f: x \to (\frac{1}{4})^x$, with domain

$$\{x: -2 \leqq x \leqq 2, x \in R\}.$$

Estimate from your graph the values of $(\frac{1}{4})^{\frac{1}{4}}$ and $(\frac{1}{4})^{-\frac{1}{4}}$.

7. The function $f: x \to a^x$, is such that $f(-2) = 25$. Find the base a, of the function.

8. The graph of the function $f: x \to (\frac{1}{4})^x$ is reflected in the y-axis.

   The image graph is defined by $g: x \to g(x)$. Find the function defined by $g(x)$.

7.3

REFLECTION IN THE LINE $y = x$

If P is any point (a, b), then on reflection in the line $y = x$,

$$P(a, b) \to P'(b, a)$$

Let $f: x \to f(x)$, $x \in R$ be a one–one correspondence and let P be any point on the graph defined by $f: x \to f(x)$, then P has coordinates $(x, f(x))$.

If the graph of f is reflected in the line $y = x$, then the point P $(x, f(x))$ maps onto the point $P'(f(x), x)$

i.e. $$P(x, f(x)) \to P'(f(x), x)$$

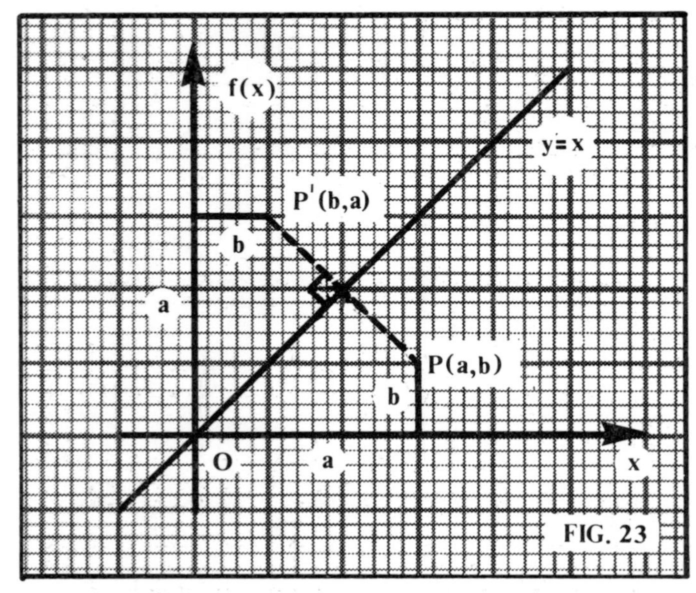

FIG. 23

Hence, any point on the image curve has coordinates $(f(x), x)$.

Let the image curve be defined by $g: x \to g(x)$ then any point on the image curve has coordinates $(x', g(x'))$

$$\therefore x' = f(x) \quad \text{and} \quad g(x') = x$$

Hence $$g(x') = g[f(x)] = x$$

and $$f(x) = f[g(x')] = x',$$

i.e. $g \circ f = I$ and $f \circ g = I$.

$\therefore$ the functions $f: x \to f(x)$ and $g: x \to g(x)$ are inverse functions.

Thus if f and f^{-1} are inverse functions, the graph of one can be obtained from the graph of the other by reflecting in the line $y = x$.

Example 1

Sketch the graphs of the functions defined by $f: x \to x^3$ and $g: x \to \sqrt[3]{x}$ on the same diagram and show their positions with respect to the line $y = x$.

Since f and g are inverse functions each graph is the image of the other on reflection in the line $y = x$.

For $f: x \to x^3$

x	0	$\frac{1}{2}$	1	$\frac{3}{2}$	2
x^3	0	$\frac{1}{8}$	1	$\frac{27}{8}$	8

For $g: x \to \sqrt[3]{x}$

x	0	$\frac{1}{8}$	1	$\frac{27}{8}$	8
$\sqrt[3]{x}$	0	$\frac{1}{2}$	1	$\frac{3}{2}$	2

Note that the domain of f is the range of g and the domain of g is the range of f.

The graph of each function is shown in figure 24.

The points $(2, 8)$ and $(8, 2)$ are corresponding points. Hence the point (a, b) on one graph and the point (p, q) on the other graph are corresponding points $\Leftrightarrow a = q$ and $b = p$.

From the above table write down 2 other sets of corresponding points.

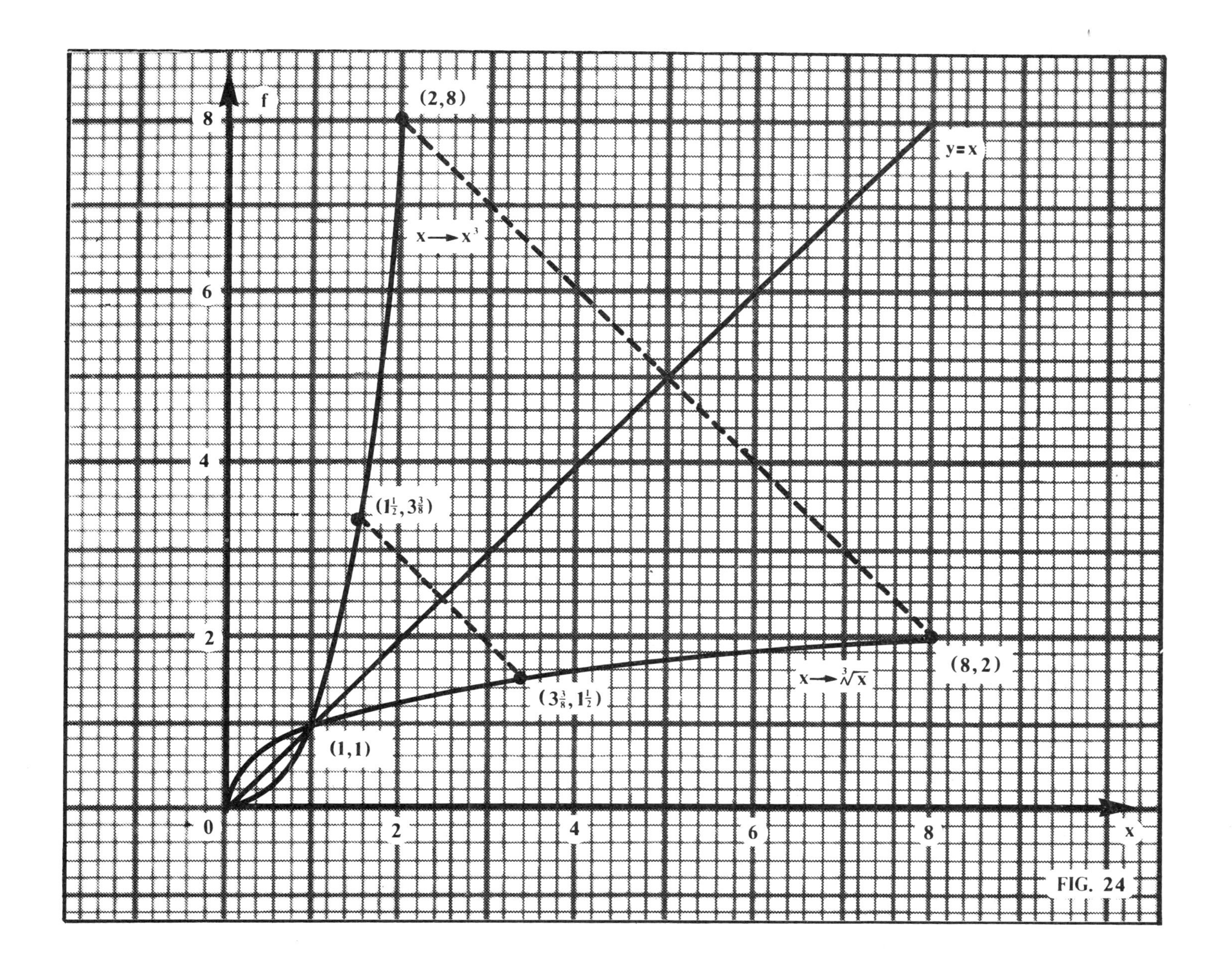
f
8
6
4
2
0
2
4
6
8
x
(2,8)
y = x
x ⟶ x³
(1½, 3⅜)
(8,2)
x ⟶ ∛x
(3⅜, 1½)
(1,1)
FIG. 24

THE LOGARITHMIC FUNCTION

The function $f: x \to 2^x$ with domain $\{-2, -1, 0, 1, 2, 3\}$ is a *one to one correspondence* from the set $\{-2, -1, 0, 1, 2, 3\}$ to the set $\{\frac{1}{4}, \frac{1}{2}, 1, 2, 4, 8\}$. It therefore has an inverse function f^{-1}, such that

$$(f^{-1} \circ f)(x) = x \Rightarrow f^{-1}(2^x) = x$$

This inverse function f^{-1} is called the *logarithm to base 2* and is written $f^{-1}: x \to \log_2 x$. Its domain is therefore $\{\frac{1}{4}, \frac{1}{2}, 1, 2, 4, 8\}$ and range $\{-2, -1, 0, 1, 2, 3\}$.

If we illustrate this inverse mapping by an arrow diagram, we get,

Domain	f^{-1}	Range		Domain	f^{-1}	Range
$\frac{1}{4}$	$\longrightarrow$	$\log_2 \frac{1}{4}$		$\frac{1}{4}$	$\longrightarrow$	-2
$\frac{1}{2}$	$\longrightarrow$	$\log_2 \frac{1}{2}$		$\frac{1}{2}$	$\longrightarrow$	-1
1	$\longrightarrow$	$\log_2 1$	but	1	$\longrightarrow$	0
2	$\longrightarrow$	$\log_2 2$		2	$\longrightarrow$	1
4	$\longrightarrow$	$\log_2 4$		4	$\longrightarrow$	2
8	$\longrightarrow$	$\log_2 8$		8	$\longrightarrow$	3

FIG. 25

Hence

$$\log_2 \tfrac{1}{4} = -2 \qquad \log_2 \tfrac{1}{2} = -1$$
$$\log_2 1 = 0 \qquad \log_2 2 = 1$$
$$\log_2 4 = 2 \qquad \log_2 8 = 3$$

Note that the set of logarithms $\{-2, -1, 0, 1, 2, 3\}$ gives the exponents or indices in the mapping $f: x \to 2^x$.

Since $f: x \to 2^x$ and $f^{-1}: x \to \log_2 x$ are inverse functions, the graph of one can be obtained from the graph of the other by reflection in the line $y = x$. This is shown in figure 26.

By the symmetry of the graph, we have, on reflecting in the line $y = x$, that

$$\left.\begin{array}{lll} (4, \log_2 4) & \text{and} & (2, 2^2) \\ \text{i.e.} \quad (x, y) & \text{and} & (y, x) \end{array}\right\} \text{ are corresponding points.}$$

$\therefore 2 = \log_2 4 \Leftrightarrow 4 = 2^2.$

Hence, if $f: n \to a^n$ and $g: p \to \log_a p$, such that (n, a^n) and $(p, \log_a p)$ are corresponding points,

then $\qquad p = a^n \Leftrightarrow n = \log_a p.$

Note: (i) If $p = a$, then $n = 1$ and hence $\log_a a = 1$.

(ii) If $p = 1$, then $n = 0$ and hence $\log_a 1 = 0$.

Example 1

Find $\log_2 16$ and $\log_3 27$.

$$\log_2 16 = n \Leftrightarrow 2^n = 16 \Rightarrow n = 4, \therefore \log_2 16 = 4$$
$$\log_3 27 = n \Leftrightarrow 3^n = 27 \Rightarrow n = 3, \therefore \log_3 27 = 3.$$

Example 2

Express 3^2 and $\frac{1}{16}$ in logarithmic form.

$3^2 = 9 \quad \Leftrightarrow \log_3 9 = 2$

$\frac{1}{16} = 4^{-2} \Leftrightarrow \log_4 \frac{1}{16} = -2 \quad$ Or $\quad \frac{1}{16} = 2^{-4} \Leftrightarrow \log_2 \frac{1}{16} = -4$

ASSIGNMENT 7.3

1. In the following show that each function with domain R is the inverse of the other.

(a) $f: x \to x^3 + 1, \quad g: x \to \sqrt[3]{(x-1)}$

(b) $f: x \to 2x + 1, \quad g: x \to \frac{1}{2}(x-1)$

Sketch the graph of each function in (a) on one diagram and the graph of each function in (b) on the one diagram for values of $x \geqq 0$. Draw the line $y = x$ on each diagram.

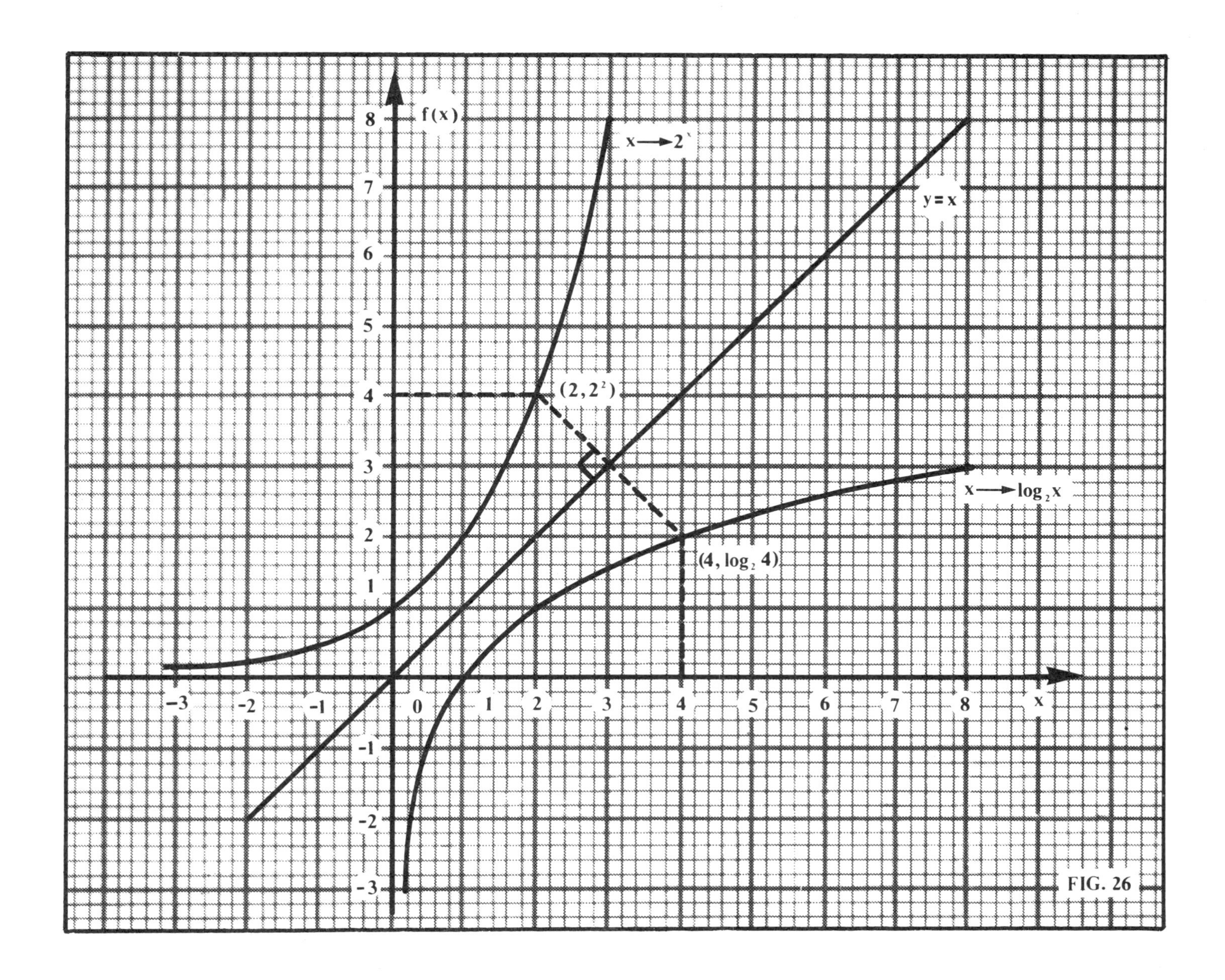

FIG. 26

Write down for each pair of functions three sets of corresponding points.

2. Write down the inverse function of each of the following functions with domain R.

   (i) $f: x \to 3^x$ (ii) $f: x \to 10^x$ (iii) $g: y \to 4^y$
   (iv) $f: z \to a^z$ (v) $f: x \to \log_2 x$ (vi) $h: x \to \log_4 x$
   (vii) $g: a \to \log_b a$

3. Construct a table of values of x and $f(x)$ for the mapping $f: x \to 10^x$ with domain $\{-2, -1, 0, 1, 2\}$ and hence write down the logarithms to base 10 of 100, 10, 1, $\frac{1}{10}$, $\frac{1}{100}$.

4. The mapping $g: x \to 3^x$ with domain $\{-2, -1, 0, 1, 2, 3\}$ has range $\{\frac{1}{9}, \frac{1}{3}, 1, 3, 9, 27\}$. Write down the logarithms to base 3 of the range of g.

5. (i) $\log_a 16 = 2 \Leftrightarrow 16 = a^2$, find a.
   (ii) $\log_3 27 = y \Leftrightarrow 27 = 3^y$, find y.
   (iii) $f: x \to \log_4 x \Leftrightarrow f^{-1}: x \to 4^x$. Is this statement true or false?

6. The mapping $f: x \to 4^x$ has domain $\{-2, -1, 0, 1, 2, 3\}$. Write down the logarithms to base 4 of 4^{-2}, 4^{-1}, 1, 4^2, 4^3.

7. A mapping $g: x \to a^x$, $a \neq 1$, has domain $\{x : x \in R\}$. Write down the inverse of this mapping and hence give the logarithms to base a of 1, a, a^{-1}, a^2, a^{-2}, a^3.

8. Express each of the following in logarithm form,

   (i) $27 = 3^3$ (ii) $64 = 4^3$ (iii) $32 = 2^5$
   (iv) $\frac{1}{9} = 3^{-2}$ (v) $\frac{1}{125} = 5^{-3}$ (vi) $4 = a^b$
   (vii) $1 = 6^0$ (viii) $p = a^n$

9. Express each of the following in exponential form,

   (i) $\log_2 8 = 3$ (ii) $3 = \log_5 125$
   (iii) $6 = \log_2 64$ (iv) $\log_2 \frac{1}{128} = -7$
   (v) $\log_a 1 = 0$ (vi) $\log_6 6 = 1$
   (vii) $-3 = \log_3 \frac{1}{27}$ (viii) $\log_b a = x$
   (ix) $\frac{1}{2} = \log_5 \sqrt{5}$

10. In the following find the value of a, b, x or y.

    (i) $x = \log_3 27$ (ii) $-a = \log_2 \frac{1}{2}$
    (iii) $-4 = \log_x 16$ (iv) $4 = \log_4 a$
    (v) $\log_4 \frac{1}{64} = y$ (vi) $-5 = \log_2 x$
    (vii) $\log_b 81 = 4$

11. Write down the value of,

    (i) $\log_2 8$ (ii) $\log_3 1$ (iii) $\log_{10} 1000$
    (iv) $\log_5 5$ (v) $\log_2 \frac{1}{16}$ (vi) $\log_3 27$
    (vii) $\log_{27} 3$

7.4

LAWS OF LOGARITHMS

We have seen that if $a^m = p \Leftrightarrow \log_a p = m$, then (m, a^m) and $(p, \log_a p)$ are corresponding points.

Also, if (n, a^n) and $(q, \log_a q)$ are corresponding points then $a^n = q \Leftrightarrow \log_a q = n$.

Hence we have,

$$a^m = p \Leftrightarrow \log_a p = m$$

and

$$a^n = q \Leftrightarrow \log_a q = n$$

From the laws of indices we know that there are relations in exponent form connecting a^m, a^n, namely,

$$a^m \times a^n = a^{m+n}, \quad a^m \div a^n = a^{m-n}, \quad (a^m)^n = a^{mn}.$$

We now wish to find the relations in logarithmic form connecting $\log_a p$ and $\log_a q$. When these index laws are expressed in their equivalent logarithmic form they are called the **Laws of Logarithms**.

1. $\log_a pq = \log_a p + \log_a q, \quad a > 0, \quad p, q > 0$

Proof: $\log_a p = m$ and $\log_a q = n \Leftrightarrow p = a^m$ and $q = a^n$

Hence $pq = a^m \times a^n = a^{m+n} \Leftrightarrow \log_a pq = m+n$

i.e. $\underline{\log_a pq = \log_a p + \log_a q}$

2. $\log_a \frac{p}{q} = \log_a p - \log_a q, \quad a > 0, \quad p, q > 0$

Proof: $\log_a p = m$ and $\log_a q = n \Leftrightarrow p = a^m$ and $q = a^n$

Hence $\frac{p}{q} = a^m \div a^n = a^{m-n} \Leftrightarrow \log_a \frac{p}{q} = m - n$

i.e. $\underline{\log_a \frac{p}{q} = \log_a p - \log_a q}$

3. $\log_a p^n = n.\log_a p, \quad a, p > 0, \quad n$ rational

Proof: $\log_a p = m \Leftrightarrow p = a^m$

$$\Rightarrow p^n = (a^m)^n = a^{mn} = a^{nm}$$

$$\Rightarrow \log_a p^n = nm = n.\log_a p$$

i.e. $\underline{\log_a p^n = n.\log_a p}$

Note: In working out questions on logarithms distinguish carefully between $\log \frac{p}{q}$ and $\frac{\log p}{\log q}$.

for $\log \frac{p}{q} = \log p - \log q$

but $\frac{\log p}{\log q} = \log p \div \log q$

Example 1

(i) If logarithms are to base 10, then

$$\log \tfrac{8}{3} = \log 8 - \log 3 = 0{\cdot}903 - 0{\cdot}477 = 0{\cdot}426$$

but $\frac{\log 8}{\log 3} = \frac{0{\cdot}903}{0{\cdot}477} = 1{\cdot}89$

(ii) If logarithms are to base 3, then

$$\log_3 \tfrac{27}{3} = \log_3 9 = \log_3 3^2 = 2\log_3 3 = 2 \text{ (since } \log_a a = 1)$$

but $\frac{\log 27}{\log 3} = \frac{\log 3^3}{\log 3} = \frac{3\log 3}{\log 3} = 3$

Example 2

Simplify $\log x^5 - 2\log x + \log \frac{1}{x^3}$

$$\log x^5 - 2\log x + \log \frac{1}{x^3} = 5\log x - 2\log x + \log x^{-3}$$
$$= 3\log x - 3\log x$$
$$= 0$$

Or

$$\log x^5 - 2\log x + \log \frac{1}{x^3} = \log x^5 - \log x^2 + \log \frac{1}{x^3}$$
$$= \log\left(\frac{x^5}{x^2}\right) + \log \frac{1}{x^3}$$
$$= \log x^3 + \log \frac{1}{x^3}$$
$$= \log\left(x^3 . \frac{1}{x^3}\right)$$
$$= \log 1$$
$$= 0 \text{ (since } \log_a 1 = 0)$$

Example 3

Evaluate without using tables, $\log_6 18 + \log_6 12 - \log_6 6$.

$$\log_6 18 + \log_6 12 - \log_6 6 = \log_6 \frac{18 \times 12}{6}$$
$$= \log_6 36 = \log_6 6^2$$
$$= 2\log_6 6$$
$$= 2 \text{ (since } \log_a a = 1)$$

Example 4

Solve the following equations,

(i) $2^{3x+1} = 128$

(ii) $2^{3x-2} = 3^{2x-3}$

(iii) $\log_3 x + \log_3 x^3 - \log_3 8x = 0$

(i) Since $128 = 2^7$, then

$$2^{3x+1} = 2^7$$
$$\Rightarrow 3x+1 = 7$$
$$\Rightarrow \quad 3x = 6 \Rightarrow x = 2$$

(ii)
$$2^{3x-2} = 3^{2x-3}$$
$$\Leftrightarrow \quad \log_{10} 2^{3x-2} = \log_{10} 3^{2x-3}$$
$$\Leftrightarrow \quad (3x-2)\log 2 = (2x-3)\log 3$$

(base 10 may be left out since there is no confusion)

$$\Leftrightarrow \quad 3x\log 2 - 2\log 2 = 2x\log 3 - 3\log 3$$
$$\Leftrightarrow \quad 3\log 3 - 2\log 2 = 2x\log 3 - 3x\log 2$$
$$\Leftrightarrow \quad \log 27 - \log 4 = x(2\log 3 - 3\log 2)$$
$$= x(\log 9 - \log 8)$$

$$\Leftrightarrow x = \frac{\log 27 - \log 4}{\log 9 - \log 8} = \frac{1{\cdot}431 - 0{\cdot}602}{0{\cdot}954 - 0{\cdot}903}$$
$$= \frac{0{\cdot}829}{0{\cdot}051} = 16{\cdot}3$$

(iii)
$$\log_3 x + \log_3 x^3 - \log_3 8x = 0$$
$$\Leftrightarrow \log_3 x + 3\log_3 x - (\log_3 8 + \log_3 x) = 0$$
$$\Leftrightarrow \quad \log_3 x + 3\log_3 x - \log_3 8 - \log_3 x = 0$$
$$\Leftrightarrow \quad 3\log_3 x = \log_3 8 = \log_3 2^3$$
$$= 3\log_3 2$$
$$\Rightarrow \quad \log_3 x = \log_3 2 \Rightarrow x = 2$$

Or
$$\log_3 x + \log_3 x^3 - \log_3 8x = 0$$
$$\Leftrightarrow \quad \log_3\left(\frac{x \cdot x^3}{8x}\right) = 0$$
$$\Leftrightarrow \quad \log_3\left(\frac{x^3}{8}\right) = 0$$
$$\Leftrightarrow \quad \frac{x^3}{8} = 3^0$$
$$\Leftrightarrow \quad \frac{x^3}{8} = 1$$
$$\Leftrightarrow \quad x^3 = 8$$
$$\Leftrightarrow \quad x = 2$$

ASSIGNMENT 7.4

1. Express each of the following in the form $\log x$.
   (i) $\log 3 + \log 8$ (ii) $\log 27 - \log 3$ (iii) $4\log 2$
   (iv) $\frac{1}{2}\log 36$ (v) $-3\log 3$

2. Write down the value of,
   (i) $\log_{10} 1000$ (ii) $\log_2 32$ (iii) $\log_3 81$
   (iv) $\log_4 \frac{1}{4}$ (v) $\log_7 1$ (vi) $\log_3 3$

3. Find the value of x or y in the following,
   (i) $\log_4 x = 2$ (ii) $\log_{10} x = \frac{1}{3}$
   (iii) $\log_7 x = -\frac{1}{2}$ (iv) $\log_8 y = -1$
   (v) $\log_3 x = 0$ (vi) $\log_{16} y = -\frac{1}{2}$
   (vii) $\log_{81} y = -\frac{1}{4}$

4. Write the following in the form $\log_b a$.
   (i) $\log_3 48 - \log_3 3$ (ii) $3\log_2 4$
   (iii) $1 + \log_5 7$ (iv) $\log_{10} 3 + 2$
   (v) $3 - \log_6 3$ (vi) $\frac{1}{3}\log_{10} 9 + 2\log_{10} 3$

5. Simplify the following (your answer may be left in the form $\log x$).

(i) $\dfrac{\log 27}{\log 3}$ (ii) $\dfrac{\log 2}{\log \frac{1}{2}}$

(iii) $\log x + \log \dfrac{1}{x}$ (iv) $\dfrac{\log \sqrt{x}}{\log x^2}$

(v) $2 \log a + \log b$ (vi) $\log x^2 - \log \dfrac{x}{y}$

(vii) $\log xy - 2 \log x$ (viii) $\dfrac{\log a^2 - \log a}{\log a^2 + \log a}$

Evaluate the following without using tables.

6. $\log_3 21 + 2 \log_3 3 - \log_3 7$ 7. $\log 625 \div \log 25$

8. $\log_6 48 - 3 \log_6 2$

9. $\frac{1}{2} \log_{10} 256 - 3 \log_{10} 2 + \frac{1}{3} \log_{10} 125$

10. $\log_3 \frac{1}{3} - \log_3 3^2 + 4 \log_3 3$

11. $\log_3 25 - \log_3 50 + \log_3 18$

12. $\log_4 8 - \log_4 \frac{1}{8}$

Solve for x the following equations,

13. $\log x + 2 \log 3 = 0$ 14. $\log x - \log 12 = \log \frac{2}{3}$

15. $3 \log x + \log \frac{1}{4} = \log x^2 - \log 3$

16. $\dfrac{\log x}{\log 2} = \dfrac{\log 4}{\log 16}$

17. Without using tables find which is greater,

$$\log_3 81 - \log_3 27 \quad \text{or} \quad \frac{\log_3 81}{\log_3 27}$$

18. If $\log_8 2 = n$ and $\log_2 8 = m$, show that $mn = 1$.

19. Simplify,

(i) $\log x^3 - 3 \log x + \log \dfrac{1}{x^2}, \quad x > 0$

(ii) $\log x^{\frac{1}{2}} + 2 \log x^{\frac{1}{4}} - \log x, \quad x > 0$

(iii) $\log_2 \sqrt{8} - \frac{1}{2} \log_2 \frac{1}{8}$

20. Solve for x the equations,

(i) $3^{2x-7} = 243$ (ii) $4^{5x} = 64$

21. Solve the following equations giving roots to three significant figures.

(i) $5^x = 12$ (ii) $2^x = 3{\cdot}2$ (iii) $3^{x-1} = 5$

(iv) $2^x = 3^{3x-1}$

22. Solve for x and y the equations,

(i) $\log_x 3 + \log_x 9 + \log_x 27 = 12$

(ii) $\log_2 xy = 8, \quad \log_2 \dfrac{x}{y} = 2$

23. Which of the following are true?

(i) $\log x + \log x^2 + \log x^3 = 6 \log x$

(ii) $\log 4 = \frac{1}{2} \log 8$

(iii) $\log_2 10 = 2 \log_2 5$

(iv) $\log \dfrac{a}{b} + \log \dfrac{b}{a} = 0$

24. If $\log_x \frac{3}{4} + 5 \log_x \frac{2}{3} - \log_x \frac{25}{81} + 2 \log_x \frac{5}{2} = 3$, find the value of x.

25. Three points $(x-h, y_1)$, (x, y_2) and $(x+h, y_3)$ are taken on the graph of the function defined by $y = 3^x$.

If $\dfrac{y_3 - y_1}{y_2} = \dfrac{3}{2}$ show that $h = \dfrac{\log 2}{\log 3}$ and hence calculate h to three significant figures.

26. Solve the equation $2^{x-1} = 3^{x+1}$ correct to three significant figures.

27. Without using tables evaluate

(i) $\log_3 \sqrt{3}$ (ii) $\log_3 \frac{1}{9}$

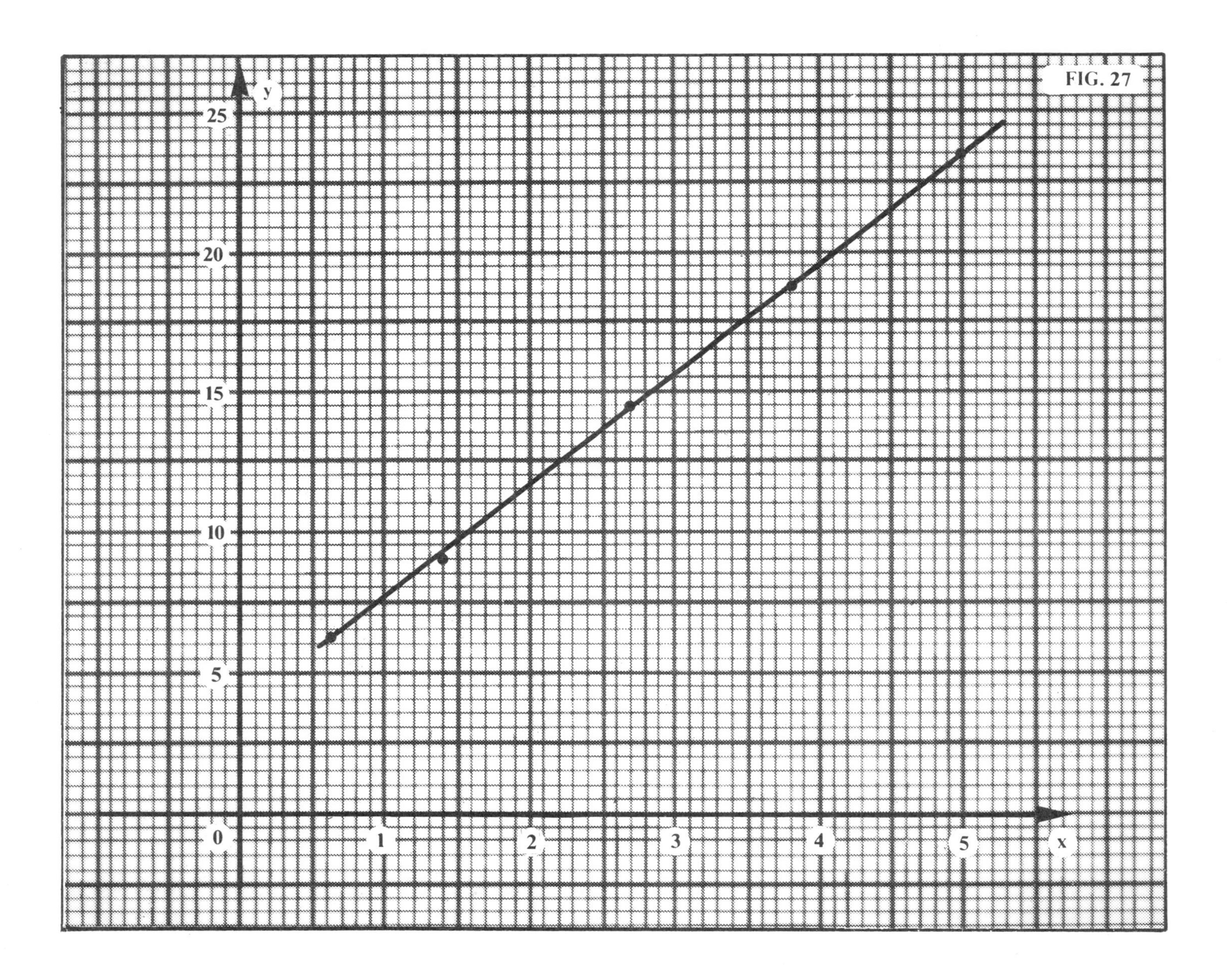
FIG. 27
y
25
20
15
10
5
0
1
2
3
4
5
x

28. Find the solution set of the equation

$$\log_7(5-2x)+\log_7(2+x) = 1.$$

DERIVATION FROM EXPERIMENTAL DATA OF A LAW OF THE FORM $y = ax^m$

If we obtain by experiment a table of corresponding values of two related variables x and y, we can show the relation between x and y on a graph. Moreover if the plotted points lie on a straight line or a straight line fits "*best*" the plotted points then the relation between x and y is of the form $y = mx+c$, where

$m =$ the slope or gradient of the line, and

$c =$ the y-coordinate of the point where the line cuts the y-axis.

Example 1

The corresponding values of two related quantities x and y obtained in an experiment are given in the table.

x	0·62	1·4	2·7	3·82	5
y	6·25	9	14·5	18·7	23·5

The relation between x and y is shown on the graph in figure 27 (opposite) where the "*best fitting*" line is drawn through the points.

Since the graph is a straight line its equation is $y = mx+c$.

To find m and c take two well separated points which lie on the line and substitute in the equation $y = mx+c$, for example $(1{\cdot}4, 9)$ and $(5, 23{\cdot}5)$. Substituting in $y = mx+c$ gives,

$$23{\cdot}5 = 5m+c \qquad (1)$$

and
$$9 = 1{\cdot}4m+c \qquad (2)$$

Subtracting gives $14{\cdot}5 = 3{\cdot}6m \Rightarrow m = \dfrac{14{\cdot}5}{3{\cdot}6} = 4{\cdot}03.$

To obtain c substitute this value of $m = 4{\cdot}03$ in equation (2)

$$9 = 1{\cdot}4m+c \Rightarrow c = 9-1{\cdot}4\times 4{\cdot}03$$
$$= 9-5{\cdot}64$$
$$= 3{\cdot}36$$

Hence the quantities x and y are related by the equation,

$$y = 4{\cdot}03x+3{\cdot}36$$

THE LAW OF THE FORM $y = ax^m$

Sometimes when we plot the corresponding values of the two related quantities x and y obtained in some experiment we do not get a straight line, but a curve whose equation we do not know or is difficult to obtain.

Example 2

The results of an experiment to find a relation between the variables y and x gave the following table.

x	1·09	1·22	1·33	1·64	1·85
y	2·34	2·88	3·47	5·25	6·61

If we plot y against x, we get a curve whose equation we do not know.

Suppose we try plotting $\log_{10} y$ against $\log_{10} x$, i.e. if $x \to \log_{10} x$, then $y \to \log_{10} y$.

$\log_{10} x$	0·038	0·085	0·125	0·215	0·266
$\log_{10} y$	0·37	0·46	0·54	0·72	0·820

The relation between $\log_{10} x$ and $\log_{10} y$ is shown on the graph in figure 28 (over) and the "*best fitting*" line is drawn through the points.

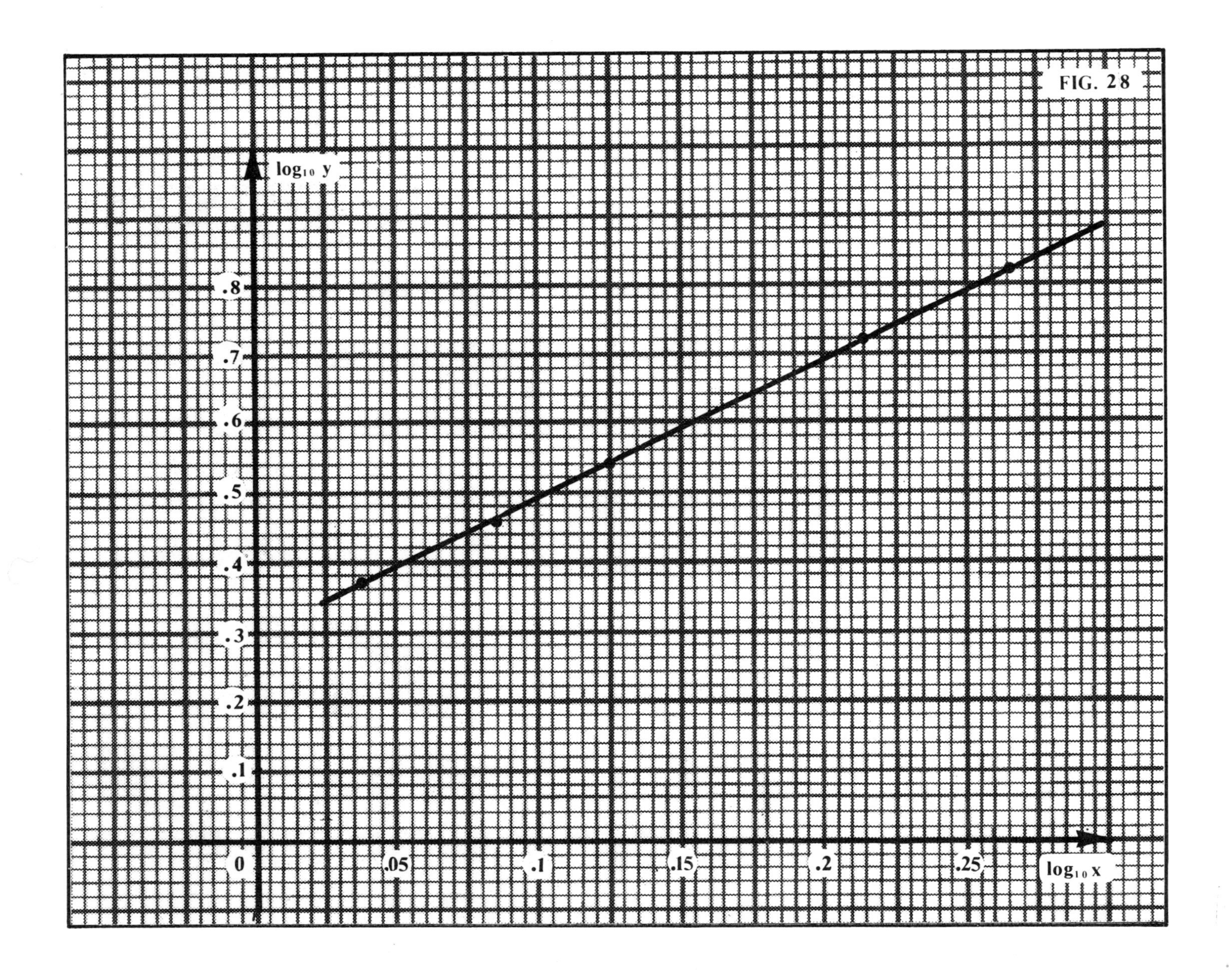
FIG. 28
$\log_{10} y$
.8
.7
.6
.5
.4
.3
.2
.1
0
.05
.1
.15
.2
.25
$\log_{10} x$

The graph is a straight line, and therefore its equation is of the form $\log_{10} y = m.\log_{10} x + c$, and writing $c = \log_{10} a$

then $\log_{10} y = m \log_{10} x + \log_{10} a \qquad (1)$

i.e. $\log_{10} y = \log_{10} x^m + \log_{10} a$

i.e. $\log_{10} y = \log_{10}(x^m a)$

i.e. $y = ax^m$, which is the formula relating x and y.

To find m and a, take two well separated points lying on the line and substitute in equation (1), for example, when

$$\left.\begin{aligned}\log_{10} x &= 0{\cdot}038\\ \log_{10} y &= 0{\cdot}37\end{aligned}\right\} \text{ and when } \left.\begin{aligned}\log_{10} x &= 0{\cdot}215\\ \log_{10} y &= 0{\cdot}72\end{aligned}\right\}$$

Hence from equation (1), $\log y = m \log x + \log a$, we get,

$$0{\cdot}37 = 0{\cdot}038m + \log a$$
$$0{\cdot}72 = 0{\cdot}215m + \log a$$

Subtracting gives $0{\cdot}35 = 0{\cdot}177m \Rightarrow m = \dfrac{0{\cdot}35}{0{\cdot}177} = 1{\cdot}98$

and $0{\cdot}72 = 0{\cdot}215m + \log a$

$$\Leftrightarrow \log a = 0{\cdot}72 - 0{\cdot}215m$$
$$= 0{\cdot}72 - 0{\cdot}215 \times 1{\cdot}98$$
$$= 0{\cdot}72 - 0{\cdot}426$$

i.e. $\log a = 0{\cdot}294$

$$\Rightarrow a = 1{\cdot}975$$

hence the quantities x and y are related by the equation

$$y = 1{\cdot}97x^{1{\cdot}98}$$

ASSIGNMENT 7.5

1. An experiment to find the relation between two variables x and y gave the table,

x	2·4	5·62	18·2	31·6	129
y	8·3	26·3	115	209	1100

Construct another table for $\log_{10} x$ and $\log_{10} y$ and by drawing the graph of $\log_{10} y$ against $\log_{10} x$, show that the relation between x and y is of the form, $y = ax^m$ and find a and m.

2. Corresponding values of two related quantities p and q are given in the table below.

p	50·1	195	501	708	1000
q	20·9	46·8	83·2	102	126

Draw the graph of $\log_{10} q$ against $\log_{10} p$. Show that $q = ap^n$ and find the value of a and n.

3. In the following plot $\log_{10} E$ against $\log_{10} p$

$\log_{10} p$	0·04	0·08	0·18	0·23	0·30
$\log_{10} E$	1·30	1·20	0·84	0·65	0·38

Show that there is a relation between p and E of the form $E = ap^n$ and find the value of a and n.

4. In the table below draw the graph of $\log_{10} E$ against $\log_{10} p$ and hence find the relation between E and p in the form $E = ap^m$ by finding the value of a and m.

$\log_{10} p$	0·08	0·142	0·2	0·26	0·31	0·36
$\log_{10} E$	0·76	0·59	0·48	0·39	0·28	0·18

5. Corresponding values of two related quantities x and y were obtained in an experiment; $\log_{10} y$ was plotted against $\log_{10} x$ and a "*best fitting*" straight line was drawn through the points as shown in figure 29. Find the formula for y in terms of x.

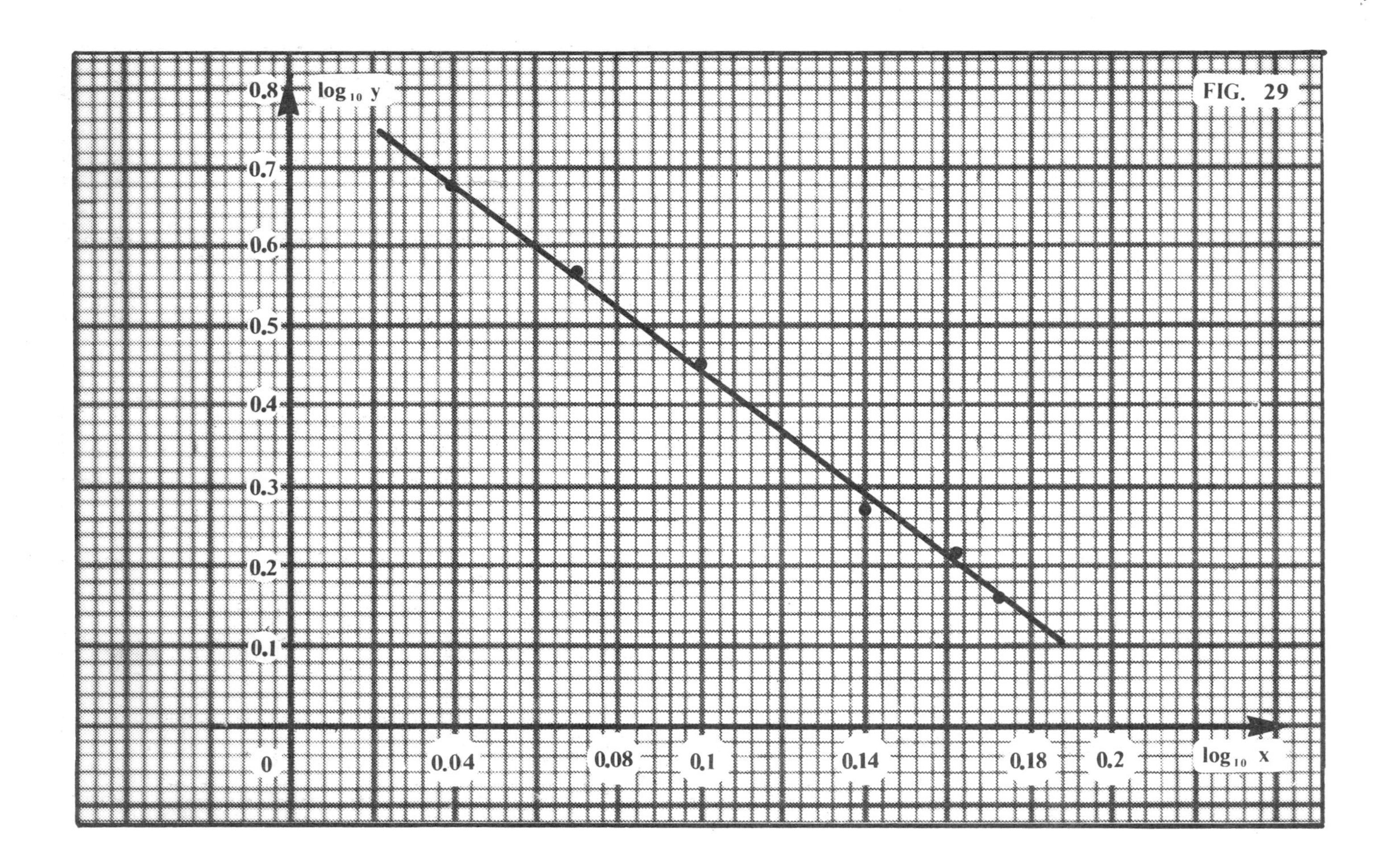

6. In an experiment the weight w grammes necessary to produce a given sag in the middle of a beam supported by two points d centimetres apart was determined for a number of values of d. When $\log_{10} w$ was plotted against $\log_{10} d$ the graph of figure 30 was obtained. Show that there is a relation of the form $w = ad^n$ between d and w, and find the value of a and n.

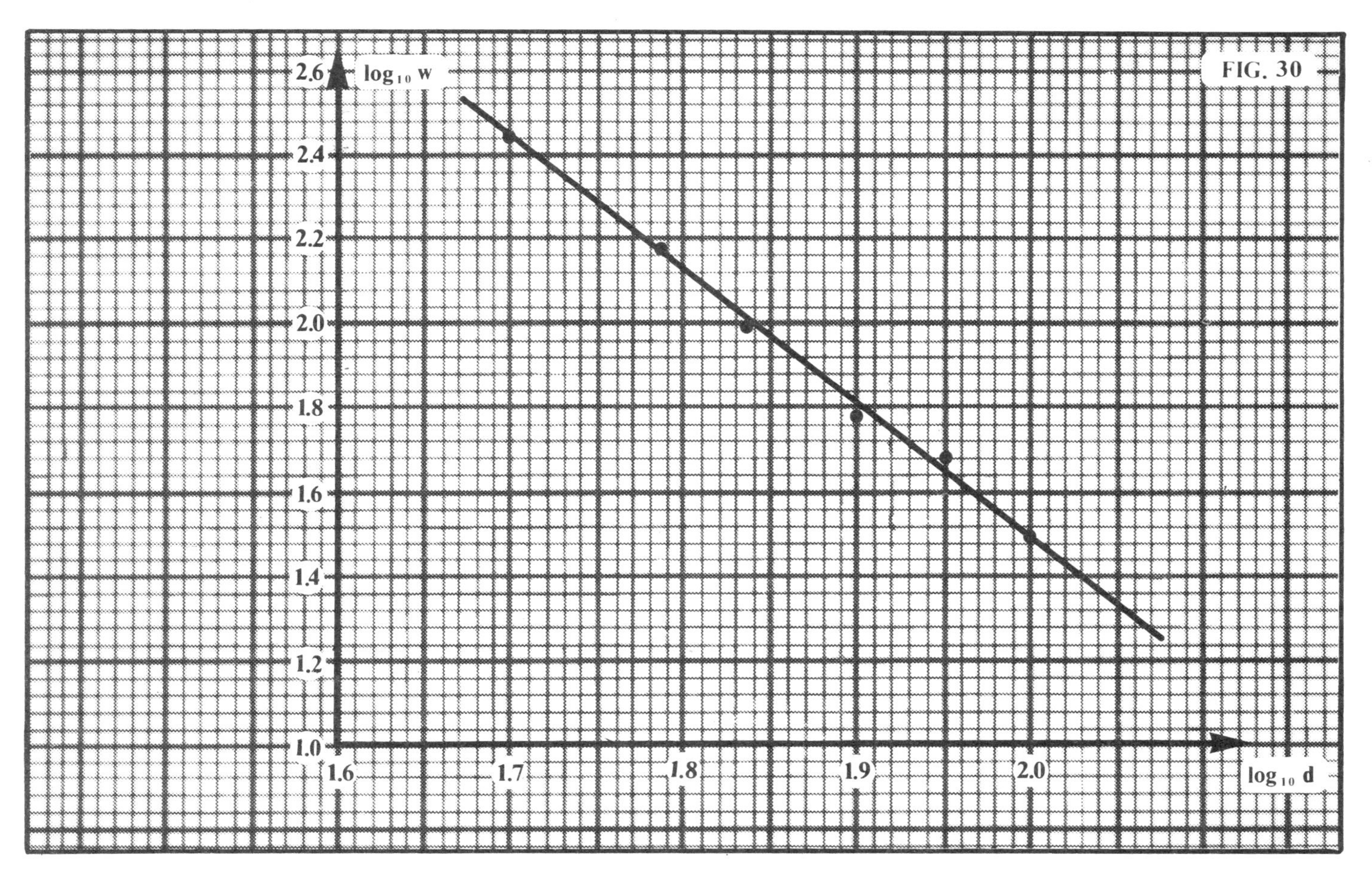

7. The table shows the time t minutes required to paint A square metres of wall. By plotting values of $\log A$ against $\log t$ express A in terms of t.

t (minutes)	2	4	5	6	8
A (square metres)	7·2	20	28	37	57

8. In an experiment to find the relation between two quantities p and v the following table was obtained.

p	40	60	70	80	90
v	10	33·5	53·5	80	114

Show that there is a relation between p and v of the form $p = kv^m$ and find the values of k and m.

7.6 CHANGE OF BASE IN LOGARITHMS

If we want to find $\log_3 2{\cdot}25$ we have no tables to base 3. We have therefore to change to base 10, and we use the formula

$$\log_b x = \frac{\log_a x}{\log_a b}$$

Proof: $\log_b x = m \Leftrightarrow x = b^m$

Hence

$$\frac{\log_a x}{\log_a b} = \frac{\log_a b^m}{\log_a b} = \frac{m \log_a b}{\log_a b}$$
$$= m$$
$$= \log_b x$$

Note: by putting $x = a$, we get $\log_b a = \dfrac{\log_a a}{\log_a b}$

Hence $\log_b a = \dfrac{1}{\log_a b}$ since $\log_a a = 1$

Or $\log_b a . \log_a b = 1$.

Example 1

Calculate $\log_3 2{\cdot}25$

$$\log_3 2{\cdot}25 = \frac{\log_{10} 2{\cdot}25}{\log_{10} 3} = \frac{0{\cdot}352}{0{\cdot}477}$$
$$= 0{\cdot}738$$

No.	log
0·352	$\bar{1}$·547
0·477	$\bar{1}$·679
0·738	$\bar{1}$·868

In many branches of mathematics logarithms are often to the base e.

The number

$$e = 1 + \frac{1}{1} + \frac{1}{1.2} + \frac{1}{1.2.3} + \frac{1}{1.2.3.4} + \ldots$$

and is obtained from the exponential series,

$$e^x = 1 + \frac{x}{1} + \frac{x^2}{1.2} + \frac{x^3}{1.2.3} + \frac{x^4}{1.2.3.4} + \ldots \text{ with } x = 1$$

Logarithms to this base e are called *natural logarithms* or *Naperian logarithms* named after Napier their inventor.

From the series for e we can approximate its value to any required degree of accuracy, to 8 places of decimals it is 2·71828182 and to 3 significant figures it is 2·72 which is the approximation we use.

Example 2

Calculate $\log_e 12$ where $e \doteqdot 2{\cdot}72$

$$\log_e 12 = \frac{\log_{10} 12}{\log_{10} e} = \frac{\log_{10} 12}{\log_{10} 2{\cdot}72}$$
$$= \frac{1{\cdot}079}{0{\cdot}435}$$
$$= 2{\cdot}48$$

No.	log
1·079	0·033
0·435	$\bar{1}$·638
2·48	0·395

ASSIGNMENT 7.6

1. Calculate the following correct to two significant figures.
   (i) $\log_3 8$ (ii) $\log_2 9$ (iii) $\log_5 21$
2. Calculate the following correct to two significant figures.
   (take $e = 2{\cdot}72$ and $\pi = 3{\cdot}14$)
   (i) $\log_3 e$ (ii) $\log_e 7$ (iii) $\log_2 \pi$ (iv) $\log_e \pi$
3. If $\log_x 10 = \log_{20} 16$, find the value of x.
4. If $\log_a 8 = \log_8 a$, find the value of a.
5. Solve for x the equation, $\log_3 x + 3\log_x 3 = 4$.
6. If $\log_a x + \log_{a^2} x = k$ show that $x^3 = a^{2k}$.
7. Prove that $\log_3 2, \log_3 4, \log_3 8, \log_3 16, \ldots$ is an arithmetic sequence. Write down the common difference in logarithmic form and calculate its value to two significant figures.
8. Prove that $\log_b a . \log_c b . \log_a c = 1$.
9. Solve for x the equation $\log_x 5 = \frac{1}{4}$.
10. Find the value of a such that $\log_a 9 = \log_{100} 3$.
11. Evaluate $\log_e 4{\cdot}75$ where $e = 2{\cdot}72$, correct to 3 significant figures.
12. Prove that
    (i) $\log_y x \log_z y \log_x z = 1$.
    (ii) $\dfrac{1}{\log_b a} + \dfrac{1}{\log_a b} = \log_b a + \log_a b$.

EXPONENTIAL GROWTH OR DECAY

Any function y which varies according to the formula $y \propto a^{px}$, $p \in R$ is said to grow or decay exponentially.

If $y \propto a^{px}$ then $y = ka^{px}$, $a > 0$, $a \neq 1$, $p \in R$, $p \neq 0$. Examples of this type of growth or decay are:

1. Certain living organisms.
2. Certain rates of chemical reactions.
3. The disintegration of radioactive elements.

In the formula $y = ka^{px}$, a is called the base and most formulae have $e = 2{\cdot}7183$ as the base.

$\therefore$ The *General Formula* is $y = ke^{px}$, where $e = 2{\cdot}72$ to three significant figures.

For $k > 0$ and $p > 0$ the formula shows growth.
For $k > 0$ and $p < 0$ the formula shows decay.

Example 1

A radioactive substance decays according to the formula $m_t = m_o e^{-kt}$ where m_o is the initial mass of the substance, m_t the mass remaining after t days and k is a constant. If the substance loses $\frac{1}{3}$ of its mass in 15 days, determine,

(i) the value of the constant k.
(ii) the half-life of the substance (i.e. the time taken for the substance to lose $\frac{1}{2}$ of its mass).

(i) After 15 days the mass remaining is $m_t = \frac{2}{3}m_o$. Hence from

$$m_t = m_o e^{-kt}$$
$$\tfrac{2}{3}m_o = m_o e^{-15k}$$
$$\Rightarrow \quad \tfrac{2}{3} = e^{-15k}$$
$$\Rightarrow \quad \tfrac{3}{2} = e^{15k}$$
$$\Rightarrow \log_{10} 1{\cdot}5 = 15k \log_{10} e$$

$$\Rightarrow \quad k = \frac{\log 1{\cdot}5}{15.\log 2{\cdot}72} \quad (e = 2{\cdot}72)$$

$$= \frac{0{\cdot}176}{15 \times 0{\cdot}435}$$

$$= 0{\cdot}027$$

(ii) We have $m_t = m_o e^{-0{\cdot}027t}$. For the substance to lose $\frac{1}{2}$ of its mass, then in time t $m_t = \frac{1}{2} m_o$

$$\therefore \quad \tfrac{1}{2} m_o = m_o e^{-0{\cdot}027t}$$

$$\Rightarrow \quad \tfrac{1}{2} = e^{-0{\cdot}027t}$$

$$\Rightarrow \quad 2 = e^{0{\cdot}027t}$$

$$\Rightarrow \log 2 = 0{\cdot}027t \log 2{\cdot}72$$

$$\Rightarrow \quad t = \frac{\log 2}{0{\cdot}027 \times \log 2{\cdot}72} = \frac{0{\cdot}301}{0{\cdot}027 \times 0{\cdot}435}$$

$$= 25{\cdot}7 \text{ days.}$$

i.e. the substance loses $\frac{1}{2}$ its mass in 25·7 days.

ASSIGNMENT 7.7

1. The distance s metres travelled by a moving body in a time t seconds is given by the formula $s = e^{-kt}$. If the body travels a distance of $\frac{3}{4}$ of a metre in 1 second determine the value of k.

2. A radio-active substance decays at a rate given by the formula $m_t = m_o e^{-0{\cdot}02t}$ where m_o is the initial mass and m_t is the mass remaining after t years.

   (i) If a mass of 500 mg disintegrates for 20 years, how much remains after that time?

   (ii) How many years are required for half the mass to disintegrate?

3. The current c_t flowing in a wire is given by the equation $c_t = c_o e^{-0{\cdot}27t}$ where c_o is the initial strength of the current and c_t the strength of the current after t seconds. How long will it take the current to be reduced to half its original strength?

4. Atmospheric pressure p_h at a height h km above the earth's surface is given by $p_h = p_o e^{-kh}$ where p_o is the pressure on the earth's surface. At heights h_1 and h_2 $(h_2 > h_1)$ the pressure is p_1 and p_2. Show that the difference in heights h_2 and h_1 is given by,

$$h_2 - h_1 = \frac{\log_e p_1 - \log_e p_2}{k}$$

5. A radioactive element decays according to the formula $m_t = m_o e^{-0{\cdot}02t}$ where m_o is the initial mass and m_t the mass remaining after t days.

   (i) If the substance loses $\frac{1}{2}$ its mass in T days, show that $T = 50 \log_e 2$.

   (ii) If a mass of 250 mg disintegrates, find the mass remaining after 10 days.

6. The intensity I_x of a source of light diminishes on passing through a glass medium according to the law $I_x = I_o e^{-\alpha x}$ where I_o is the initial intensity, I_x the intensity after passing through x cm of the glass medium and α is a constant.

   (i) If the intensity is reduced by one third on passing through 5 cm of the glass, calculate to 2 significant figures the value of the constant α.

   (ii) Calculate how far through the medium the light has travelled if the intensity is reduced by half its original value.

7. Water flows from a reservoir according to the law $h_t = h_o 2^{-kt}$, where h_o is the original height of the water level and h_t the height of the water level after t days.

   (i) If the water level decreases by half its original height in T days, show that $kT = 1$.

(ii) If the water level decreases by one fifth of its original height in 10 days, show that,

$$k = \frac{\log_{10} 1{\cdot}25}{10 . \log_{10} 2}$$

and hence deduce the number of days required to reduce the water level to half its original height.

8. A radioactive element disintegrates according to the law $m = m_o e^{-pt}$ where m_o is the initial mass of the element and m the mass remaining after t seconds. If at the end of one second half the mass has disintegrated show that $e^{-p} = \frac{1}{2}$.

   If at the end of t seconds $\frac{3}{5}$ of the mass has disintegrated show that

$$(\tfrac{1}{2})^t = \tfrac{2}{5}$$

   and hence evaluate t to three significant figures.

9. The temperature of a cooling liquid is measured at intervals of five minutes and the difference $T°$ centigrade from room temperature is given in the table

t (minutes)	0	5	10	15	20
T (°C)	45·7	24	12·6	6·9	3·8
$\log_{10} T$	1·66	1·38	1·10	0·84	0·58

   Taking 1 cm to represent 1 unit on the t axis and 4 cm to represent 1 unit on the log T axis draw the graph of log T against t.

   Is it reasonable to assume a law of variation of the form $T = Aa^t$? If so find the values of A and a.

ASSIGNMENT 7.8

Objective items testing Sections 7.1–7.7

Instructions for answering these items are given on Page 22.

1. If $\log a = 2 \log b - \log 3$ then a is equal to,

   A. $b^2 - 3$
   B. $2b - 3$
   C. $3b^2$
   D. $\frac{b^2}{3}$
   E. $\frac{2b}{3}$

2. $\log_{\sqrt{2}} 8$ equals

   A. 3
   B. $\frac{1}{2}$
   C. $\frac{1}{4}$
   D. 4
   E. 6

3. The value of $(27^{-\frac{1}{3}} + 16^{\frac{1}{2}})^{-1}$ is

   A. $\frac{1}{7}$
   B. 7
   C. $\frac{13}{4}$
   D. $\frac{3}{13}$
   E. $\frac{4}{13}$

4. Which one of the following gives the solutions of the equation $\log_2(x^2-14x) = 5$?

   A. $7 \pm \sqrt{74}$

   B. $16, 16$

   C. $7 \pm \sqrt{59}$

   D. $-2, 16$

   E. None of these.

5. Which one of the following statements is **false**?

   A. The function $f: x \to a^x$, $a > 1$ has domain the set of all real numbers.

   B. $\log_a x = 1 \Rightarrow x = a$

   C. $1 + \log_3 2 = \log_3 6$

   D. $\dfrac{\log_2 9}{\log_2 3} = 3$

   E. $\log_3 2 . \log_2 3 = 1$

6. A radio-active substance obeys the law $m = m_o 10^{-kt}$ An expression for t in terms of m, m_o and k would be

   A. $k(\log m - \log m_o)$

   B. $\dfrac{1}{k}\log_{10}\dfrac{m}{m_o}$

   C. $\dfrac{1}{k}(\log m_o + \log m)$

   D. $\dfrac{1}{k}\log_{10}\dfrac{m_o}{m}$

   E. some other expression involving m, m_o and k.

7. If $\log_{10} x = p - \log_{10} y$, which one of the following is true?

   A. $x = \dfrac{p}{y}$

   B. $x = y10^p$

   C. $x = p - 10^y$

   D. $x = x^{-p}$

   E. $x = \dfrac{10^p}{y}$

8. The solution set of the equation

   $$\log_{10} x^2 - 4\log_{10} x + 6 = 0,\ x > 0, \text{ is}$$

   A. $\{10\}$

   B. $\{100, 10\}$

   C. $\{1000\}$

   D. $\{1, 10\}$

   E. none of these values.

9. If $p = q^x$ then the value of x is given by

   (1) $\log p - \log q$

   (2) $\dfrac{\log p}{\log q}$

   (3) $\log_q p$

10. Consider the following statements.

   (1) $\log\dfrac{a}{b} - \log\dfrac{b}{a} = 2\log\dfrac{a}{b}$

   (2) $\log 2a = 2\log a$

   (3) $\log\frac{1}{4} + \log 4 = 0$

11. (1) $9^x = 243$ (2) $\log_9 243 = x$.

12. (1) $a + b = 2$ (2) $\log a + \log b = \log 2$.

UNIT 8: DEDUCTIVE REASONING

DEDUCTIVE REASONING: INTRODUCTION

In deductive reasoning we use statements which are called **premises.** These statements are *known to be true* or are *taken for granted.* From these statements or premises a conclusion is drawn.

Example 1

	All big cars are expensive:	First premise
	A Rover is a big car:	Second premise
∴	A Rover is an expensive car:	Conclusion

Note 1: If the first two statements are true, then the conclusion is true.

Note 2: The premise above can be represented by a **Venn diagram.**

Let $E = \{\text{all cars}\}$, $B = \{\text{big cars}\}$, $X = \{\text{expensive cars}\}$, then,

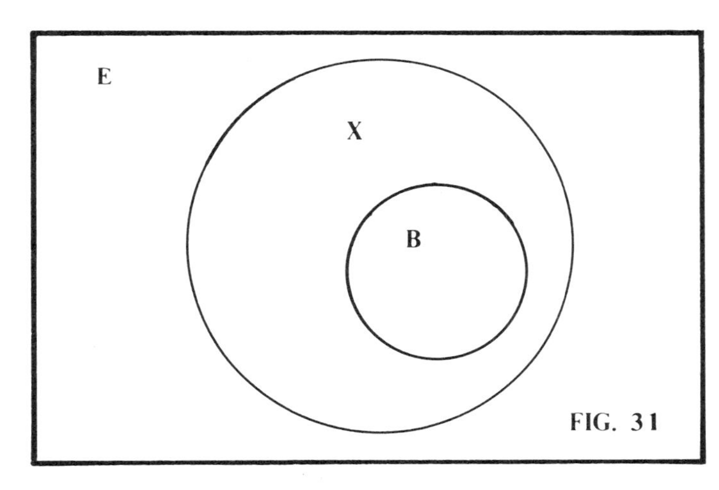

FIG. 31

From the premises or the Venn diagram of figure 31, we see that the statement,

(i) if a car is big it is expensive, is True;

(ii) if a car is expensive it is big, is False;

(iii) if a car is not big it is not expensive, is False;

(iv) if a car is not expensive it is not big, is True.

Example 2

	All scientists have a knowledge of physics.	First premise
	Some scientists are engineers.	Second premise
∴	Some engineers have a knowledge of physics.	Conclusion

This is the valid conclusion that can be drawn from the first two statements.

Let $P = \{$people who have a knowledge of physics$\}$
$S = \{$scientists$\}$
$G = \{$engineers$\}$
then this can be illustrated by a Venn diagram, (figure 32), where
$E = \{$all people$\}$

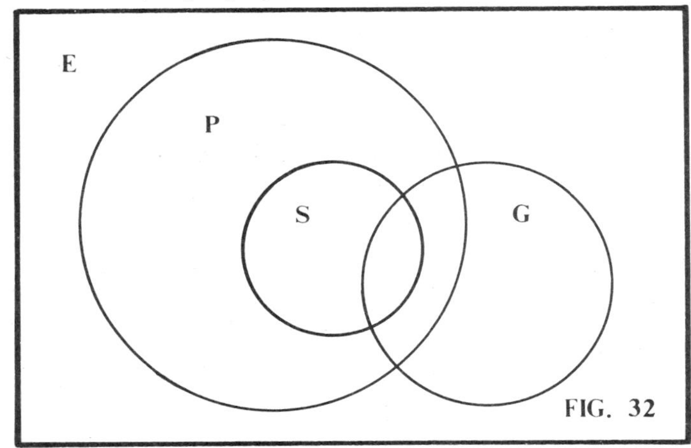

FIG. 32

Example 3

"All successful politicians are noted for their tact."
"All members of parliament are successful politicians."

From these two statements one or more of the following conclusions is/are valid.

I. Politicians who are not tactful are not successful.
II. Politicians not successful are not members of parliament.
III. Politicians not members of parliament are not tactful.
IV. Politicians not tactful are not members of parliament.

To answer this question it is best to draw a Venn diagram.

Let $T = \{$tactful people$\}$
$S = \{$successful politicians$\}$, then $S \subset T$
$M = \{$members of parliament$\}$, then $M \subset S$

Hence the Venn diagram:

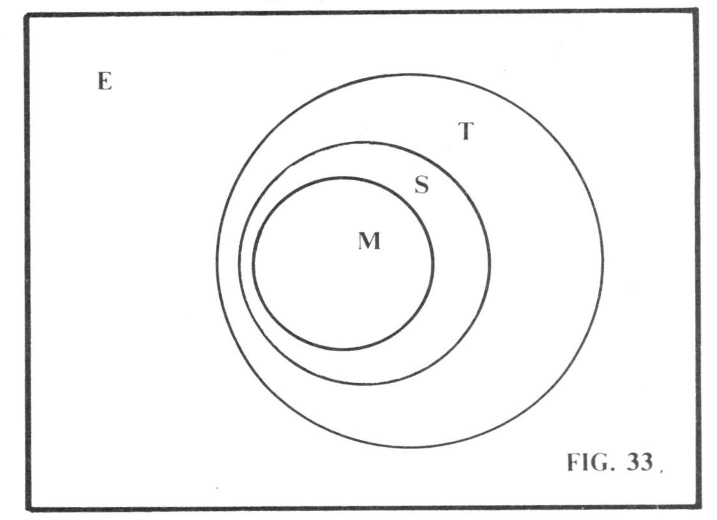

FIG. 33.

From the diagram, we see that:

Statement I, "Politicians who are not tactful are not successful," is valid.
Statement II, "Politicians not successful are not members of parliament," is valid.
Statement III, "Politicians not members of parliament are not tactful," is not valid.
Statement IV, "Politicians not tactful are not members of parliament," is valid.

ASSIGNMENT 8.1

Deduce where possible a valid conclusion from each set of premises in questions 1–4.

1. All biologists have studied botany.
   Miss Brown is a biologist.

2. All quadrilaterals have four and only four sides.
   ABCDE has five sides.

3. Every rhombus has two axes of symmetry.
   Every rhombus is a kite.
   Square ABCD is also a rhombus.

4. All plants use energy from the sun.
   A truffle uses energy from the sun.

5. Every civil servant has passed his O levels.
   Some civil servants have passed their A levels.
   What can be said about,
   (i) Mr. Smith who is a civil servant?
   (ii) Miss Green who has passed her O levels?
   (iii) Mr. Jones who has passed his A levels?
   (iv) Miss James who has not passed her O levels?

6. Only persons whose names are on the voter's roll may vote. In order to vote a person must reside in Britain.
   What can you deduce with certainty about the present circumstances of:
   (i) John Sim who lives in Britain?
   (ii) Mr. Scott whose name is on the voter's roll?
   (iii) Miss Hill whose name is on the voter's roll and lives in Britain?
   (iv) Mary Blythe who voted at the last election?

7. Examine and draw a Venn diagram to illustrate the validity of the following argument,
   "All Polonians are blue-eyed."
   "All blue-eyed people have red hair."
   Hence all Polonians have red hair.

8. "All tall buildings are more than eight storeys high."
   "All skyscrapers are tall."
   From these two statements, which of the following are valid conclusions?
   I. Buildings which are not tall are not skyscrapers.
   II. Buildings which are not skyscrapers are not more than eight storeys high.
   III. Buildings not more than eight storeys high are not skyscrapers.
   IV. Buildings not more than eight storeys high are not tall.

9. "All old paintings are valuable."
   "All Rembrandts are old paintings."
   Which of the following statements are valid conclusions?
   I. All Rembrandts are valuable.
   II. Some Rembrandts are not old paintings.
   III. No Rembrandt is valuable.
   IV. Some Rembrandts are not valuable.

10. "All forces are vectors."
    "All torques are forces."
    Which of the following statements are valid conclusions?
    I. No vector is a torque.
    II. Some torques are vectors.
    III. All torques are vectors.
    IV. All vectors are torques.

STATEMENTS AND NEGATION OF STATEMENTS 8.2

A **statement** in mathematics is a sentence which is meaningful. It must be either *true* or *false* but not both true *and* false.

Example 1

(i) 17 is a prime number.
There are seven days in a week.
21 is greater than 19.
These are all true statements.

(ii) 24 is a prime number.

9 is less than 6.
There are eleven months in a year.

These are all false statements.

(iii) There are a thousand in a kilogramme.
I am older than Tom.
$5x-2 = 13$.
x is greater than 8.

Such sentences are called **open sentences** since we cannot say whether they are true or false. To make these sentences true or false, we must define the thousand objects in a kilogramme, my age and Tom's age, and state a replacement for x.

For example, "9 is greater than 8" is true
but "7 is greater than 8" is false.

NEGATION

Given a statement, we can make another statement which denies the given statement, and the negation may be said in several ways.

Example 2

The statement "The girl has red hair" may be negated thus:
The girl does not have red hair.
It is not true that the girl has red hair.
It is false that the girl has red hair.

Note 1: "The girl has brown hair," is not the negation of the statement "the girl has red hair," for both statements may be false.

Note 2: A statement and its negation must exhaust all possibilities. For example, "The girl has red hair" or "The girl does not have red hair," ensures there is no other possibility.

Example 3

Write down the negation of the following statements,

(i) John is taller than Tom.

(ii) $x < y$

(iii) This coat is not blue.

Negations,

(i) John is not taller than Tom.

(ii) $x \nless y$

(iii) This coat is blue.

ASSIGNMENT 8.2

Which of the following are true statements?

1. The area of a rectangle is length times breadth.
2. There are seven days in a week.
3. Every number which is divisible by 2 is divisible by 4.
4. If $n \in N$ then $2n$ is even.
5. $5x-1 = 9$
6. $(2n)^3 = 8n^3$
7. If a quadrilateral has 4 equal sides it is a square.
8. -9 is less than -3.
9. Jane is younger than Mary.
10. For all $x \in R$, $(x-2)^2 \geqq 0$.
11. If 3 is a factor of n and 2 is a factor of n, then 6 is a factor of n.

Write down the negation of the following.

12. Tom is 16 years old.
13. $\triangle$ABC is not an isosceles triangle.
14. James Scott is a good footballer.
15. That hat is brown.
16. Mary Jones does not wear spectacles.
17. It is not true that 337 is a prime number.
18. It is false that James is not older than Paul.
19. $a^m \times a^n = a^{mn}$.
20. This boy has not blue eyes.

QUANTIFIED STATEMENTS AND THEIR NEGATION

8.3

In example 2, section 8.1,

"All scientists have a knowledge of physics",
"Some scientists are engineers", we used the words *All* and *Some*.

These words are called **quantifiers.**

All is called the **universal quantifier** and includes every member of a set.

Some is called the **existential quantifier** and means *at least one.*

For example, "Some scientists are engineers." This is taken to mean that there exists, or there is at least, one scientist who is an engineer.

Illustrating these by means of a Venn diagram,

let $E = \{\text{all people}\}$
$P = \{\text{people who have a knowledge of physics}\}$
$S = \{\text{scientists}\}$
$G = \{\text{engineers}\}$

Then "All scientists have a knowledge of physics" means $S \subset P$; (see figure 34)

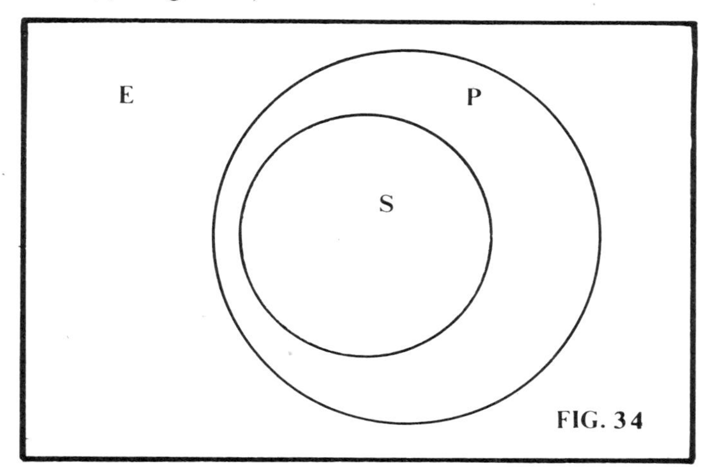

FIG. 34

and "Some scientists are engineers" means $S \cap G \neq \varnothing$; (see figure 35).

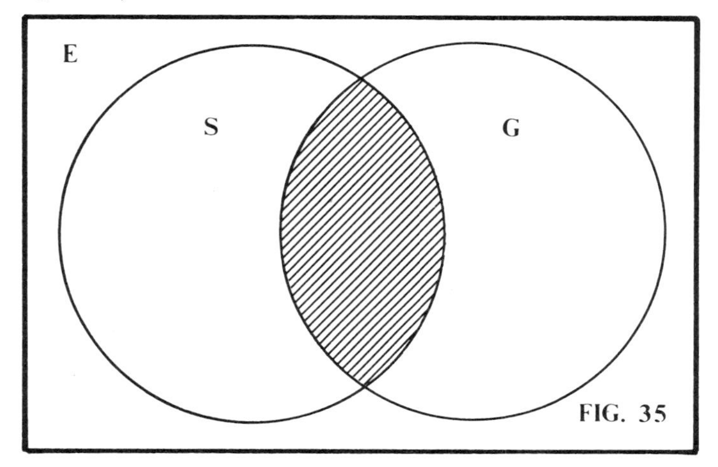

FIG. 35

Note that *some* means *one or more* and does not exclude the possibility of *all.*

For example, "Some men are mortal," does not exclude the possibility that "All men are mortal."

Again, consider the equations

(i) $x^2-9=0$

(ii) $x^2-4x+4=(x-2)^2$

The first equation is true for the replacements $x=3$ and $x=-3$.

Hence, for some x (or there exists an x such that) $x^2-9=0$, $x \in R$, is true.

The second equation $x^2-4x+4=(x-2)^2$ is true for *all* values of $x \in R$.

Hence for *all* x, $x^2-4x+4=(x-2)^2$, $x \in R$ is true, and we say that $x^2-4x+4=(x-2)^2$, $x \in R$, is an **identity.**

NEGATION OF QUANTIFIED STATEMENTS

To negate a statement we deny the truth of the statement. Hence to negate "All scientists have a knowledge of physics", there must be some (at least one) who have no knowledge of physics.

Hence the negation of "All scientists have a knowledge of physics" is "Some scientists do not have a knowledge of physics".

Again, since "Some scientists are engineers" implies at least one scientist is an engineer, then to negate this we say "No scientist is an engineer."

Hence the negation of "Some scientists are engineers" is "No scientist is an engineer," or "All scientists are not engineers."

Example 1

Negate the following quantified statements.

(i) All dogs hate cats.
(ii) Some triangles are isosceles.
(iii) No schoolboy likes football.
(iv) Some girls do not like spectacles.

Negations are:

(i) Some dogs do not hate cats.
(ii) All triangles are not isosceles or no triangle is isosceles.
(iii) Some schoolboys like football.
(iv) All girls like spectacles.

Hence the summary for statements and their negations:

Statement	*Negation*
All Ps are *Qs*.	*Some Ps* are *not Qs*.
Some Ps are *Qs*.	*All Ps* are *not Qs* or *No Ps* are *Qs*.
All Ps are *not Qs*. or *No Ps* are *Qs*.	*Some Ps* are *Qs*.
Some Ps are *not Qs*.	*All Ps* are *Qs*.

Example 2

What valid conclusion can be drawn from the statements,

"All Swiss people speak French."
"Some Swiss people speak German."?

Let $F=$ {French speaking people}
$S=$ {Swiss people}, then $S \subset F$.
$G=$ {German speaking people}, then $S \cap G \neq \varnothing$.

Hence the Venn diagram (figure 36).

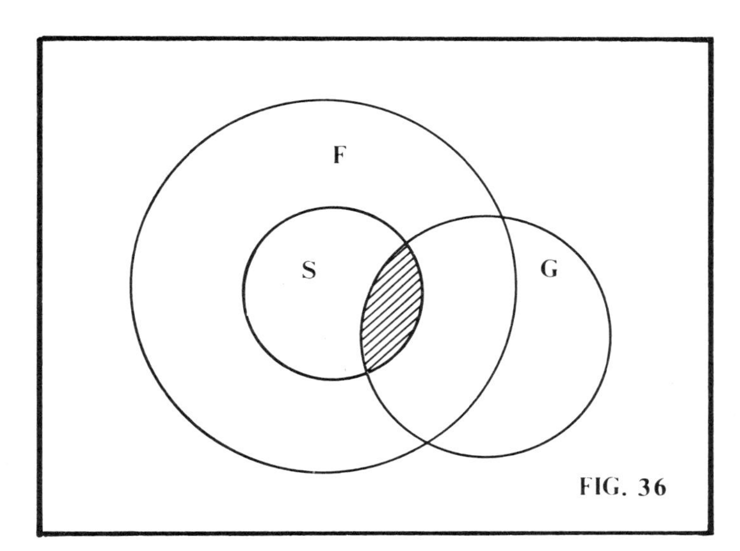

FIG. 36

and the valid conclusion is,
"Some Swiss people speak both French and German."

Example 3

From the statements,

"No boy dislikes football."
"Some rugby fans are boys."

say which of the following are valid conclusions.

I. Some rugby fans like football.
II. No rugby fans like football.
III. All rugby fans dislike football.
IV. All boys who like rugby like football.

Draw a Venn diagram with,

$F = \{\text{people who like football}\}$

$B = \{\text{boys}\}$, then $B \subset F$

$R = \{\text{rugby fans}\}$, then $B \cap R \neq \varnothing$

Hence diagram (figure 37).

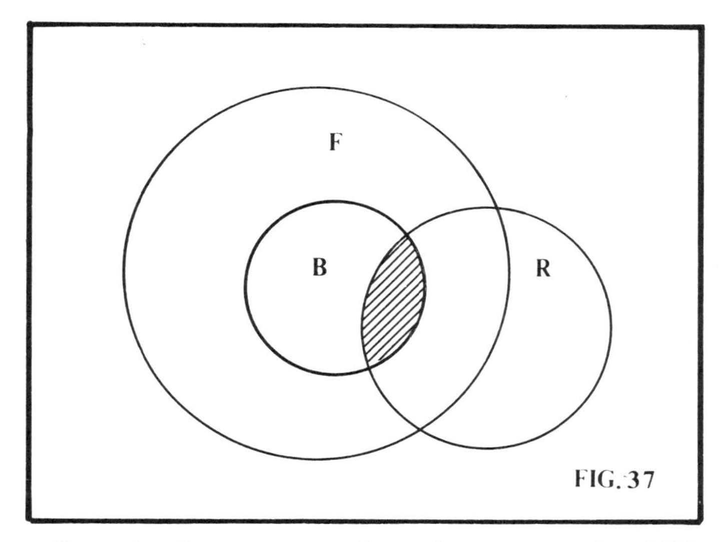

FIG. 37

From the diagram we see that only statements I and IV are valid conclusions.

ASSIGNMENT 8.3

Draw Venn diagrams to illustrate the following quantified statements. What valid conclusion can you make from each?

1. All cats are four legged animals.
   Some cats enjoy milk.

   Let $C = \{\text{cats}\}$
   $E = \{\text{animals}\}$
   $F = \{\text{four legged animals}\}$
   $A = \{\text{animals which enjoy milk}\}$

2. All sine functions are periodic functions. Some sine functions do not have a period of 2π.

Let $E = \{\text{all functions}\}$
$T = \{\text{functions having a period of } 2\pi\}$
$S = \{\text{sine functions}\}$
$P = \{\text{periodic functions}\}$

3. All men are sports fans.
Some men like music.

Let $M = \{\text{men}\}$
$S = \{\text{sports fans}\}$
$R = \{\text{people who like music}\}$

and let all be sub-sets of the universal set E, the set of all people.

4. All trees are flowering-plants.
Some trees are evergreen plants.

Let $T = \{\text{trees}\}$
$F = \{\text{flowering plants}\}$
$G = \{\text{evergreen plants}\}$

and let all be subsets of the universal set E, the set of plants.

5. Negate the following quantified statements,
   (i) All dogs hate cats.
   (ii) Some dogs hate cats.
   (iii) Some dogs do not hate cats.
   (iv) All dogs do not hate cats.

6. Negate the following,
   (i) All trees are plants.
   (ii) Some men are not sports fans.
   (iii) Some girls like hockey.
   (iv) No man has blond hair.
   (v) No scientist is not an engineer.
   (vi) All boys are not good at mathematics.

7. Negate the following and say whether the statement or the negation is true.
   (i) No triangle is isosceles.
   (ii) Some matrices are commutative under multiplication.
   (iii) All numbers of the form $6n-1$ are prime, $n \in N$.
   (iv) All mappings are not functions.
   (v) There exists a value of θ such that $\cos\theta > 1$.
   (vi) All functions have inverses.

In questions 8–12, say whether the conclusion drawn from the premises is true or false. Draw a Venn diagram to check your answers.

8. Some books are expensive.
All first editions are expensive.
Therefore some books are first editions.

9. All musicians are artistes.
Some men are musicians.
Hence some men are artistes.

10. All biologists are scientists.
Some women are biologists.
Hence some women scientists are not biologists.

11. Some men are football fans.
Some men are rugby fans.
Therefore some rugby fans are football fans.

12. Boys are either wild or quiet.
Boys are either football fans or rugby fans.
Therefore some quiet boys are football fans.

State a valid conclusion which may be drawn from each set of given premises in questions 13–15.

13. No sensible person has bad manners.
    Some sensible people have quick tempers.

14. Some antiques are beautiful.
    Beautiful things are a joy for ever.

15. Some kites are not rhombuses.
    All squares are rhombuses.

16. All boys like football.
    Some cricket fans are boys.

    From these two statements say which of the following conclusions is/are valid.

    I. All cricket fans like football.
    II. Some cricket fans are football fans.
    III. Some football fans like cricket.
    IV. No boy likes football and cricket.

IMPLICATIONS AND CONVERSES—EQUIVALENCE

8.4

An **implication** is a statement made up of two parts so that if the first part is true then the second part is also true. We use the words, if . . . then . . .

For example, "If I study then I shall pass the examination," or simply, "If I study, I shall pass the examination."

We say the first part implies the second. "If I study" implies "I shall pass the examination." The symbol $\Rightarrow$ is used for implies.

Examples of implications are:

(i) John crosses a busy road haphazardly $\Rightarrow$ he will be knocked down.

(ii) A child plays with fire $\Rightarrow$ it will burn itself.

(iii) $x = 4 \Rightarrow x^2 = 16$.

(iv) $\triangle ABC$ has two equal sides $\Rightarrow$ $\triangle ABC$ has two equal angles.

(v) $\cos\theta = \frac{1}{2} \Rightarrow \theta = 60°$ or $300°$, for $0 \leqq \theta \leqq 360$.

(vi) n is divisible by $9 \Rightarrow n$ is divisible by 3.

If we interchange the parts of an implication we obtain the **Converse** of the implication, which may or may not be true.

From the above examples,

From (iv)

Implication: $\triangle ABC$ has two equal sides $\Rightarrow$ $\triangle ABC$ has two equal angles—True.
Converse: $\triangle ABC$ has two equal angles $\Rightarrow$ $\triangle ABC$ has two equal sides—True.

From (v)

Implication: $\cos\theta = \frac{1}{2} \Rightarrow \theta = 60°$ or $300°$ for $0 \leqq \theta \leqq 360$—True.
Converse: $\theta = 60°$ or $300° \Rightarrow \cos\theta = \frac{1}{2}$ for $0 \leqq \theta \leqq 360$—True.

From (iii)

Implication: $x = 4 \Rightarrow x^2 = 16$—True.
Converse: $x^2 = 16 \Rightarrow x = 4$—False.

but

Implication: $x = 4$ or $-4 \Rightarrow x^2 = 16$—True.

and

Converse: $x^2 = 16 \Rightarrow x = 4$ or -4—True.

From (vi)

Implication: n is divisible by $9 \Rightarrow n$ is divisible by 3—True.
Converse: n is divisible by $3 \Rightarrow n$ is divisible by 9—False.

If an implication and its converse are both true then the truth of one implies the truth of the other. We say that one statement is **equivalent** to the other, i.e. both implication and converse are equivalent statements. For this we use the symbol $\Leftrightarrow$ which means,

"Implies and is implied by,"
or "If and only if,"
or "If one is true then the other is true."

Thus implication (iv) and its converse may be written

$\triangle$ABC has two equal sides $\Leftrightarrow$ $\triangle$ABC has two equal angles

also $x = 4$ or $-4 \Leftrightarrow x^2 = 16$

These are sometimes called two-way implications.

Example 1

What two true statements are contained in the two-way implication,

"Point P(a, b) lies on the circle $x^2+y^2 = r^2 \Leftrightarrow a^2+b^2 = r^2$."?

The two statements are,

"If P(a, b) lies on the circle $x^2+y^2 = r^2$ then $a^2+b^2 = r^2$,"

and "If $a^2+b^2 = r^2$ then the point P(a, b) lies on the circle $x^2+y^2 = r^2$."

Example 2

Write as a two-way implication the statements

(i) "n is even $\Rightarrow$ n is divisible by 2."
"n is divisible by 2 $\Rightarrow$ n is even."

(ii) "Two straight lines of gradient m_1 and m_2 are parallel $\Rightarrow m_1 = m_2$."
"$m_1 = m_2 \Rightarrow$ the lines are parallel."

The two-way implications are,

(i) "n is even $\Leftrightarrow$ n is divisible by 2."

(ii) "Two straight lines with gradient m_1 and m_2 are parallel $\Leftrightarrow m_1 = m_2$."

THE USE OF A COUNTER-EXAMPLE

A counter-example may be used to disprove a statement which claims to be true for all cases.

Example 3

Every number of the form $6n-1$, $n \in N$, is a prime number.

The statement is true for $n = 1, 2, 3, 4, 5$.

But $n = 6$ gives $6n-1 = 35$ which is not prime. Hence we have disproved the statement by producing the counter-example $n = 6$.

For what value of n does the statement break down again?

ASSIGNMENT 8.4

Complete the following mathematical implications.

1. If two triangles are equiangular, then...
2. If the sum of the lengths of two sides of a triangle is 11 units, then the length of the third side...
3. If $(x-\alpha)$ is a factor of $f(x)$, then...
4. $\triangle$ABC has 3 equal sides $\Rightarrow$ $\triangle$ABC has...

Examine the statements in questions 5–9 and determine whether the conclusion is valid.

5. If Mr. Smith gets a rise in salary, he will buy a new car.
Mr. Smith has received a rise in salary.
Therefore, Mr. Smith has bought a new car.
6. If Mr. Smith gets a salary rise, he will buy a new car.
Mr. Smith has bought a new car.
Therefore, Mr. Smith has received a rise in salary.
7. If it is raining, I shall get wet.
It is not raining.
Therefore, I shall not get wet.

8. If butter costs 30p a pound then butter is expensive.
   Butter is expensive.
   Therefore, butter costs 30p a pound.

9. (i) All novelists have a good command of English.
   Henry Maris is a novelist.
   Therefore, Henry Maris has a good command of English.

   (ii) All novelists have a good command of English.
   James Ennis has a good command of English.
   Therefore, James Ennis is a novelist.

In the following questions, write down the converse of each implication. State whether the converse is true and whether the implication can be replaced by a two-way implication. If it can rewrite the implication and converse using the equivalence sign.

10. If $\triangle ABC$ is equilateral, it is isosceles.

11. n divides $x \Rightarrow n$ divides x^2.

12. $x > y \Rightarrow -x < -y$.

13. $2^n = p \Rightarrow \log_2 p = n$.

14. If a car is big then it is expensive.

15. $a > 3 \Rightarrow a^2 > 9$, for all $a \in R$.

16. $A \cap B = A \Rightarrow A \subset B$.

Use a counter-example to disprove any of the following statements you think are false.

17. $a > b \Rightarrow a^2 > b^2$, $a, b \in R$.

18. Thirty days have September, April, June and November. All the rest have thirty-one.

19. All simultaneous equations have a solution.

20. If $x \in R$ then $x^2 > x$.

21. If n divides ab, then n divides a or n divides b, $(a, b, n \in N)$.

22. m odd and n odd $\Rightarrow mn$ is odd.

23. No triangle has only two axes of symmetry.

ASSIGNMENT 8.5

Objective items testing Sections 8.1–8.4

Instructions for answering these items are given on Page 22.

1. The negation of the statement "All teachers are human" is
   A. All teachers are not human.
   B. Some teachers are not human.
   C. No teachers are human.
   D. No teachers are not human.
   E. Some humans are not teachers.

2. Which of the following is a valid deduction from the statement "All pupils who walk to school are active"?
   A. All active pupils walk to school.
   B. All pupils who do not walk to school are not active.
   C. No pupil who is inactive walks to school.
   D. Some pupils who are active do not walk to school.
   E. Some pupils who are not active walk to school.

3. Which of the following is a valid deduction from the statement, "Every old painting is valuable"?
   A. All valuable paintings are old.
   B. If a painting is not old it is not valuable.
   C. No valuable paintings are old.
   D. If a painting is valuable it is old.
   E. No old paintings are not valuable.

4. In which of the following can $\Rightarrow$ be replaced by the equivalence sign?
   A. In triangle ABC, $A = B \Rightarrow \sin 2A = \sin 2B$.
   B. P the point (a, b) lies on the curve $y^2 = x \Rightarrow b^2 = a$.
   C. $x = \cos 30°$ and $y = \sin 30° \Rightarrow x^2 + y^2 = 1$.
   D. $f(x) = x^2 - 4$ and $g(x) = x - 2 \Rightarrow f \circ g(x) = x^2 - 4x$.
   E. ABCD is a rhombus $\Rightarrow$ ABCD is a kite.

5. Which of the following is a valid deduction from the statements,
   "All diplomats are tactful people"
   "Some diplomats are politicians."?
   A. All tactful people are diplomats.
   B. All politicians are diplomats.
   C. No politician is tactful.
   D. No tactful person is a politician.
   E. Some politicians are tactful.

6. The negation of the statement "No boys dislike examinations" is
   A. No boys like examinations.
   B. Some boys like examinations.
   C. All boys like examinations.
   D. Some boys do not like examinations.
   E. All boys do not like examinations.

7. x is a real number and $-1 < x < 0$. We can deduce that
   A. $x^2 < -1$
   B. $-1 < x^2 < 0$
   C. $x^2 < 0$
   D. $0 < x^2 < 1$
   E. $x^2 > 1$

8. p and q are two real numbers such that $p > 2q$. Which one or more of the following statements must be true?
   1. $q > \frac{1}{2}p$
   2. $q < 2p$
   3. $-p < -2q$
   4. $p^2 > 4q^2$

   A. 3 only
   B. 1, 2 and 4 only
   C. 1, 3 and 4 only
   D. 3 and 4 only
   E. Some other combination.

9. "All actuaries are mathematicians." Which of the following statements is/are valid conclusions?
   (1) If a person is a mathematician he is an actuary.
   (2) If a person is not an actuary he is not a mathematician.
   (3) If a person is not a mathematician he is not an actuary.

10. Which of the following have true converses?
    (1) If $x > 0$ then $2^x > 0$, $x \in R$.
    (2) If $0 < k < 1$ then $\log_{10} k < 0$.
    (3) If ABCD is a rhombus then ABCD has only two axes of symmetry.

11. (1) a and b are odd integers.
    (2) $a + b$ is even.

12. (1) My first edition is valuable.
    (2) All first editions are valuable.

ASSIGNMENT 8.6

Supplementary Examples

1. The roots of the equation $px^2+5x+27=0$ are in the ratio 3:2. Find the value of p.

2. (a) If $5^x = 20$ prove that $x = \dfrac{1+\log_{10} 2}{1-\log_{10} 2}$

   (b) Without using tables evaluate

   (i) $\log_3 12 - 2\log_3 6$ (ii) $\log_2 \frac{4}{15} + \log_2 \frac{3}{40} - \log_2 \frac{8}{25}$

3. Simplify $\left(\dfrac{p^4}{r^{\frac{2}{3}}s}\right)^{\frac{3}{2}} \div \left(\dfrac{s^{\frac{3}{4}}x\sqrt{p^3}}{p^{-3}x\sqrt{r}}\right)^{-2}$

4. The sum of the roots of a quadratic equation is 3 and the sum of the squares of the roots is 19. Find the equation.

5. Simplify $\dfrac{2}{1+\sqrt{3}} + \dfrac{3}{3-2\sqrt{3}} - \dfrac{\sqrt{3}}{4+2\sqrt{3}}$

6. If $\log_x 9{\cdot}32 = 2{\cdot}4$ find the value of x to 3 significant figures.

7. Determine the values of $\sin\theta$ for which the equation $x^2+4x\sin\theta+1=0$ has equal roots and find the values of θ in the interval $0 \leqq \theta \leqq 90$ for which the equation has real roots.

8. Find the value of x such that $\log_x(10x^2-31x+30) = 3$.

9. (i) The equation $k(2x+1)(x+2) = x$ has equal roots. Find the value of k.

   (ii) If k lies between these values state with reasons the nature of the roots of the equation.

10. If $\log_a p = 2+3\log_a q$, express p in terms of q and a.

11. If $\dfrac{4\sqrt{3}}{1+\sqrt{3}}, p, \dfrac{9-5\sqrt{3}}{7-4\sqrt{3}}$ are three consecutive terms of a geometric sequence, find p in its simplest form.

12. Simplify $\dfrac{a^{\frac{2}{3}}b^{\frac{1}{3}}+a^{\frac{1}{3}}b^{\frac{2}{3}}}{a^{\frac{4}{3}}b^{-\frac{1}{3}}+a}$ when $a=8b$.

13. Write down the condition that the equation $ax^2+bx+c=0$ will have no real roots. If $k = \dfrac{x^2+9}{x}$ has no real roots in x show that $-6 < k < 6$.

14. In a chemical experiment the quantity in grams of a substance t minutes after the start of the experiment is given by the relation $m = m_0 e^{-pt}$, where p is a constant and m_0 grams is the quantity of the substance at the beginning of the experiment.

    If at the beginning of the experiment there are 10 grams of the substance and after 30 minutes there are 5 grams find the time required to reduce the substance to 2 grams.

15. Find the quadratic equation whose roots are $5+2\sqrt{3}$ and $5-2\sqrt{3}$.

16. Find the values of $y+\dfrac{3}{y}$ and $y^2+\dfrac{9}{y^2}$ when $y = 3+\sqrt{6}$.

17. Express the following in equivalent logarithmic form,

    (i) $9 = 3^2$ (ii) $1000 = 10^3$

    (iii) $2 = 8^{\frac{1}{3}}$ (iv) $2^{-2} = \frac{1}{4}$ (v) $16^{\frac{1}{4}} = 2$.

18. Express the following in equivalent index form,

    (i) $\log_2 4 = 2$ (ii) $\log_{10} 1000 = 3$

    (iii) $\log_2 \frac{1}{4} = -2$ (iv) $\log_3 81 = 4$

    (v) $\log_x p = n$.

19. On the same diagram draw the graphs of $f: x \rightarrow 3^x$ and $g: x \rightarrow 3^{-x}$ for $\{x: -3 \leqq x \leqq 3, x \in R\}$.

    (a) Write down the coordinates of the point of intersection of each curve with the y-axis.

(b) Show that if the point (a, b) lies on one curve then the point $(-a, b)$ lies on the other curve for $\{(a, b): -3 \leqq a \leqq 3, \frac{1}{27} \leqq b \leqq 27, a, b \in R\}$.

Solve the system of equations in questions 20–26.

20. $x+2y+z = -6$
$2x-y-4z = 1$
$2x-2y+3z = -21$

21. $2x-2y-z = 3$
$4x-3y-2z = 8$
$6x-4y+3z = -23$

22. $x+y+z = 4$
$3x+2y+4z = 11$
$2x+3y-5z = -3$

23. $y = (x-2)(x-5)$
$y = x-2$

24. $(x-3y)(2x-y) = 12$
$x+2y = 4$

25. $3x^2+8xy+5y^2 = 60$
$3x-y = 6$

26. $x^2-3xy+5y^2-6x = 0$
$x-3y = 4$

27. Find the equation connecting x and y when the parameter t is eliminated in each of the following pairs of equations.

(i) $x = t-1,\ y = 3t+2$

(ii) $\log_2 t = x+y,\ \log_2 x+\log_2 y = t$

(iii) $y = t^2 = t,\ x = t+1$

(iv) $x = t+\log_{10} t,\ y = t-\log_{10} t$

(v) $x \sin^2 t = 1+\cos t,\ y = 1-\cos t$.

28. With reference to rectangular axes OX and OY, the coordinates (x, y) of a point on a curve are given by: $x = 4 \cos \theta°$, $y = 2 \sin \theta°$, where θ is a variable. Tabulate values of x and y for $\theta = 0$, 30, 45, 60, 90, 120, 135, 150, 180 degrees and hence sketch the part of the curve for $0 \leqq \theta \leqq 180$, $\theta \in R$.

Deduce the sketch of the curve for $0 \leqq \theta \leqq 360$, $\theta \in R$. By eliminating θ find the cartesian equation of the curve.

29. The intensity I_0 of a beam of light passing through a coloured filter x millimetres thick emerges with an intensity I given by the formula $I = I_0 e^{-px}$, where p is a constant. If a filter 4 millimetres thick decreases the intensity by 50 percent of the original intensity I_0, calculate by what percentage of I_0 a filter of 8 millimetres would decrease the intensity.

30. "All record breaking athletes have been trained rigorously." From this statement what, if anything, can you deduce about,

I. Donald Smythe who has been trained rigorously.
II. Douglas Myles who held the world record for the 100 metres.
III. John Falls who has never been trained rigorously.
IV. Graham Mann who has never broken a world record?

31. "All squares are rhombuses."
"ABCD is not a rhombus."
Make, if possible, any valid conclusion about ABCD.

32. "All rational numbers are real numbers."
"All surds are real numbers."
From *these two statements*, which of the following is/are valid conclusions?

I. Some rational numbers are surds.
II. Some surds are rational numbers.
III. Some real numbers are both rational and surds.
IV. Every real number is either a rational or a surd.

33. If $\dfrac{\log_p 3}{\log_q 3} = x$, show that $q = p^x$. Hence or otherwise show that if $\dfrac{\log_p 2}{\log_q 2} = y$, then $x = y$.

GEOMETRY

UNIT 1: REFLECTION

REFLECTION IN AN AXIS

1.1

Let a point P reflected in an axis m have image P_1, as in figure 1.

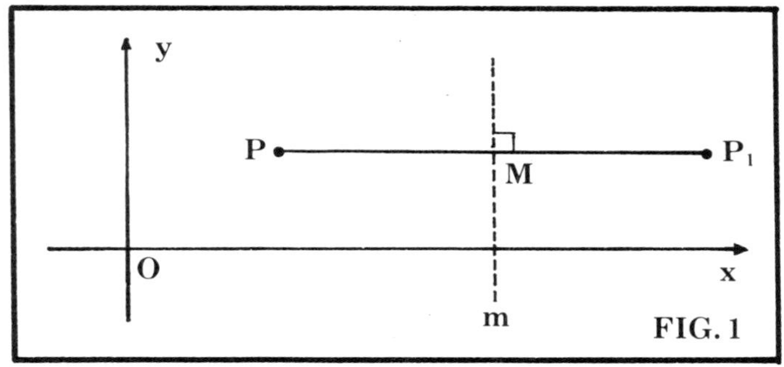

FIG. 1

Let PP_1 cut the axis m at M, then $PM = MP_1$, and $\vec{PP_1} = 2\vec{PM}$.

(i) If P is the point (a, b), and m is the line $x = h$, then $\vec{OP}$ represents $\begin{pmatrix} a \\ b \end{pmatrix}$, $\vec{OM}$ represents $\begin{pmatrix} h \\ b \end{pmatrix}$, and $\vec{PM}$ represents

$$\begin{pmatrix} h-a \\ 0 \end{pmatrix}$$

Hence $\vec{PP_1} = 2\vec{PM}$ represents

$$\begin{pmatrix} 2h-2a \\ 0 \end{pmatrix}$$

and $\vec{OP_1} = \vec{OP} + \vec{PP_1}$ represents

$$\begin{pmatrix} a \\ b \end{pmatrix} + \begin{pmatrix} 2h-2a \\ b \end{pmatrix} = \begin{pmatrix} 2h-a \\ b \end{pmatrix}$$

i.e. P_1 has coordinates $(2h-a, b)$,

i.e. the image of the point (a, b) under reflection in the line $x = h$ is the point $(2h - a, b)$.

(ii) If P is the point (a, b) and m the line $y = k$, it follows that $\vec{OP_1}$ represents

$$\begin{pmatrix} a \\ 2k - b \end{pmatrix}$$

i.e. P_1 has coordinates $(a, 2k - b)$.

REFLECTION IN (OR HALF-TURN ABOUT) A POINT

It follows from 1.1 that if the point $P(a, b)$ is reflected in the point $M(h, k)$ then, $\vec{OP_1}$ represents

$$\begin{pmatrix} 2h - a \\ 2k - b \end{pmatrix}$$

and therefore P_1 has coordinates $(2h - a, 2k - b)$.

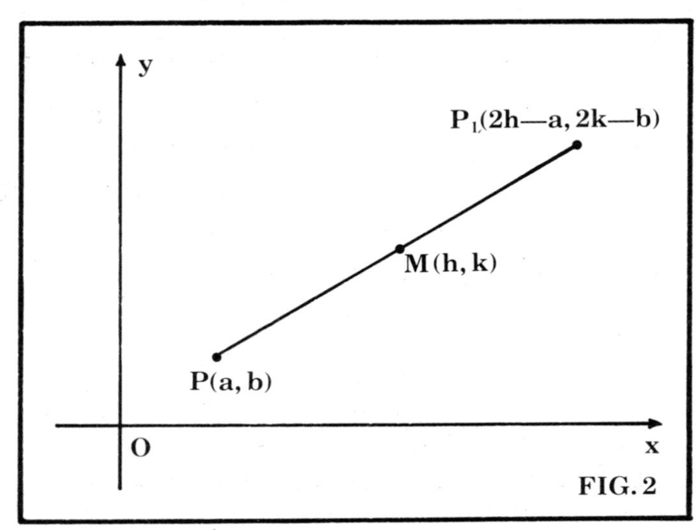

FIG. 2

Example 1

Find the image A′ of A(5, 2) and the image B′ of B(−1, 4) under a reflection in the point $(7, 4\frac{1}{2})$ and show that ABA′B′ is a parallelogram.

The image of $P(a, b)$ under reflection in $M(h, k)$ is $P'(2h - a, 2k - b)$, hence the image of A(5, 2) under reflection in $(7, 4\frac{1}{2})$ is A′(14−5, 9−2), i.e. A′(9, 7), and the image of B(−1, 4) under reflection in $(7, 4\frac{1}{2})$ is B′(14+1, 9−4), i.e. B′(15, 5).

Also

$$\vec{AB} = \vec{OB} - \vec{OA} \text{ represents } \begin{pmatrix} -1 \\ 4 \end{pmatrix} - \begin{pmatrix} 5 \\ 2 \end{pmatrix} = \begin{pmatrix} -6 \\ 2 \end{pmatrix}$$

$$\text{and } \vec{A'B'} = \vec{OB'} - \vec{OA'} \text{ represents } \begin{pmatrix} 15 \\ 5 \end{pmatrix} - \begin{pmatrix} 9 \\ 7 \end{pmatrix} = \begin{pmatrix} 6 \\ -2 \end{pmatrix}$$

$$\Rightarrow \vec{B'A'} \text{ represents } \begin{pmatrix} -6 \\ 2 \end{pmatrix}$$

i.e. $\vec{AB}$ represents $\begin{pmatrix} -6 \\ 2 \end{pmatrix}$ and $\vec{B'A'}$ represents $\begin{pmatrix} -6 \\ 2 \end{pmatrix}$

⇒ ABA′B′ is a parallelogram.

Example 2

K is the mid-point of the line joining the points P and Q and R is any point not in the same straight line as PQ. L is the image of P under reflection in R. M is the image of L under reflection in K, and N is the image of Q under reflection in R. Prove that

(i) N, P and M are collinear,

(ii) P is the mid-point of MN.

Note: PK = KQ
PR = RL
LK = KM
QR = RN

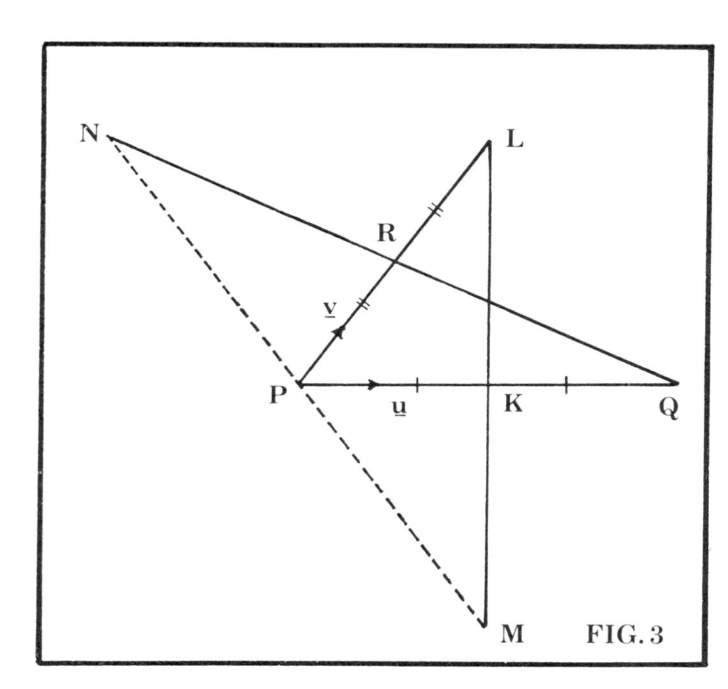

FIG. 3

Let $\vec{PK}$ represent $\boldsymbol{u}$ and $\vec{PR}$ represent $\boldsymbol{v}$.

(i) $\vec{PK}$ represents $\boldsymbol{u}$ and K the mid-point of PQ $\Rightarrow \vec{PQ}$ represents $2\boldsymbol{u}$; $\vec{PR}$ represents $\boldsymbol{v}$ and R the mid-point of PL $\Rightarrow \vec{PL}$ represents $2\boldsymbol{v}$; $\vec{LK} = \vec{PK} - \vec{PL}$ represents $\boldsymbol{u} - 2\boldsymbol{v} \Rightarrow \vec{LM}$ represents $2(\boldsymbol{u} - 2\boldsymbol{v})$.

Hence $\vec{PM} = \vec{PL} + \vec{LM}$ represents

$$2\boldsymbol{v} + 2(\boldsymbol{u} - 2\boldsymbol{v}) = 2\boldsymbol{u} - 2\boldsymbol{v}$$

Again $\vec{QR} = \vec{PR} - \vec{PQ}$ represents $\boldsymbol{v} - 2\boldsymbol{u}$

$$\Rightarrow \vec{QN} \text{ represents } 2(\boldsymbol{v} - 2\boldsymbol{u})$$

hence $\vec{PN} = \vec{PQ} + \vec{QN}$ represents

$$2\boldsymbol{u} + 2(\boldsymbol{v} - 2\boldsymbol{u}) = 2\boldsymbol{v} - 2\boldsymbol{u}$$

Hence $\vec{PN}$ represents $2\boldsymbol{v} - 2\boldsymbol{u}$ and $\vec{PM}$ represents $2\boldsymbol{u} - 2\boldsymbol{v}$ $\Big\} \Rightarrow \vec{PN} = -\vec{PM}$

$\Rightarrow \vec{PN} = \vec{MP}$

$\Rightarrow$ N, P and M are collinear

(ii) Since $\vec{PN}$ represents $2(\boldsymbol{v} - \boldsymbol{u})$ and $\vec{PM}$ represents

$$2(\boldsymbol{u} - \boldsymbol{v}) = -2(\boldsymbol{v} - \boldsymbol{u})$$

it follows that $|\vec{PN}| = |\vec{PM}|$, i.e. P is the mid-point of MN.

REFLECTION IN TWO PARALLEL AXES

1.2

(i) Let a point P reflected in an axis m_1 have image P_1, and P_1 reflected in an axis m_2 have image P_2. Let PP_1 cut axis m_1 at M_1 and axis m_2 at M_2 at right angles.

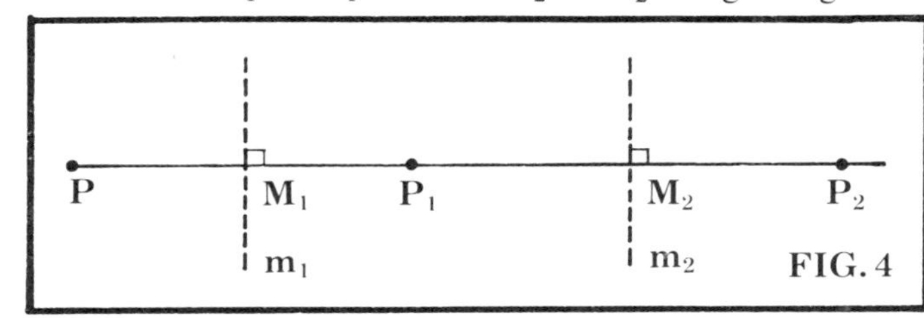

FIG. 4

P_1 the image of P in $m_1 \Rightarrow PM_1 = M_1P_1$

P_2 the image of P_1 in $m_2 \Rightarrow P_1M_2 = M_2P_2$

but $\vec{PP_2} = \vec{PP_1} + \vec{P_1P_2} = 2\vec{M_1P_1} + 2\vec{P_1M_2}$

$= 2(\vec{M_1P_1} + \vec{P_1M_2})$

$= 2\vec{M_1M_2}$

i.e. the distance between the point P and its image after the second reflection is twice the separation distance between the axes. *This separation distance is independent of the position of* P.

Hence $\quad \overrightarrow{PP_2} = 2\overrightarrow{M_1M_2}$ and if $\overrightarrow{M_1M_2}$ represents $\boldsymbol{u}$

then $\quad \overrightarrow{PP_2}$ represents $2\boldsymbol{u}$

(ii) Let $P(a, b)$ be reflected in the line $x = h_1$ and its image in the line $x = h_2$, so that $P \to P_1 \to P_2$.

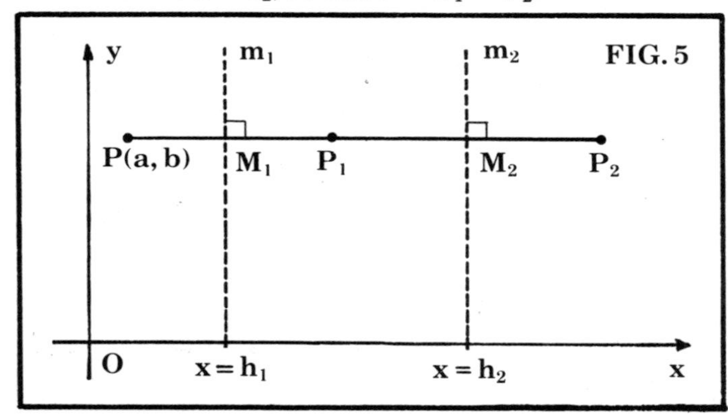

Hence

$\overrightarrow{OP}$ represents $\begin{pmatrix} a \\ b \end{pmatrix}$, $\quad \overrightarrow{M_1M_2}$ represents $\begin{pmatrix} h_2 - h_1 \\ 0 \end{pmatrix}$

and $\quad \overrightarrow{PP_2} = 2\overrightarrow{M_1M_2}$ represents $\begin{pmatrix} 2(h_2 - h_1) \\ 0 \end{pmatrix}$

also $\overrightarrow{OP_2} = \overrightarrow{OP_1} + \overrightarrow{PP_2}$ represents

$$\begin{pmatrix} a \\ b \end{pmatrix} + \begin{pmatrix} 2(h_2 - h_1) \\ 0 \end{pmatrix} = \begin{pmatrix} 2(h_2 - h_1) + a \\ b \end{pmatrix}$$

i.e. P_2 has coordinates $[2(h_2 - h_1) + a, b]$.

Similarly if $P(a, b)$ is reflected in the line $y = k_1$ and its image is reflected in the line $y = k_2$, it follows that,

$$\overrightarrow{OP_2} \text{ represents } \begin{pmatrix} a \\ 2(k_2 - k_1) + b \end{pmatrix}$$

i.e. P_2 has coordinates $[a, 2(k_2 - k_1) + b]$.

REFLECTION IN TWO POINTS

It follows from 1.2(ii) that if $P(a, b)$ has image P_1 under reflection in $M_1(h_1, k_1)$ and P_1 has image P_2 under reflection in $M_2(h_2, k_2)$, then,

$$\overrightarrow{OP_2} \text{ represents } \begin{pmatrix} 2(h_2 - h_1) + a \\ 2(k_2 - k_1) + b \end{pmatrix}$$

i.e. P_2 has coordinates $[2(h_2 - h_1) + a, 2(k_2 - k_1) + b]$.

Example 1

Two circles, centres A and B, intersect at P and Q. Two parallel lines through P and Q cut the circle with centre A at R and S respectively and the circle with centre B at T and V respectively. Prove that RTVS is a parallelogram.

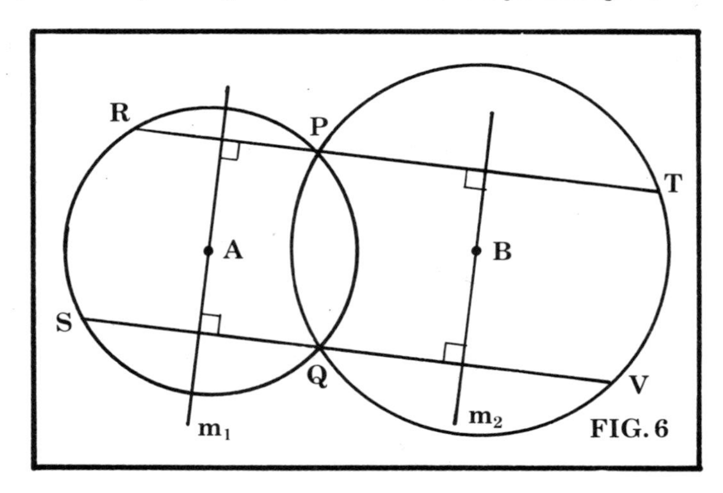

Since chords RP and SQ are parallel the straight line through centre A perpendicular to RP and SQ passes through the mid-point of RP and SQ.

Similarly the parallel straight line through centre B passes through the mid-point of PT and QV.

Denote these two parallel lines through A and B by m_1 and m_2.

Hence $R \to P$ under reflection in m_1 and $P \to T$ under reflection in m_2

$$\Rightarrow \overrightarrow{RT} \text{ represents } 2\boldsymbol{u}$$

where $\boldsymbol{u}$ denotes the separation vector between m_1 and m_2.

Similarly $\overrightarrow{SV}$ represents $2\boldsymbol{u}$, hence $\overrightarrow{RT} = \overrightarrow{SV} \Rightarrow$ RTVS is a parallelogram.

REFLECTION IN TWO INTERSECTING AXES

(a) *Consider first the resultant of two rotations (anticlockwise), about the same point.*

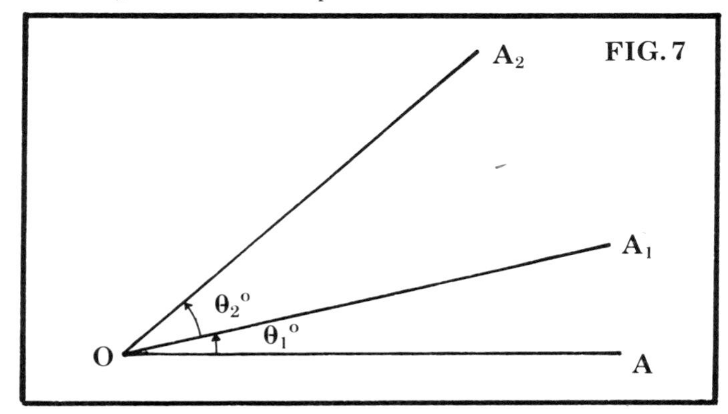

Under a rotation about O through an angle $\theta_1°$, $OA \to OA_1$.

Under a rotation about O through an angle $\theta_2°$, $OA_1 \to OA_2$.

Hence $OA \to OA_2$ under a rotation about O and through an angle $\theta_1° + \theta_2°$, i.e. the resultant of two rotations about the same point is the same as a single rotation about the point, and the angle of the single rotation is the sum of the two separate rotations.

(b) *Reflection in two intersecting axes.*

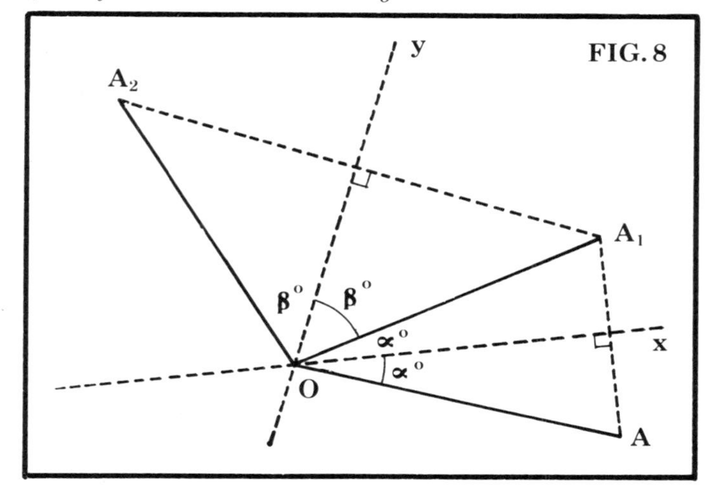

Let the axes OX and OY make an angle $\theta°$, i.e. $\angle XOY = \theta°$.

Let $\angle AOX = \alpha°$, then under a reflection in OX, $A \to A_1$, $OA \to OA_1$ and therefore $\angle A_1OX = \alpha°$.

Let $\angle A_1OY = \beta°$, then under a reflection in OY, $A_1 \to A_2$, $OA_1 \to OA_2$ and therefore $\angle A_2OY = \beta°$.

Also $OA \to OA_2$ under a rotation about O through an angle $2\alpha° + 2\beta°$,

i.e. $\quad \angle AOA_2 = 2\alpha° + 2\beta° = 2(\alpha° + \beta°)$

But $\quad \angle XOY = \theta°$

Hence $\quad \angle AOA_2 = 2\theta°$

i.e. The composition of reflection in two intersecting axes making an angle $\theta°$ is equivalent to a rotation about the point of intersection of the axes and through an angle $2\theta°$ (i.e. twice the angle between the axes).

(c) *Reflection in two perpendicular axes*

It follows that if the axes OX and OY are perpendicular

then $\theta = 90$ and reflection in OX followed by reflection in OY is the same as a rotation of 180° about O.

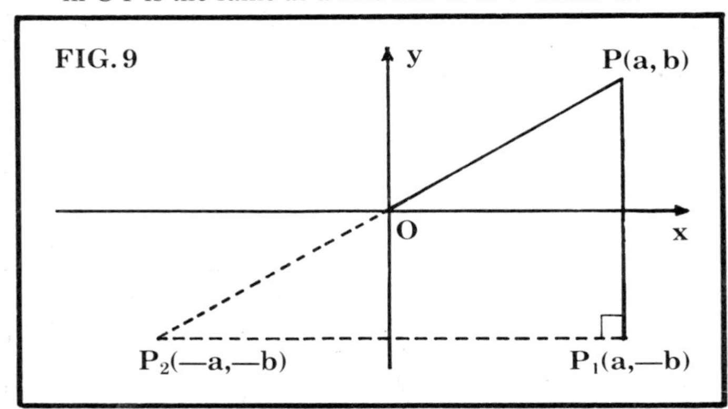

Again if OX and OY are the coordinate axes then a rotation of 180° is the same as a half-turn about (or reflection in) O.

From the figure $P(a, b) \to P_1(a, -b)$ under reflection in OX and $P_1(a, -b) \to P_2(-a, -b)$ under reflection in OY.

Hence the composition of reflection in two intersecting perpendicular axes is equivalent to a half-turn about (or reflection in) the point of intersection of the axes.

1.4 **Example 1**

Angle AOB is the angle at the centre and angle ACB is the angle at the circumference standing on the same arc AB of a circle centre O. Prove that angle AOB is twice the size of angle ACB.

Lines X′OX and Y′OY are the perpendicular bisectors of the chords AC and BC respectively.

$$OA \to OC \text{ under reflection in } X'OX$$
$$OC \to OB \text{ under reflection in } Y'OY$$

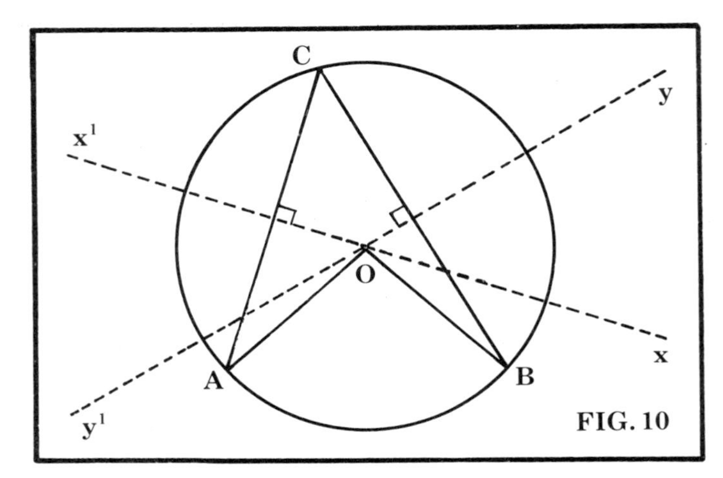

FIG. 10

$\Rightarrow OA \to OB$ under a rotation about O and through an angle twice the size of the angle between the axes X′OX and Y′OY.

$\Rightarrow$ reflex angle $AOB = 2\angle YOX'$

and smaller angle $AOB = 2\angle XOY$.

$$\left.\begin{array}{ll}\text{But} & \angle XOY + \angle YOX' = 180^\circ \\ \text{and} & \angle ACB + \angle YOX' = 180^\circ\end{array}\right\} \Rightarrow \angle XOY = \angle ACB$$

hence $\angle AOB = 2\angle ACB$.

THE GLIDE REFLECTION

Under a reflection M in XY,

$$\triangle ABC \to \triangle A_1B_1C_1$$

Under a translation T parallel to XY,

$$\triangle A_1B_1C_1 \to \triangle A_2B_2C_2$$

Such a transformation, namely a reflection in an axis followed by a translation parallel to the axis is called a **Glide Reflection.**

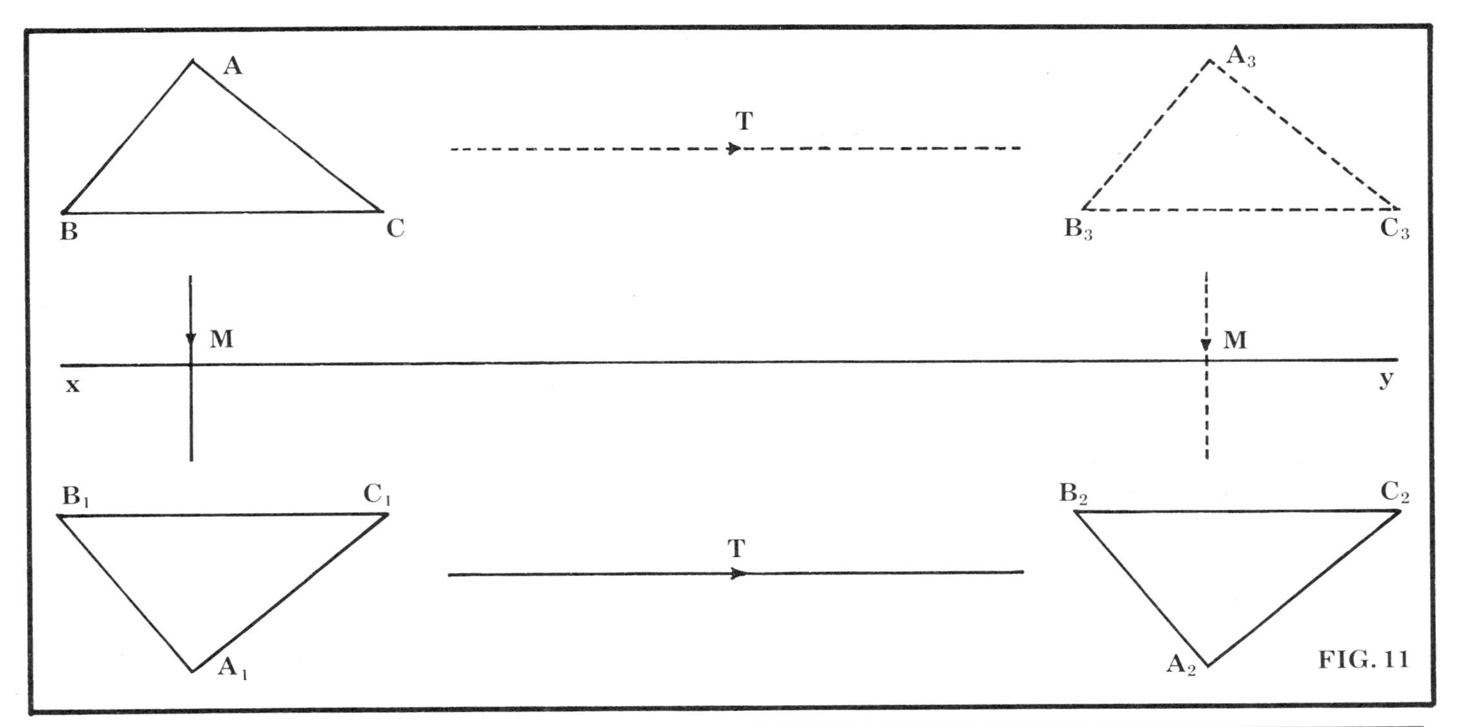

FIG. 11

M maps $\triangle ABC \to \triangle A_1B_1C_1$
and T maps $\triangle A_1B_1C_1 \to \triangle A_2B_2C_2$
then $T \circ M(\triangle ABC) \to \triangle A_2B_2C_2$

(where the notation $T \circ M$ is used to indicate the composition of two transformations and the order is to be taken T *after* M).

Also, since T maps $\triangle ABC \to \triangle A_3B_3C_3$
and M maps $\triangle A_3B_3C_3 \to \triangle A_2B_2C_2$
then $M \circ T(\triangle ABC) \to \triangle A_2B_2C_2$

Hence $M \circ T = T \circ M$, i.e. the glide reflection is commutative.

SUMMARY OF THESE TRANSFORMATIONS

1.5

(a) (i) Reflection in an axis

$$x = h \Rightarrow P(a, b) \to P'(2h - a, b)$$

Reflection in an axis

$$y = k \Rightarrow P(a, b) \to P'(a, 2k - b)$$

(ii) Reflection in a point

$$(h, k) \Rightarrow P(a, b) \to P'(2h - a, 2k - b)$$

(b) (i) Reflection in two parallel axes, $P \to P_2$ and PP_2 cuts the axes in M_1 and $M_2 \Rightarrow \overrightarrow{PP_2} = 2\overrightarrow{M_1M_2}$.

Thus the resultant of the composition of reflections

in two parallel axes whose separation is denoted by $\boldsymbol{u}$ is a translation $2\boldsymbol{u}$.

(ii) Reflection in two parallel axes $x = h_1$ and $x = h_2$

$$\Rightarrow P(a, b) \to P'[2(h_2 - h_1) + a, b)]$$

Reflection in two parallel axes $y = k_1$ and $y = k_2$

$$\Rightarrow P(a, b) \to P'[a, 2(k_2 - k_1) + b]$$

(iii) Reflection in two points (h_1, k_1) and (h_2, k_2)

$$\Rightarrow P(a, b) \to P'[2(h_2 - h_1) + a, 2(k_2 - k_1) + b]$$

(c) (i) The composition of reflections in two intersecting axes making an angle θ is equivalent to a rotation of 2θ about the point of intersection of the axes.

(ii) The composition of reflections in two perpendicular axes is equivalent to a half-turn about the point of intersection of the axes.

(d) A reflection M in an axis followed by a translation T parallel to the axis is called a Glide Reflection, and

$$M \circ T = T \circ M$$

ASSIGNMENT 1.1

1. Write down the coordinates of the image of each of the following points under reflection in the given lines:

(i) $(3, 4)$; line $x = 5$ (ii) $(-3, 9)$; line $x = -2$

(iii) $(4, -2)$; line $x = -7$ (iv) $(-2, -3)$; line $x = 7$

(v) $(-2, 1)$; line $y = 2$ (vi) $(-1, 4)$; line $y = -7$

(vii) $(2, -5)$; line $y = -5$ (viii) $(-2, -7)$; line $y = 3$

2. Write down the coordinates of the image of each of the following points under reflection in (or a half-turn about) the given points:

(i) $(5, 7)$ reflected in $(0, 0)$

(ii) $(8, -1)$ reflected in $(3, 2)$

(iii) $(-5, 2)$ reflected in $(3, -1)$

(iv) $(-5, -4)$ reflected in $(-2, 5)$
(v) $(-2, -3)$ reflected in $(4, 7)$
(vi) $(-4, -7)$ reflected in $(-2, 4)$
(vii) $(-4, -2)$ reflected in $(-3, -3)$
(viii) $(5, 7)$ reflected in $(-1, -6)$

3. The point $P(a, b)$ maps onto the point $P_1(a_1, b_1)$ under reflection in the point (s, t). Write down the coordinates a_1 and b_1 in terms of a, b, s and t.

4. Write down the coordinates of the image of each of the following points under successive reflection in the given lines:
(i) $(5, 2)$ reflected in $x = 4$ and $x = 7$
(ii) $(5, 2)$ reflected in $x = 7$ and $x = 4$
(iii) $(3, -4)$ reflected in $x = 4$ and $x = -3$
(iv) $(-2, -5)$ reflected in $x = -2$ and $x = -6$
(v) $(-9, 4)$ reflected in $x = s$ and $x = t$
(vi) $(7, 2)$ reflected in $y = 4$ and $y = -6$
(vii) $(-7, -2)$ reflected in $y = -4$ and $y = 6$
(viii) $(5, 2)$ reflected in $y = -3$ and $y = 0$
(ix) $(-2, 4)$ reflected in $y = a$ and $y = b$
(x) (a, b) reflected in $y = s$ and $y = t$

5. Write down the image of each of the following points under successive reflection in the given points:
(i) $(1, 7)$ reflected in $(2, 4)$ and then in $(3, 7)$
(ii) $(1, 7)$ reflected in $(3, 7)$ and then in $(2, 4)$
(iii) $(-2, -4)$ reflected in $(-2, -1)$ and then in $(1, 3)$
(iv) $(-2, 3)$ reflected in $(2, 4)$ and then in $(-2, 4)$

6. $P(a, b)$ is reflected in the line $y = s$ and then in the line $y = t$, and $P(a, b)$ is reflected in the line $y = t$ and then in the line $y = s$. Are the images the same?

7. If M_1 denotes reflection in the point (p, q) and M_2 denotes reflection in the point (r, s), does $M_1 \circ M_2 = M_2 \circ M_1$? Test by applying $M_1 \circ M_2$ and $M_2 \circ M_1$ to the point (x, y) and finding the image point in each case.

8. With rectangular axes OX and OY draw the line l with equation $y = 2x+2$. Draw l_1 the image of l under reflection in the line $x = 0$ and write down the equation of l_1. Draw the image l_2 of l_1 under reflection in the line $x = 3$ and write down the equation of l_2. What single translation would map l onto l_2?

9. P, Q and R are three points (x, y), (x_1, y_1) and (x_2, y_2) respectively. P maps onto P′ under reflection in Q followed by reflection in R. Write down the coordinates of P′ in terms of the coordinates of Q and R.
What single translation would map P onto P′?
Write down the relation between $\overrightarrow{PP'}$ and $\overrightarrow{QR}$.
PQRS and QRTV are two parallelograms on the opposite side of QR. The diagonals of PQRS intersect at X and the diagonals of QRTV intersect at Y. By successive reflections in X and Y prove that PVTS is a parallelogram.

10. On a coordinate diagram with origin O mark the point P(4, 2). Give the coordinates of,
(i) P_1, the image of P under reflection in the x-axis followed by reflection in the line $y = x$.
(ii) P_2, the image of P under reflection in the line $y = x$ followed by reflection in the x-axis.
(iii) P_3, the image of P under reflection in the y-axis followed by reflection in the line $y = x$.
(iv) P_4, the image of P under reflection in the line $y = x$ followed by reflection in the y-axis.
If X denotes the operation of reflection in the x-axis,
Y denotes the operation of reflection in the y-axis,

M denotes the operation of reflection in the line $y = x$,

Which of the following operations are equal: $Y \circ M$, $M \circ Y$, $X \circ M$ and $M \circ X$?

11. PQRS is a rectangle in which PK and SK bisect the angles SPQ and PSR respectively. Under successive reflections in PK and SK the image of P and S are P′ and S′ respectively.

    Prove that,

    (i) P′ lies on SR and S′ lies on PQ

    (ii) P′S′ is equal and parallel to RQ

    Would the same results hold if PQRS was a parallelogram but not a rectangle?

12. ABCD is a square whose diagonals AC and BD intersect at O. The straight line POQ cuts BC at P and AD at Q. XY is an axis through O and parallel to AD. P_1 is the image of P under reflection in BD and P_2 is the image of P_1 under reflection in XY. Under the same successive reflections in BD and XY, $Q \rightarrow Q_1 \rightarrow Q_2$.

    (i) Show that P_1 lies on AB.

    (ii) What single transformation would map P onto P_2?

    (iii) Prove that PP_2QQ_2 is a square.

13. O is the centre of a circle circumscribing a triangle ABC in which angle BAC is 45°. The composite reflection in OB followed by reflection in OC maps triangle ABC onto the triangle $A_1B_1C_1$. Draw a sketch of triangle ABC and its image $A_1B_1C_1$ and prove that the two triangles are congruent.

    Why are the arcs AA_1, BB_1 and CC_1 equal in length?

14. A point P has coordinates (5, 35°). The point P is reflected in the line $\theta = 60°$ and its image P_1 is reflected in the line $\theta = 140°$ to obtain an image P_2. Find the polar coordinates of P_2.

UNIT 2: TRANSFORMATIONS

TRANSFORMATIONS USING COORDINATES AND MATRICES

2.1

By using coordinates we can map a point or set of points onto a second point or second set of points according to some rule, that is under a certain transformation or mapping the point $P(a, b)$ is mapped onto the point $P'(a', b')$ where a' and b' are related to a and b.

Example 1

Under a certain transformation $P(a, b) \rightarrow P'(a', b')$ where $a' = 2a$ and $b' = -2b$.

(i) Find the image of the points $(1, 3)$, $(4, -2)$, $(0, 0)$, $(-2, -2)$.

(ii) Find the equation of the image of the line $y = x+2$.

(i) $(a, b) \rightarrow (a', b')$, where $a' = 2a$ and $b' = -2b$ can be written

$$(a, b) \rightarrow (2a, -2b)$$

hence

$$(1, 3) \rightarrow (2, -6)$$

$$(4, -2) \rightarrow (8, 4)$$

$$(0, 0) \rightarrow (0, 0)$$

$$(-2, -2) \rightarrow (-4, 4)$$

(ii) $a' = 2a \Rightarrow a = \dfrac{a'}{2}$ and $b' = -2b \Rightarrow b = -\dfrac{b'}{2}$.

If (a, b) lies on the line $y = x+2$ then $b = a+2$, i.e. (a, b) lies on the line

$$y = x+2 \Leftrightarrow \quad b = a+2$$
$$\Leftrightarrow -\frac{b'}{2} = \frac{a'}{2}+2$$
$$\Leftrightarrow \quad -b' = a'+4$$
$$\Leftrightarrow \quad b' = -a'-4$$

i.e. the point (a', b') lies on the line $y = -x-4$ and $y = -x-4$ is therefore the image of the line $y = x+2$ under the given transformation.

Example 2

Under a composite mapping

$$P(x, y) \to P_1(x_1, y_1) \to P_2(x_2, y_2)$$

where, $\quad x_1 = x, \quad x_2 = x_1+2$

$\quad y_1 = -y, \quad y_2 = y_1-2$

(i) State the geometrical transformation effected by this composite mapping.

(ii) Find the image of the set of points

$$\{(2, 2), (-2, -3), (4, 6), (0, 0)\}$$

under this composite mapping.

(iii) Find the equation of the image of the line $y = 4x$.

(i) $(x, y) \to (x_1, y_1)$ where $x_1 = x$ and $y_1 = -y$, means $(x, y) \to (x, -y)$ and this is a reflection in the x-axis

$(x_1, y_1) \to (x_2, y_2)$ where $x_2 = x_1+2, y_2 = y_1-2$ can be written as

$$(x_1, y_1) \to (x_1+2, y_1-2)$$

and this is a translation $\begin{pmatrix} 2 \\ -2 \end{pmatrix}$.

Hence the composite mapping

$$(x, y) \to (x_1, y_1) \to (x_2, y_2)$$

is a reflection in the x-axis followed by a translation

$$\begin{pmatrix} 2 \\ -2 \end{pmatrix}$$

(ii) Again $\quad (x, y) \to (x_1, y_1) \to (x_2, y_2)$

$$\Rightarrow (x, y) \to (x, -y) \to (x_1+2, y_1-2)$$
$$\Rightarrow (x, y) \to (x+2, -y-2)$$

Hence $\quad (2, 2) \to (2+2, -2-2) = (4, -4)$

$$(-2, -3) \to (-2+2, 3-2) = (0, 1)$$
$$(4, 6) \to (4+2, -6-2) = (6, -8)$$
$$(0, 0) \to (0+2, -0-2) = (2, -2)$$

(iii) $\quad x_2 = x_1+2 \Rightarrow x_1 = x_2-2$

and $\quad y_2 = y_1-2 \Rightarrow y_1 = y_2+2$

Also $\quad x_1 = x \Rightarrow x = x_2-2$

and $\quad y_1 = -y \Rightarrow y = -y_1 = -y_2-2$

(x, y) lies on the line $y = 4x \Leftrightarrow \quad y = 4x$

$$\Leftrightarrow (-y_2-2) = 4(x_2-2)$$
$$\Leftrightarrow \quad -y_2 = 4x_2-8+2$$
$$\Leftrightarrow \quad y_2 = -4x_2+6$$

i.e. the point (x_2, y_2) lies on the line $y = -4x+6$ and $y = -4x+6$ is therefore the image of the line $y = 4x$ under the composite mapping.

ASSIGNMENT 2.1

1. Under a certain mapping $P(x, y) \to P'(x', y')$ where $x' = -x$ and $y' = y$.
   (i) Find the image of the set
   $$\{(2, 1), (3, -5), (0, 0), (-2, -6)\}$$
   (ii) Find the equation of the image of the line $y = -2x$ under this mapping.
   (iii) State the geometric transformation represented by this mapping.

2. Under a certain mapping $P(a, b) \to P'(a', b')$ where $a' = 2a - b$ and $b' = -2b$.
   (i) Express a and b in terms of a' and b'.
   (ii) Give the images of the points O(0, 0), A(2, 0), B(2, 4) and show the triangle OAB and its image in a diagram.
   (iii) Show that the line with equation $y - x = 1$ maps onto the line with equation $y + 2x = -4$.

3. Under a certain mapping in which $P(x, y) \to P'(x', y')$, $x' = x + 3y$, $y' = y$.
   (i) A square OABC has vertices (0, 0), (2, 0), (2, 2) and (0, 2) respectively. Give the images of the vertices under this transformation and show the square and its image on a diagram.
   (ii) Is area preserved by this mapping? [i.e. is the area of the square OABC equal to the area of its image?].
   (iii) Show that the centre of the square maps onto the point of intersection of the diagonals of the image figure.

4. Under a composite mapping
   $$P(x, y) \to P_1(x_1, y_1) \to P_2(x_2, y_2)$$
   where
   $$x_1 = x + 2, \quad y_1 = y - 2$$
   $$x_2 = -x_1, \quad y_2 = y_1$$
   (i) State the geometrical transformation effected by this composite transformation.
   (ii) Show that the triangle OAB with vertices (0, 0), (4, 0) and (2, 2) respectively maps onto a triangle O′A′B′ of equal area.
   (iii) Check that in this mapping, distance is also preserved [i.e. show that OA = O′A′, AB = A′B′ and OB = O′B′].

5. Under a certain transformation $P(a, b) \to P'(a', b')$, where $a' = \sqrt{3}a + b$ and $b' = a - \sqrt{3}b$.
   (i) Show that OP′ = 2OP.
   (ii) Under this transformation find the images of the square OABC, where O is (0, 0), A(1, 0), B(1, 1) and C(0, 1) and show the square OABC and its image O′A′B′C′ on the same diagram.
   (iii) Compare the areas of the square OABC and its image.

6. Under a certain transformation $P(x, y) \to P'(x', y')$, where $x' = x - y$ and $y' = x + y$. Show that any point with coordinates (k, k) maps onto the point $(0, 2k)$ and hence or otherwise deduce that the transformation maps the line $y = x$ onto the y-axis.

MATRICES ASSOCIATED WITH TRANSFORMATIONS 2.2

Suppose under a certain mapping $P(x, y) \to P'(x', y')$ where $x' = x + 2y$ and $y' = -2y$, then

$$P(x, y) \to P'(x', y') \to P'(x + 2y, -2y)$$

We can write this as

$$x' = x+2y = 1.x+2.y$$
$$y' = \quad -2y \; = 0.x+(-2)y$$

Which again may be written as

$$\begin{pmatrix} x' \\ y' \end{pmatrix} = \begin{pmatrix} 1 & 2 \\ 0 & -2 \end{pmatrix} \begin{pmatrix} x \\ y \end{pmatrix}$$

Thus the matrix

$$\begin{pmatrix} 1 & 2 \\ 0 & -2 \end{pmatrix}$$

maps $P(x, y)$ onto $P'(x', y')$ and

$$\begin{pmatrix} 1 & 2 \\ 0 & -2 \end{pmatrix} \begin{pmatrix} x \\ y \end{pmatrix} = \begin{pmatrix} x+2y \\ -2y \end{pmatrix}$$

We have thus changed the point (x, y) into the point $(x+2y, -2y)$ by means of this operation. This is called a **mapping** and by means of the matrix operator

$$\begin{pmatrix} 1 & 2 \\ 0 & -2 \end{pmatrix}$$

every point of the plane has been mapped onto another point of the plane except the point $(0, 0)$ which maps onto itself.

$$\begin{pmatrix} 1 & 2 \\ 0 & -2 \end{pmatrix}$$

is said to be the matrix associated with the mapping or the transformation.

Example 1

Under a certain mapping $P(x, y) \rightarrow P_1(x_1, y_1)$ where

$$x_1 = x \quad \text{and} \quad y_1 = 2x-y$$

(i) Find the matrix M of the mapping and show that M maps the point $(3, -2)$ onto the point $(3, 8)$.

(ii) Find the set of points which map onto themselves under the operation of M.

(iii) Show that the inverse M^{-1} of M is M itself and hence check that the point $(3, 8)$ maps back onto the point $(3, -2)$ under M.

(i) $$P(x, y) \rightarrow P_1(x_1, y_1)$$

where $$x_1 = x \quad \text{and} \quad y_1 = 2x-y$$

We can write the above as

$$x_1 = 1.x+0.y$$
$$y_1 = 2.x+(-1)y$$

i.e. $$\begin{pmatrix} x_1 \\ y_1 \end{pmatrix} = \begin{pmatrix} 1 & 0 \\ 2 & -1 \end{pmatrix} \begin{pmatrix} x \\ y \end{pmatrix}$$

i.e. $M = \begin{pmatrix} 1 & 0 \\ 2 & -1 \end{pmatrix}$ is the matrix of the mapping

and $$\begin{pmatrix} 1 & 0 \\ 2 & -1 \end{pmatrix} \begin{pmatrix} 3 \\ -2 \end{pmatrix} = \begin{pmatrix} 3+0 \\ 6+2 \end{pmatrix} = \begin{pmatrix} 3 \\ 8 \end{pmatrix}$$

i.e. $(3, -2) \rightarrow (3, 8)$ under M.

(ii) Apply M to the point (a, b)

$$\begin{pmatrix} 1 & 0 \\ 2 & -1 \end{pmatrix} \begin{pmatrix} a \\ b \end{pmatrix} = \begin{pmatrix} a \\ 2a-b \end{pmatrix}$$

Hence the point (a, b) maps onto the point

$$(a, b) \Leftrightarrow \; b = 2a-b$$
$$\Leftrightarrow 2b = 2a$$
$$\Leftrightarrow \; b = a$$

hence $$\begin{pmatrix} 1 & 0 \\ 2 & -1 \end{pmatrix} \begin{pmatrix} a \\ a \end{pmatrix} = \begin{pmatrix} a \\ a \end{pmatrix}$$

i.e. the set of points $\{(k, k) : k \in R\}$ map onto themselves under the operation of M.

(iii) $M = \begin{pmatrix} 1 & 0 \\ 2 & -1 \end{pmatrix}$

$$\Rightarrow M^{-1} = -1\begin{pmatrix} -1 & 0 \\ -2 & 1 \end{pmatrix} = \begin{pmatrix} 1 & 0 \\ 2 & -1 \end{pmatrix} = M$$

Thus M is its own inverse, and

$$\begin{pmatrix} 1 & 0 \\ 2 & -1 \end{pmatrix}\begin{pmatrix} 3 \\ 8 \end{pmatrix} = \begin{pmatrix} 3 \\ 6-8 \end{pmatrix} = \begin{pmatrix} 3 \\ -2 \end{pmatrix}$$

i.e. $(3, 8) \rightarrow (3, -2)$ under M and $(3, -2) \rightarrow (3, 8)$ under M.

SPECIAL MATRICES ASSOCIATED WITH TRANSFORMATIONS

(i) *The identity or 'no change' matrix.*

Since $\begin{pmatrix} 1 & 0 \\ 0 & 1 \end{pmatrix}\begin{pmatrix} x \\ y \end{pmatrix} = \begin{pmatrix} x \\ y \end{pmatrix}$

the matrix $I = \begin{pmatrix} 1 & 0 \\ 0 & 1 \end{pmatrix}$

is called the **identity** or **'no change'** matrix since it leaves every point in the plane unchanged, i.e. I maps every point onto itself.

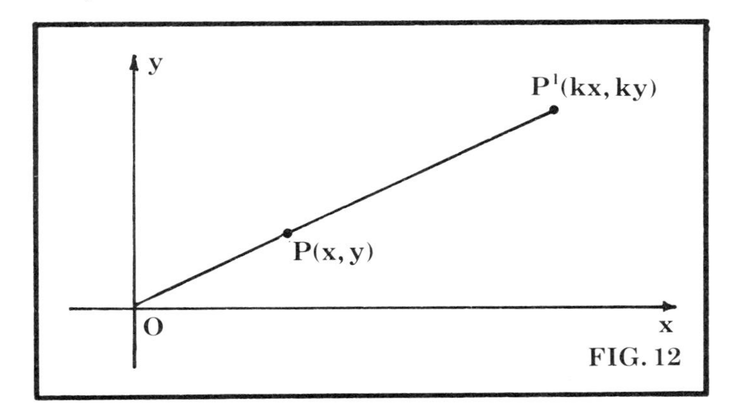

FIG. 12

(ii) *The enlargement matrix.*

Since $\begin{pmatrix} k & 0 \\ 0 & k \end{pmatrix}\begin{pmatrix} x \\ y \end{pmatrix} = \begin{pmatrix} kx \\ ky \end{pmatrix}$

the matrix $\begin{pmatrix} k & 0 \\ 0 & k \end{pmatrix}$

maps each point (x, y) onto (kx, ky).

From figure 12, $\vec{OP'} = k\vec{OP}$.

Note: If $k \in Q$, the matrix

$$\begin{pmatrix} k & 0 \\ 0 & k \end{pmatrix}$$

may be associated with a dilatation centre O and scale factor k.

(iii) *Reflection in the x-axis.*

For reflection in the x-axis $P(x, y) \rightarrow P'(x', y')$ where

$$x' = x \quad \text{and} \quad y' = -y$$

i.e. $x' = 1.x + 0.y$

$y' = 0.x + (-1)y$

or $\begin{pmatrix} x' \\ y' \end{pmatrix} = \begin{pmatrix} 1 & 0 \\ 0 & -1 \end{pmatrix}\begin{pmatrix} x \\ y \end{pmatrix}$

Hence $\begin{pmatrix} 1 & 0 \\ 0 & -1 \end{pmatrix}$

is the matrix associated with a reflection in the x-axis.

(iv) *Reflection in the y-axis.*

Similarly $\begin{pmatrix} -1 & 0 \\ 0 & 1 \end{pmatrix}$

is the matrix associated with a reflection in the y-axis.

(v) *Shearing.*

The operation which maps the square OABC onto the parallelogram OAB_1C_1 is called a **shear** parallel to the x-axis (see figure 13, page 142).

If the points C and B move a distance k parallel to

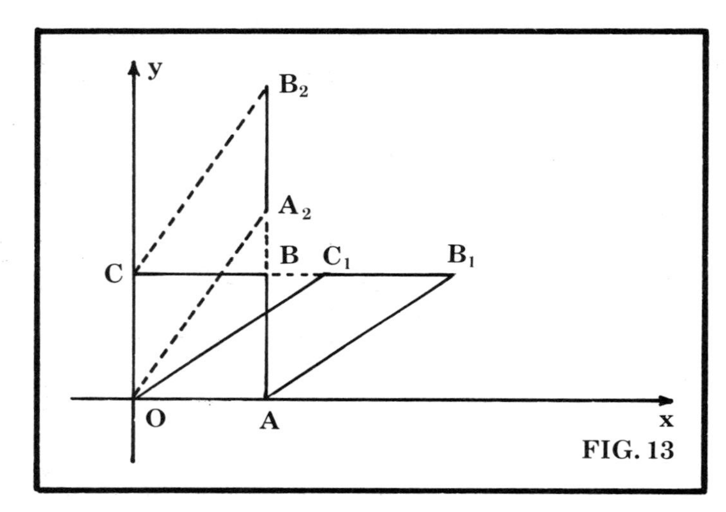

FIG. 13

the x-axis then the matrix associated with this operation is

$$\begin{pmatrix} 1 & k \\ 0 & 1 \end{pmatrix}$$

The corresponding shear parallel to the y-axis in which the square OABC maps onto the parallelogram OA_2B_2C has associated matrix

$$\begin{pmatrix} 1 & 0 \\ k & 1 \end{pmatrix}$$

Example 1

(i) With what transformations are the matrices

$$\text{(a)} \begin{pmatrix} 0 & 1 \\ 1 & 0 \end{pmatrix}, \quad \text{(b)} \begin{pmatrix} 0 & -1 \\ -1 & 0 \end{pmatrix} \quad \text{and} \quad \text{(c)} \begin{pmatrix} 0 & -1 \\ 1 & 0 \end{pmatrix}$$

associated?

(ii) Find the matrix associated with a half-turn about the origin.

(i) (a) Let $P(x, y) \rightarrow P'(x', y')$ under the matrix transformation

$$\begin{pmatrix} 0 & 1 \\ 1 & 0 \end{pmatrix}$$

Then $$\begin{pmatrix} x' \\ y' \end{pmatrix} = \begin{pmatrix} 0 & 1 \\ 1 & 0 \end{pmatrix}\begin{pmatrix} x \\ y \end{pmatrix} = \begin{pmatrix} y \\ x \end{pmatrix}$$

i.e. $P(x, y) \rightarrow P'(y, x)$ (see figure 14).

Hence the matrix

$$\begin{pmatrix} 0 & 1 \\ 1 & 0 \end{pmatrix}$$

is associated with a reflection in the line $y = x$.

(b) Again let $P(x, y) \rightarrow P'(x', y')$ then

$$\begin{pmatrix} x' \\ y' \end{pmatrix} = \begin{pmatrix} 0 & -1 \\ -1 & 0 \end{pmatrix}\begin{pmatrix} x \\ y \end{pmatrix} = \begin{pmatrix} -y \\ -x \end{pmatrix}$$

i.e. $P(x, y) \rightarrow P'(-y, -x)$ (see figure 15).

Hence the matrix

$$\begin{pmatrix} 0 & -1 \\ -1 & 0 \end{pmatrix}$$

is associated with a reflection in the line $y = -x$.

(c) Again let $P(x, y) \rightarrow P'(x', y')$ then

$$\begin{pmatrix} x' \\ y' \end{pmatrix} = \begin{pmatrix} 0 & -1 \\ 1 & 0 \end{pmatrix}\begin{pmatrix} x \\ y \end{pmatrix} = \begin{pmatrix} -y \\ x \end{pmatrix}$$

i.e. $P(x, y) \rightarrow P'(-y, x)$ (see figure 16).

Hence the matrix

$$\begin{pmatrix} 0 & -1 \\ 1 & 0 \end{pmatrix}$$

is associated with a quarter turn about the origin, i.e. with a rotation of 90° anticlockwise about the origin.

(ii) Since a half turn is equivalent to a reflection in the x-axis followed by a reflection in the y-axis, then

$$P(x, y) \rightarrow P_1(x_1, y_1)$$

by reflection in the x-axis and

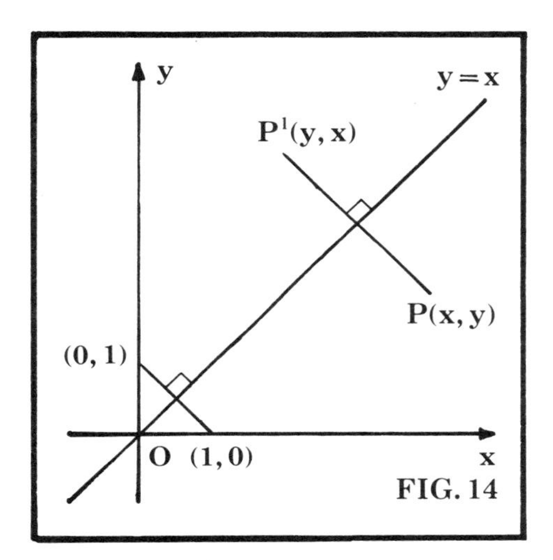

FIG. 14

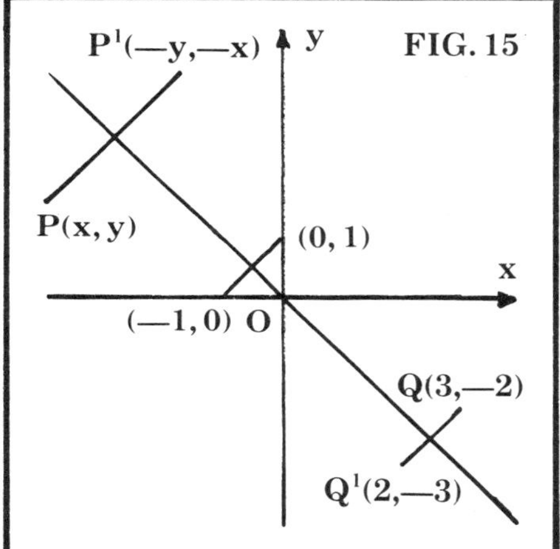

FIG. 15

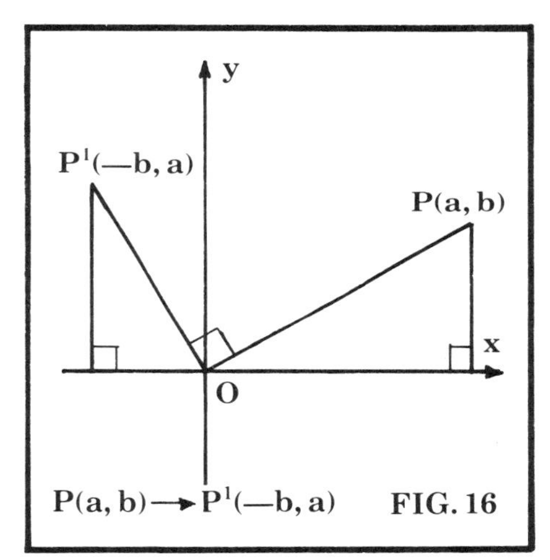

FIG. 16

$$P_1(x_1, y_1) \rightarrow P_2(x_2, y_2)$$

by reflection in the y-axis.

i.e. $$\begin{pmatrix} x_1 \\ y_1 \end{pmatrix} = \begin{pmatrix} 1 & 0 \\ 0 & -1 \end{pmatrix}\begin{pmatrix} x \\ y \end{pmatrix}$$

and $$\begin{pmatrix} x_2 \\ y_2 \end{pmatrix} = \begin{pmatrix} -1 & 0 \\ 0 & 1 \end{pmatrix}\begin{pmatrix} x_1 \\ y_1 \end{pmatrix}$$

Hence

$$\begin{pmatrix} x_2 \\ y_2 \end{pmatrix} = \begin{pmatrix} -1 & 0 \\ 0 & 1 \end{pmatrix}\begin{pmatrix} x_1 \\ y_1 \end{pmatrix} = \begin{pmatrix} -1 & 0 \\ 0 & 1 \end{pmatrix}\begin{pmatrix} 1 & 0 \\ 0 & -1 \end{pmatrix}\begin{pmatrix} x \\ y \end{pmatrix}$$

But $$\begin{pmatrix} -1 & 0 \\ 0 & 1 \end{pmatrix}\begin{pmatrix} 1 & 0 \\ 0 & -1 \end{pmatrix} = \begin{pmatrix} -1 & 0 \\ 0 & -1 \end{pmatrix}$$

Hence $$\begin{pmatrix} x_2 \\ y_2 \end{pmatrix} = \begin{pmatrix} -1 & 0 \\ 0 & -1 \end{pmatrix}\begin{pmatrix} x \\ y \end{pmatrix}$$

and therefore $$\begin{pmatrix} -1 & 0 \\ 0 & -1 \end{pmatrix}$$

is the matrix associated with a half-turn about the origin.

Note:

(1) If X denotes the operation of reflection in the x-axis and Y denotes the operation of reflection in the y-axis then $Y \text{ o } X$ denotes the operation of reflection in the x-axis followed by the operation of reflection in the y-axis.

(2) If $$A = \begin{pmatrix} 1 & 0 \\ 0 & -1 \end{pmatrix}$$

is the matrix associated with the transformation X,

and $$B = \begin{pmatrix} -1 & 0 \\ 0 & 1 \end{pmatrix}$$

is the matrix associated with the transformation Y,

then $AB\begin{pmatrix} x \\ y \end{pmatrix} = \begin{pmatrix} x_2 \\ y_2 \end{pmatrix}$ and $BA\begin{pmatrix} x \\ y \end{pmatrix} = \begin{pmatrix} x_2 \\ y_2 \end{pmatrix}$

i.e. reflection in the y-axis followed by reflection in the x-axis is also equivalent to a half-turn about the origin. i.e. $X \text{ o } Y = Y \text{ o } X$.

2.4 THE ROTATION MATRIX

It is shown in Trigonometry, section 2.3 that the matrix

$$R = \begin{pmatrix} \cos\theta & -\sin\theta \\ \sin\theta & \cos\theta \end{pmatrix}$$

is the matrix associated with the rotation of the plane anti-clockwise through an angle θ about the origin.

Example 1

With reference to rectangular axes OX and OY, the point A has coordinates (x, y). Under an anticlockwise rotation of θ radians about O, OA maps onto OB where B is the point (x', y').

(i) Express x' and y' the coordinates of B in terms of x, y and θ.

(ii) Express x and y the coordinates of A in terms of x', y' and θ.

(iii) If the line l has equation $y = mx$, where $m = \tan\alpha$, show that l maps onto the line l' with equation $y = Mx$, where $M = \tan(\alpha + \theta)$.

(i) Under a rotation of θ radians $A(x, y) \to B(x', y')$.

Hence $\begin{pmatrix} x' \\ y' \end{pmatrix} = R\begin{pmatrix} x \\ y \end{pmatrix}$ where $R = \begin{pmatrix} \cos\theta & -\sin\theta \\ \sin\theta & \cos\theta \end{pmatrix}$

i.e. $\begin{pmatrix} x' \\ y' \end{pmatrix} = \begin{pmatrix} \cos\theta & -\sin\theta \\ \sin\theta & \cos\theta \end{pmatrix}\begin{pmatrix} x \\ y \end{pmatrix} = \begin{pmatrix} x\cos\theta - y\sin\theta \\ x\sin\theta + y\cos\theta \end{pmatrix}$

i.e. $x' = x\cos\theta - y\sin\theta$

$y' = x\sin\theta + y\cos\theta$

(ii) $$\begin{pmatrix} x' \\ y' \end{pmatrix} = R\begin{pmatrix} x \\ y \end{pmatrix} \Rightarrow \begin{pmatrix} x \\ y \end{pmatrix} = R^{-1}\begin{pmatrix} x' \\ y' \end{pmatrix}$$

where R^{-1} is the inverse of the matrix R.

But $R = \begin{pmatrix} \cos\theta & -\sin\theta \\ \sin\theta & \cos\theta \end{pmatrix} \Rightarrow R^{-1} = \begin{pmatrix} \cos\theta & \sin\theta \\ -\sin\theta & \cos\theta \end{pmatrix}$

Hence

$$\begin{pmatrix} x \\ y \end{pmatrix} = \begin{pmatrix} \cos\theta & \sin\theta \\ -\sin\theta & \cos\theta \end{pmatrix}\begin{pmatrix} x' \\ y' \end{pmatrix} = \begin{pmatrix} x'\cos\theta + y'\sin\theta \\ -x'\sin\theta + y'\cos\theta \end{pmatrix}$$

i.e. $x = x'\cos\theta + y'\sin\theta$

$y = -x'\sin\theta + y'\cos\theta$

(iii) Under the given rotation, line l with equation

$$y = mx \to -x'\sin\theta + y'\cos\theta = m(x'\cos\theta + y'\sin\theta)$$
$$\to y'(\cos\theta - m\sin\theta) = x'(m\cos\theta + \sin\theta)$$

but $m = \tan\alpha = \dfrac{\sin\alpha}{\cos\alpha}$

hence

$$y = mx \to y'\left(\cos\theta - \frac{\sin\alpha}{\cos\alpha}\sin\theta\right)$$
$$= x'\left(\frac{\sin\alpha}{\cos\alpha}\cos\theta + \sin\theta\right)$$
$$\to y'(\cos\theta\cos\alpha - \sin\alpha\sin\theta)$$
$$= x'(\sin\alpha\cos\theta + \cos\alpha\sin\theta)$$
$$\to y'\cos(\alpha + \theta) = x'\sin(\alpha + \theta)$$
$$\to y' = \frac{\sin(\alpha + \theta)}{\cos(\alpha + \theta)}x'$$
$$\to y' = \tan(\alpha + \theta)x'$$

i.e. the point (x', y') lies on the line $y = Mx$, where $M = \tan(\alpha + \theta)$ and $y = Mx$ is therefore the image l' of l under this rotation.

Example 2

P is mapped onto P′ by a rotation through θ radians anticlockwise about the origin. P′ is mapped onto P″ by reflection in the x-axis. Find the matrix which maps P onto

P″. Show that this matrix maps the circle with equation $x^2+y^2 = 1$ onto itself.

Let P be the point (x, y), P′ the point (x', y') and P″ the point (x'', y''). The matrix associated with a rotation of θ radians anticlockwise about the origin is

$$R = \begin{pmatrix} \cos\theta & -\sin\theta \\ \sin\theta & \cos\theta \end{pmatrix} \quad \text{and} \quad P \to P'$$

Thus
$$\begin{pmatrix} x' \\ y' \end{pmatrix} = R\begin{pmatrix} x \\ y \end{pmatrix}$$

The matrix associated with a reflection in the x-axis is

$$A = \begin{pmatrix} 1 & 0 \\ 0 & -1 \end{pmatrix} \quad \text{and} \quad P' \to P''$$

Thus
$$\begin{pmatrix} x'' \\ y'' \end{pmatrix} = A\begin{pmatrix} x' \\ y' \end{pmatrix} = AR\begin{pmatrix} x \\ y \end{pmatrix}$$

hence the matrix AR maps P → P″ where

$$AR = \begin{pmatrix} 1 & 0 \\ 0 & -1 \end{pmatrix}\begin{pmatrix} \cos\theta & -\sin\theta \\ \sin\theta & \cos\theta \end{pmatrix} = \begin{pmatrix} \cos\theta & -\sin\theta \\ -\sin\theta & -\cos\theta \end{pmatrix}$$

To show the circle $x^2+y^2 = 1$ maps onto itself under the transformation associated with the matrix AR.

Let the matrix AR map $P(x, y)$ onto $P_1(x_1, y_1)$.

Hence
$$\begin{pmatrix} x_1 \\ y_1 \end{pmatrix} = \begin{pmatrix} \cos\theta & -\sin\theta \\ -\sin\theta & -\cos\theta \end{pmatrix}\begin{pmatrix} x \\ y \end{pmatrix}$$

i.e.
$$\begin{pmatrix} x_1 \\ y_1 \end{pmatrix} = AR\begin{pmatrix} x \\ y \end{pmatrix}$$

$$\Rightarrow \begin{pmatrix} x \\ y \end{pmatrix} = (AR)^{-1}\begin{pmatrix} x_1 \\ y_1 \end{pmatrix} = -1\begin{pmatrix} -\cos\theta & \sin\theta \\ \sin\theta & \cos\theta \end{pmatrix}\begin{pmatrix} x_1 \\ y_1 \end{pmatrix}$$

$$= \begin{pmatrix} \cos\theta & -\sin\theta \\ -\sin\theta & -\cos\theta \end{pmatrix}\begin{pmatrix} x_1 \\ y_1 \end{pmatrix}$$

i.e.
$$x = \cos\theta \,.\, x_1 - \sin\theta \,.\, y_1$$
$$y = -\sin\theta \,.\, x_1 - \cos\theta \,.\, y_1$$

Thus the circle

$$x^2+y^2 = 1$$
$$\to (x_1\cos\theta - y_1\sin\theta)^2 + (-x_1\sin\theta - y_1\cos\theta)^2 = 1$$
$$\to {x_1}^2\cos^2\theta - 2x_1y_1\cos\theta\sin\theta + {y_1}^2\sin^2\theta$$
$$+ {x_1}^2\sin^2\theta + 2x_1y_1\cos\theta\sin\theta + {y_1}^2\cos^2\theta = 1$$
$$\to (\cos^2\theta + \sin^2\theta){x_1}^2 + (\sin^2\theta + \cos^2\theta){y_1}^2 = 1$$
$$\to {x_1}^2 + {y_1}^2 = 1$$

i.e. the image point (x_1, y_1) satisfies the equation $x^2+y^2 = 1$ and therefore lies on the circle with equation $x^2+y^2 = 1$.

Thus the matrix AR maps the circle $x^2+y^2 = 1$ onto itself.

2.5 SUMMARY OF TRANSFORMATIONS AND THEIR ASSOCIATED MATRICES

(i)
$$I = \begin{pmatrix} 1 & 0 \\ 0 & 1 \end{pmatrix}$$

—the identity matrix.

I maps $P(x, y)$ onto $P'(x', y')$, where

$$x' = x \quad \text{and} \quad y' = y$$

(ii)
$$M = \begin{pmatrix} k & 0 \\ 0 & k \end{pmatrix}$$

—the enlargement matrix or the matrix associated with a dilatation $[0, k]$, where 0 is the origin and k the scale factor.

M maps $P(x, y)$ onto $P'(x', y')$, where

$$x' = kx \quad \text{and} \quad y' = ky$$

(iii)
$$A = \begin{pmatrix} 1 & 0 \\ 0 & -1 \end{pmatrix}$$

—the matrix associated with a reflection in the x-axis.

A maps $P(x, y)$ onto $P'(x', y')$, where

$$x' = x \quad \text{and} \quad y' = -y$$

(iv) $$B = \begin{pmatrix} -1 & 0 \\ 0 & 1 \end{pmatrix}$$

—the matrix associated with a reflection in the y-axis.

B maps $P(x, y)$ onto $P'(x', y')$, where

$$x' = -x \quad \text{and} \quad y' = y$$

(v) $$S_1 = \begin{pmatrix} 1 & k \\ 0 & 1 \end{pmatrix}$$

—the matrix associated with a shear parallel to the x-axis.

S_1 maps $P(x, y)$ onto $P'(x', y')$, where

$$x' = x + ky \quad \text{and} \quad y' = y$$

$$S_2 = \begin{pmatrix} 1 & 0 \\ k & 1 \end{pmatrix}$$

—the matrix associated with a shear parallel to the y-axis.

S_2 maps $P(x, y)$ onto $P'(x', y')$, where

$$x' = x \quad \text{and} \quad y' = kx + y$$

(vi) $$C = \begin{pmatrix} 0 & 1 \\ 1 & 0 \end{pmatrix}$$

—the matrix associated with a reflection in the line $y = x$.

C maps $P(x, y)$ onto $P'(x', y')$, where

$$x' = y \quad \text{and} \quad y' = x$$

(vii) $$D = \begin{pmatrix} 0 & -1 \\ -1 & 0 \end{pmatrix}$$

—the matrix associated with a reflection in the line $y = -x$.

D maps $P(x, y)$ onto $P'(x', y')$, where

$$x' = -y \quad \text{and} \quad y' = -x$$

(viii) $$R = \begin{pmatrix} \cos\theta & -\sin\theta \\ \sin\theta & \cos\theta \end{pmatrix}$$

—the matrix associated with a rotation of the plane anticlockwise through an angle θ.

R maps $P(x, y)$ onto $P'(x', y')$, where

$$x' = x\cos\theta - y\sin\theta \quad \text{and} \quad y' = x\sin\theta + y\cos\theta$$

(ix) $$R_1 = \begin{pmatrix} 0 & -1 \\ 1 & 0 \end{pmatrix}$$

—the matrix associated with a rotation of the plane through an angle of $+90°$ (i.e. a quarter turn) which may be obtained by putting $\theta = 90°$ in R.

R_1 maps $P(x, y)$ onto $P'(x', y')$, where

$$x' = -y \quad \text{and} \quad y' = x$$

(x) $$R_2 = \begin{pmatrix} -1 & 0 \\ 0 & -1 \end{pmatrix}$$

—the matrix associated with a rotation of the plane through an angle of $+180°$ (i.e. a half-turn about the origin) which may be obtained by putting $\theta = 180°$ in R.

R_2 maps $P(x, y)$ onto $P'(x', y')$, where

$$x' = -x \quad \text{and} \quad y' = -y$$

R_2 may also be obtained by reflection in the x-axis followed by reflection in the y-axis, or reflection in the y-axis followed by reflection in the x-axis.

i.e. $$R_2 = BA = AB$$

ASSIGNMENT 2.2

1. Under a certain mapping $P(x, y) \to P(x', y')$, where $x' = x + 2y$ and $y' = y$. Write this mapping in matrix form and state the matrix of the mapping.
2. Under a certain mapping $P(x, y) \to P_1(x_1, y_1)$, where

$x_1 = x$ and $y_1 = 2x+y$. Write down the mapping in the form

$$\begin{pmatrix} x_1 \\ y_1 \end{pmatrix} = A\begin{pmatrix} x \\ y \end{pmatrix}$$

where A is a 2×2 matrix.

State the matrix associated with this mapping and use this matrix to find the image of the set of points $\{(0,0), (1,0), (-1,-1), (0,-1)\}$.

Show this set of points and the set of image points on a diagram.

3. Write down the matrix associated with a reflection in the line $y = x$ and use this matrix to find the coordinates of the images on reflection in the line $y = x$ of the points O(0, 0), P(3, 0), Q(3, 2) and R(0, 2).

   Show the figure OPQR and its image on a diagram.

4. Under the transformation with matrix

$$\begin{pmatrix} 1 & 3 \\ 2 & 6 \end{pmatrix}$$

   find the image of the set of points

$$\{(1,-1), (0,-2), (3,5), (a,b)\}$$

   Show that this transformation maps the plane onto the line with equation $y = 2x$.

5. The matrix $\begin{pmatrix} \cos\theta & -\sin\theta \\ \sin\theta & \cos\theta \end{pmatrix}$

   is the matrix associated with a rotation of θ radians or degrees anticlockwise about the origin. Evaluate the matrices associated with an anticlockwise rotation of

   (i) $90°, 45°, 30°, 120°$.

   (ii) $\frac{\pi}{3}, \pi, \frac{5\pi}{6}$ radians.

   (iii) $-\frac{\pi}{3}, -\frac{\pi}{6}, -\frac{\pi}{2}$ radians.

   (iv) $0°$. Explain the geometrical significance of this rotation of $0°$.

6. P(x, y) is mapped onto P$_1(x_1, y_1)$ by a reflection in the x-axis followed by a reflection in the origin. Find the matrix which maps P onto P$_1$ and identify geometrically this matrix.

7. P(x, y) is mapped onto P$_1(x_1, y_1)$ by reflection in the y-axis and P$_1$ is mapped onto P$_2$ by rotation anticlockwise of $90°$ about the origin. Find the matrix which maps P onto P$_2$ and identify this matrix.

   Show that this matrix maps the circle $x^2+y^2 = 1$ onto itself and maps the line $y = 2x+1$ onto the line $x = 2y-1$.

In questions 8 and 9,

X denotes the operation of reflection in the x-axis with associated matrix A.

Y denotes the operation of reflection in the y-axis with associated matrix B.

M denotes the operation of reflection in the line $y = x$ with associated matrix P.

N denotes the operation of reflection in the line $y = -x$ with associated matrix Q.

8. (i) To what single transformation is $X \circ Y$ equivalent?

   (ii) What is the matrix associated with the transformation $X \circ Y$?

   (iii) Is the transformation $Y \circ X$ equivalent to the transformation $X \circ Y$? Check by finding the matrix products AB and BA.

9. (i) Write down the matrices P and Q associated with the reflections in the lines $y = x$ and $y = -x$ respectively.

   (ii) What single transformation is associated with $M \circ N$?

(iii) What is the matrix associated with the transformation $M \circ N$?

(iv) Is the transformation $N \circ M$ equivalent to the transformation $M \circ N$? Check by finding the matrix products PQ and QP.

(v) Can you say that

$$X \circ Y = Y \circ X = M \circ N = N \circ M?$$

10. Matrix A represents a transformation of reflection of every point $P(x, y)$ in the y-axis and matrix B represents the rotation through 90° anticlockwise about the origin.

(i) Find the matrices A and B and calculate the products AB and BA.

(ii) Indicate by diagrams the images of $P(x, y)$ under the transformations AB and BA.

(iii) Find a matrix X such that $XAB = BA$.

11. Matrix A represents the reflection of every point in the x-axis and matrix B represents the rotation of the plane through 90° anticlockwise about the origin.

(i) Give the 2×2 matrices A and B and find the matrix products AB and BA.

(ii) Indicate on a diagram the positions P_1 and P_2 of the images of $P(x, y)$ under the transformations given by AB and BA respectively.

(iii) Find the matrix X such that $ABX = BA$ and state the transformation represented by X.

12. Describe the geometrical transformations associated with the matrices

$$\text{(i) } A = \begin{pmatrix} 0 & -1 \\ 1 & 0 \end{pmatrix} \quad \text{and} \quad \text{(ii) } B = \begin{pmatrix} -1 & 0 \\ 0 & 1 \end{pmatrix}$$

If $X = AB$, calculate X and describe the geometrical transformation with which it is associated.

Calculate X^{-1} the inverse of X.

13. Identify the transformations represented by the matrices

$$P = \begin{pmatrix} 0 & 1 \\ 1 & 0 \end{pmatrix} \quad \text{and} \quad Q = \begin{pmatrix} 0 & -1 \\ -1 & 0 \end{pmatrix}$$

Show by multiplying these matrices that $PQ = QP$ and interpret geometrically this result.

If I is the 2×2 identity matrix and $X = PQ$ complete the following table for matrix multiplication,

·	I	P	Q	X
I				
P			X	
Q				
X		Q		

14. If matrix $P = \begin{pmatrix} p & q \\ 1 & 1 \end{pmatrix}$, and matrix $Q = \begin{pmatrix} 1 & 2 \\ 1 & 0 \end{pmatrix}$ and the product $PQ = \begin{pmatrix} 0 & 4 \\ 2 & 2 \end{pmatrix}$ find the values of p and q and calculate the product QP.

If X is another 2×2 matrix such that $PQX = QP$ find the matrix X and state the geometrical transformation with which it is associated.

Hence find the equation of the image of the line $x - 3y = 0$ under the transformation associated with matrix X.

15. A is the matrix associated with reflection in the x-axis and B the matrix associated with a reflection in the line $y = x$.

Write down matrix A and matrix B.

If $C = BA$, calculate C and describe concisely the transformation associated with C.

16. A is the matrix associated with a rotation of 90° anticlockwise about the origin and B the matrix associated with a reflection in the line $y = x$.

Find the matrices A and B and calculate the matrix products AB and BA. What transformations are associated with AB and BA?

If X is a matrix such that $XAB = BA$, find the matrix X and state the transformation associated with X.

Hence or otherwise find the image of the line $y = 3x + 2$ under the transformation associated with X.

17. Write down the 2×2 matrix which represents a rotation of θ radians anticlockwise about the origin.

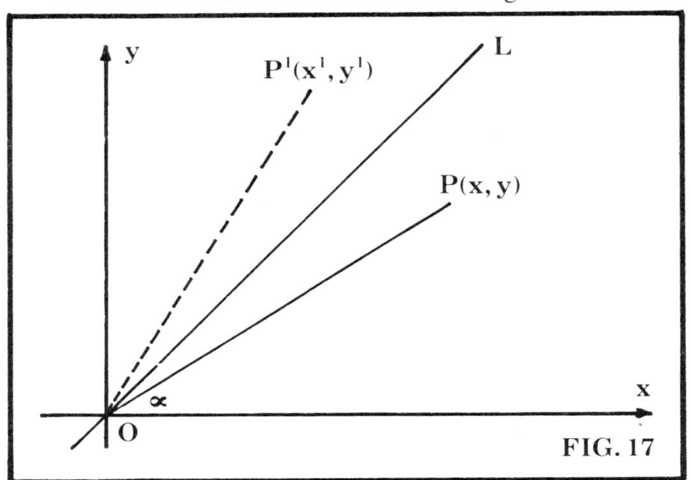

FIG. 17

In figure 17, OL is the line with equation $y = x$. P is the point (x, y) such that angle XOP $= \alpha$ radians.

If $P'(x', y')$ is the image of P under reflection in OL, show that

$$\text{angle POP}' = \frac{\pi}{2} - 2\alpha$$

If $P(x, y)$ maps onto P′ by a rotation of

$$\left(\frac{\pi}{2} - 2\alpha\right)$$

anticlockwise about O, write down the 2×2 matrix R such that

$$\begin{pmatrix} x' \\ y' \end{pmatrix} = R \begin{pmatrix} x \\ y \end{pmatrix}$$

writing R in its simplest form and hence express the coordinates of P′ in terms of x, y and α.

18. Show that any point on the line $y = x$ is mapped onto itself by the transformation whose matrix is

$$N = \begin{pmatrix} 2 & -1 \\ 1 & 0 \end{pmatrix}$$

If any point on the line $y = x$ is also mapped onto itself by the transformation whose matrix is

$$\begin{pmatrix} p & 0 \\ 1 & q \end{pmatrix}$$

calculate the values of p and q.

19. Under a certain transformation $p(x, y) \to P'(x', y')$, where

$$x' = x - y \quad \text{and} \quad y' = x + y$$

Write this transformation in the form

$$\begin{pmatrix} x' \\ y' \end{pmatrix} = P \begin{pmatrix} x \\ y \end{pmatrix}$$

where P is the matrix associated with the transformation. If

$$Q = \begin{pmatrix} 2 & 0 \\ 0 & 2 \end{pmatrix}$$

identify the transformation associated with Q and find the 2×2 matrix R such that $RQ = P$.

20. T is the transformation with associated matrix

$$A = \begin{pmatrix} -1 & 1 \\ 1 & 1 \end{pmatrix}$$

(i) Show that T maps the point (p, q) onto the point (p', q') where $p' = q - p$ and $q' = q + p$ and hence or otherwise show that T maps points on the y-axis to points on the line $y-x=0$.

(ii) If the transformation T is equivalent to a transformation with associated matrix B followed by a reflection in the line $y=-x$ with associated matrix C, find the 2×2 matrix C and hence find the matrix B such that $CB=A$.

ASSIGNMENT 2.3

Objective items testing sections 2.1–2.5

Instructions for answering these items are given on page 152.

1. Under a composite mapping

$$P(x,y)\rightarrow P_1(x_1,y_1)\rightarrow P_2(x_2,y_2)$$

where $\quad x_1=-2x \quad x_2=-x_1-2$

$\quad y_1=1-2y \quad y_2=y_1+3$

The image of the point $(-2,4)$ is

A. $(0,-13)$

B. $(-6,-4)$

C. $(6,8)$

D. $(12,-1)$

E. $(-4,-6)$

2. O is the origin and P the point $(4,7)$. After an anticlockwise rotation of $90°$ about O followed by a reflection in the y-axis, the image of P is

A. $(-7,-4)$

B. $(7,-4)$

C. $(-7,4)$

D. $(-4,-7)$

E. $(7,4)$

3. Under a certain mapping $P(a,b)\rightarrow P'(a',b')$, where

$$a'=2a-b \quad \text{and} \quad b'=3a-2b$$

Under the inverse mapping $P'(a',b')\rightarrow P(a,b)$, a and b are given by

A. $a=-2a'+b'$ and $b=-3a'+2b'$

B. $a=a'-2b$ and $b=a'-2b'$

C. $a=2b'-a'$ and $b=2b'-3a'$

D. $a=2a'-b'$ and $b=3a'-2b'$

E. none of these.

4. The transformation represented by the matrix

$$\begin{pmatrix}1 & 0\\ 2 & 1\end{pmatrix}$$

maps the line $y=4x$ onto the line with equation

A. $y=-4x$

B. $4y=x$

C. $y=6x$

D. $6y=x$

E. $y=4x$

5. P reflected in the x-axis maps onto P′ and P′ rotated through an angle of $90°$ anticlockwise about the origin maps onto P″. The matrix which maps P onto P″ is

A. $\begin{pmatrix}0 & -1\\ -1 & 0\end{pmatrix}$

B. $\begin{pmatrix}0 & 1\\ 1 & 0\end{pmatrix}$

C. $\begin{pmatrix}-1 & 0\\ 0 & 1\end{pmatrix}$

D. $\begin{pmatrix}-1 & 0\\ 0 & -1\end{pmatrix}$

E. $\begin{pmatrix} 1 & 0 \\ 0 & -1 \end{pmatrix}$

6. A transformation T maps $P(x, y) \to P'(x', y')$ such that $x' = 2x$ and $y' = 2y$. Which one of the following statements is false?

   A. The matrix of the transformation is $\begin{pmatrix} 2 & 0 \\ 0 & 2 \end{pmatrix}$.

   B. T maps the circle $x^2 + y^2 = 1$ onto the circle $x^2 + y^2 = 2$.

   C. T is the dilatation $[O, 2]$, where O is the origin.

   D. The transformation in which $P'(x', y') \to P(x, y)$ such that $x = \frac{1}{2}x'$ and $y = \frac{1}{2}y'$ is the inverse mapping of T.

   E. O(0, 0) is the only point which maps onto itself under T.

7. The transformation with matrix

$$\begin{pmatrix} -2 & 0 \\ 4 & 1 \end{pmatrix}$$

maps P onto P_1 and the transformation with matrix

$$\begin{pmatrix} 1 & -1 \\ 3 & 0 \end{pmatrix}$$

maps P_1 onto P_2. If P is the point $(2, -1)$ then P_2 has coordinates

   A. $(-4, 7)$

   B. $(-11, -12)$

   C. $(3, 6)$

   D. $(-6, 18)$

   E. $(-11, -5)$

8. A point P is reflected in the line $y = x$ and its image is reflected in the line $y = -x$. Which one of the following statements is not equivalent to this composite transformation?

   A. Reflection in the x-axis followed by reflection in the y-axis.

   B. A rotation of 180° anticlockwise about O.

   C. A dilatation centre the origin and with scale factor -1.

   D. The transformation with matrix $\begin{pmatrix} -1 & 0 \\ 0 & -1 \end{pmatrix}$.

   E. The transformation with matrix $\begin{pmatrix} 0 & -1 \\ -1 & 0 \end{pmatrix}$.

9. In figure 18 the centre of the square is at the origin and its sides are parallel to the x- and y-axes.

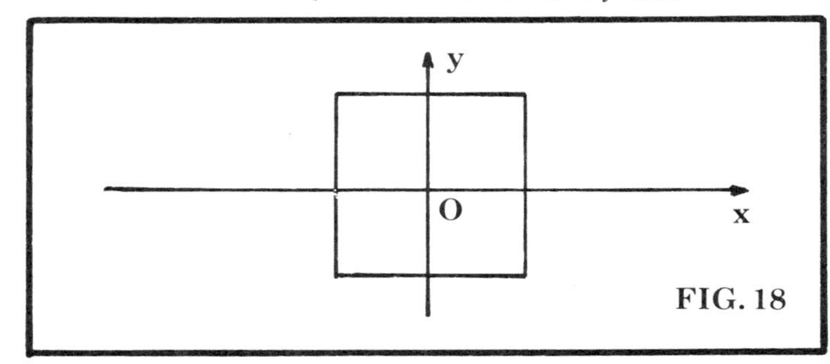

FIG. 18

X denotes reflection in the x-axis
Y denotes reflection in the y-axis
H denotes reflection in the origin
R denotes rotation anticlockwise through an angle 90° about O.

Which of the following transformations is/are equivalent to I the identity transformation?

   (1) $X \circ R \circ H$

   (2) $Y \circ X \circ H$

   (3) $R \circ H \circ R$

10. The transformation T with associated matrix

$$\begin{pmatrix} k & 0 \\ 0 & k \end{pmatrix}$$

maps P onto P′ such that $\vec{OP'} = -3\vec{OP}$, where O is the origin.

(1) $k = -3$

(2) T is the dilatation $[0. -3]$

(3) $|\vec{OP'}| = 3|\vec{OP}|$

11. (1) $\begin{pmatrix} 0 & -1 \\ 1 & 0 \end{pmatrix}$ is the matrix associated with a rotation of $\theta°$, $0 \leqq \theta \leqq 360$

(2) $\theta = 90$

12. (1) P denotes reflection in the x-axis, Q denotes reflection in the y-axis.

(2) $P \circ Q = H$, where H denotes reflection in the origin.

Objective test items

In the objective test items situated at the end of each unit, answer the items as follows (unless otherwise stated).

(i) In questions 1–8 the correct answer is given by one of the options A, B, C, D or E.

(ii) In questions 9 and 10 one or more of the three statements is/are correct.

Answer A if statement (1) only is correct.
B if statement (2) only is correct.
C if statement (3) only is correct.
D if statements (1), (2) and (3) are correct.
E if some other combination of the given statements is correct.

(iii) In questions 11 and 12 two statements are numbered (1) and (2).

Answer A if (1) implies (2) but (2) does not imply (1).
B if (2) implies (1) but (1) does not imply (2).
C if (1) implies (2) and (2) implies (1).
D if (1) denies (2) or (2) denies (1).
E if none of the above relationships hold.

UNIT 3: VECTORS

DEFINITION OF VECTORS IN TWO DIMENSIONS

3.1

A **Scalar** is a quantity that can be completely specified by a real number, for example, time, speed, temperature, distance.

A **Vector** is a quantity that requires a *magnitude* and *direction* for its complete specification, for example, force, velocity, acceleration, displacement.

Thus a vector must have three things to specify it completely,

(i) a magnitude or length

(ii) a direction in space

(iii) a sense. This sense gives a more precise idea of direction, thus

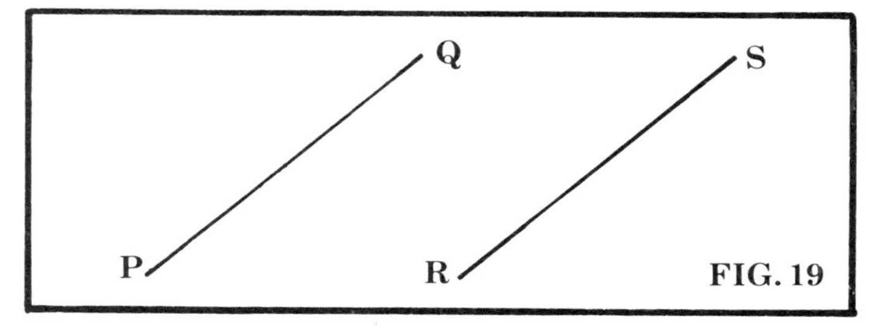

FIG. 19

PQ and RS are parallel in direction and in the same sense.
PQ and SR are parallel in direction but in the opposite sense.

3.2 3.4

NOTATION

We can denote vectors in two ways,

(a) If P and Q are two points in space, the line segment joining P and Q represents a vector. This directed line segment is denoted by $\overrightarrow{PQ}$.

(i) The magnitude of the vector is the distance between P and Q and is denoted by $|\overrightarrow{PQ}|$.

(ii) The direction of the vector is that of the line PQ and its sense is from P to Q.

Thus $$|\overrightarrow{PQ}| = |\overrightarrow{QP}|$$

and $\overrightarrow{PQ}$ and $\overrightarrow{QP}$ are in the same direction but in the opposite sense.

3.5

Hence $$\overrightarrow{PQ} = -\overrightarrow{QP}$$

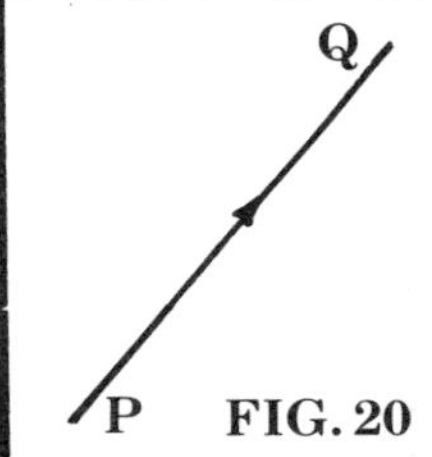

FIG. 20

(b) We sometimes use single letters $a, b, c, \ldots$ to denote vectors. In print these letters are usually shown in bold or heavy type; in writing on paper it is often difficult to distinguish them from other letters so we underline them thus, $\underline{a}, \underline{b}, \underline{c}, \ldots$, so that we may write $\overrightarrow{AB} = \mathbf{a}$ or $\overrightarrow{AB} = \underline{a}$ for convenience.

$|\mathbf{a}|$ or $|\underline{a}|$ denotes the magnitude of the vector $\mathbf{a}$ or $\underline{a}$.

3.3 3.6

FURTHER DEFINITIONS

(a) *Negative vectors*
The vector $-\mathbf{a}$, called the **negative** of $\mathbf{a}$ is a vector which has the same magnitude and direction as $\mathbf{a}$ but opposite sense.

(b) *The zero vector*
The **zero vector** is denoted by $\mathbf{0}$ (or $\underline{0}$). It has zero magnitude but its direction and sense are not defined.

ADDITION OF VECTORS

If $\mathbf{a}$ and $\mathbf{b}$ are two vectors represented by the line segments OP and PQ, then the vector $\mathbf{a}+\mathbf{b}$ is defined to be represented by $\overrightarrow{OQ}$. If $\overrightarrow{OQ}$ represents $\mathbf{c}$ then $\overrightarrow{OQ} = \overrightarrow{OP} + \overrightarrow{PQ}$,

i.e. $$\mathbf{c} = \mathbf{a}+\mathbf{b} \quad \text{(see figure 21)}$$

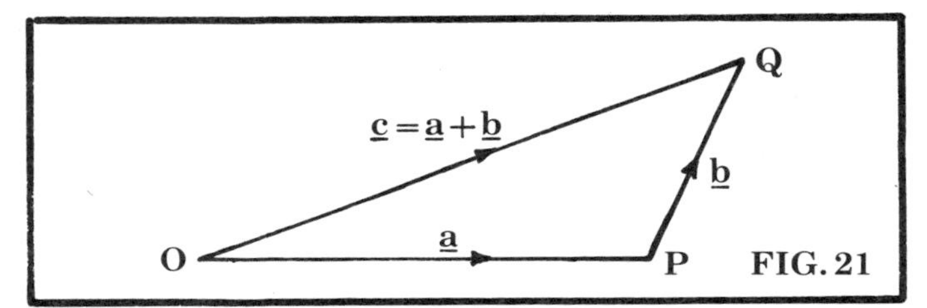

FIG. 21

MULTIPLICATION BY A SCALAR OR NUMBER *k*

If $\mathbf{a}$ is a vector and k a non-zero number then $k\mathbf{a}$ is a vector with the same direction as $\mathbf{a}$ and with k times its magnitude.

(i) If $k > 0$ the sense of $k\mathbf{a}$ is the same as $\mathbf{a}$.

(ii) If $k < 0$ the sense of $k\mathbf{a}$ is opposite to the sense of $\mathbf{a}$.

Parallel vectors
Two or more non-zero vectors are said to be parallel if they are in the same direction.

Hence two vectors $\mathbf{a}$ and $\mathbf{b}$ are parallel $\Leftrightarrow \mathbf{b} = k\mathbf{a}$ for some non-zero number k.

ALGEBRAIC LAWS

(a) for scalars k and m

(i) $k(\mathbf{a}+\mathbf{b}) = k\mathbf{a}+k\mathbf{b}$ —distributive law

(ii) $(k+m)\mathbf{a} = k\mathbf{a}+m\mathbf{a}$ —distributive law

(iii) $k(m\mathbf{a}) = (km)\mathbf{a}$

(b) for vectors $\boldsymbol{a}$, $\boldsymbol{b}$, $\boldsymbol{c}$

(i) $\boldsymbol{a}+\boldsymbol{b} = \boldsymbol{b}+\boldsymbol{a}$	—commutative law for addition
(ii) $(\boldsymbol{a}+\boldsymbol{b})+\boldsymbol{c} = \boldsymbol{a}+(\boldsymbol{b}+\boldsymbol{c})$	—associative law for addition
(iii) $\boldsymbol{a}+\boldsymbol{0} = \boldsymbol{0}+\boldsymbol{a} = \boldsymbol{a}$	—additive identity
(iv) $\boldsymbol{a}+(-\boldsymbol{a}) = (-\boldsymbol{a})+\boldsymbol{a} = \boldsymbol{0}$	—additive inverse
(v) $0\boldsymbol{a} = \boldsymbol{0}$	—zero times a vector $\boldsymbol{a}$ gives the zero vector

Example 1

If P and Q are the mid-points of AB and AC of triangle ABC (figure 22(a)), prove that $\overrightarrow{PQ} = \frac{1}{2}\overrightarrow{BC}$.

From quadrilateral PQCB:

$$\overrightarrow{PQ}+\overrightarrow{QC}+\overrightarrow{CB}+\overrightarrow{BP} = \boldsymbol{0} \qquad (1)$$

but $\quad \overrightarrow{BP} = \frac{1}{2}\overrightarrow{BA} \quad$ and $\quad \overrightarrow{QC} = \frac{1}{2}\overrightarrow{AC}$

hence $\quad \overrightarrow{PQ}+\frac{1}{2}\overrightarrow{AC}+\overrightarrow{CB}+\frac{1}{2}\overrightarrow{BA} = \boldsymbol{0}$ from (1)

$\Rightarrow \overrightarrow{PQ}+\frac{1}{2}(\overrightarrow{BA}+\overrightarrow{AC})+\overrightarrow{CB} = \boldsymbol{0}$

$\Rightarrow \overrightarrow{PQ}+\frac{1}{2}\overrightarrow{BC}+\overrightarrow{CB} = \boldsymbol{0} \quad$ since $\quad \overrightarrow{BA}+\overrightarrow{AC} = \overrightarrow{BC}$

$\Rightarrow \overrightarrow{PQ}-\frac{1}{2}\overrightarrow{BC} = \boldsymbol{0} \quad$ since $\quad \overrightarrow{CB} = -\overrightarrow{BC}$

$\Rightarrow \overrightarrow{PQ} = \frac{1}{2}\overrightarrow{BC}$

OR In figure 22(b) let $\overrightarrow{AB}$ represent the vector $\boldsymbol{u}$, $\Rightarrow \overrightarrow{AP}$ represents $\frac{1}{2}\boldsymbol{u}$, let $\overrightarrow{AC}$ represent the vector $\boldsymbol{v}$, $\Rightarrow \overrightarrow{AQ}$ represents $\frac{1}{2}\boldsymbol{v}$ and let $\overrightarrow{PQ}$ represent the vector $\boldsymbol{p}$.

From $\quad \triangle APQ: \overrightarrow{AP}+\overrightarrow{PQ} = \overrightarrow{AQ} \Rightarrow \frac{1}{2}\boldsymbol{u}+\boldsymbol{p} = \frac{1}{2}\boldsymbol{v}$

$\Rightarrow \boldsymbol{p} = \frac{1}{2}(\boldsymbol{v}-\boldsymbol{u})$

Let $\overrightarrow{BC}$ represent the vector $\boldsymbol{q}$ then

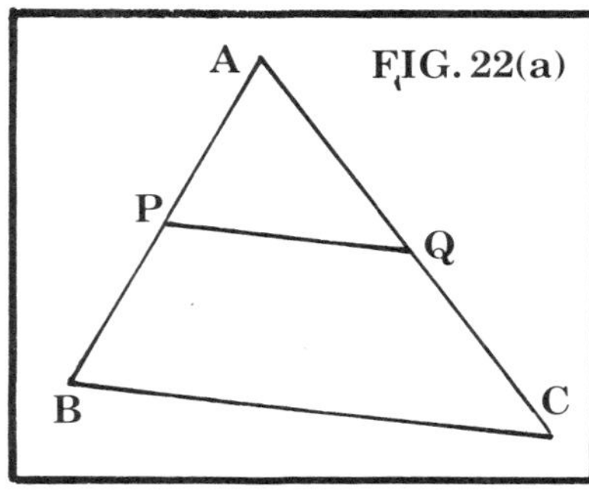

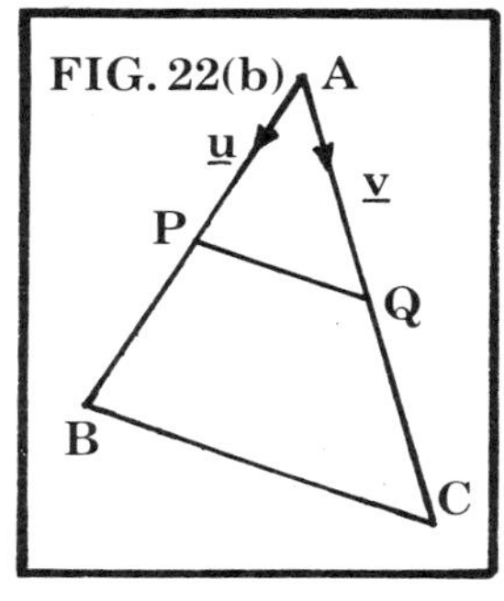

from $\quad \triangle ABC: \overrightarrow{AB}+\overrightarrow{BC} = \overrightarrow{AC} \Rightarrow \boldsymbol{u}+\boldsymbol{q} = \boldsymbol{v}$

$\Rightarrow \quad \boldsymbol{q} = \boldsymbol{v}-\boldsymbol{u}$

$\Rightarrow \quad \boldsymbol{p} = \frac{1}{2}\boldsymbol{q}$

Hence $\quad \overrightarrow{PQ} = \frac{1}{2}\overrightarrow{BC}$

ASSIGNMENT 3.1

1. In figure 23, PQRS is a quadrilateral. $\overrightarrow{PS}$, $\overrightarrow{SR}$ and $\overrightarrow{QR}$ represent vectors $\boldsymbol{a}$, $\boldsymbol{b}$ and $\boldsymbol{c}$. Express the vector represented by
   (i) $\overrightarrow{PR}$ in terms of $\boldsymbol{a}$ and $\boldsymbol{b}$.
   (ii) $\overrightarrow{SQ}$ in terms of $\boldsymbol{b}$ and $\boldsymbol{c}$.
   (iii) $\overrightarrow{PQ}$ in terms of $\boldsymbol{a}$, $\boldsymbol{b}$ and $\boldsymbol{c}$.

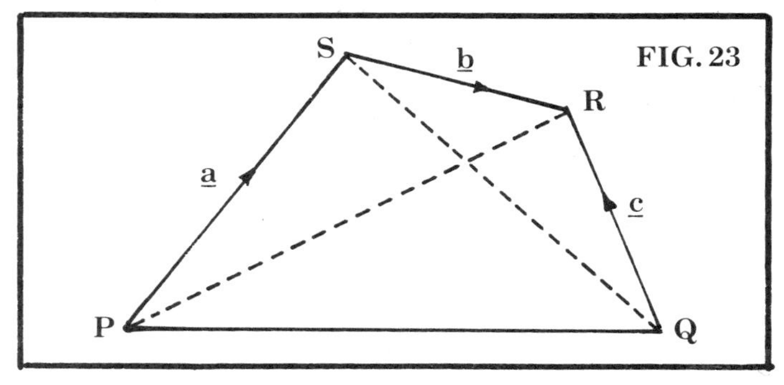

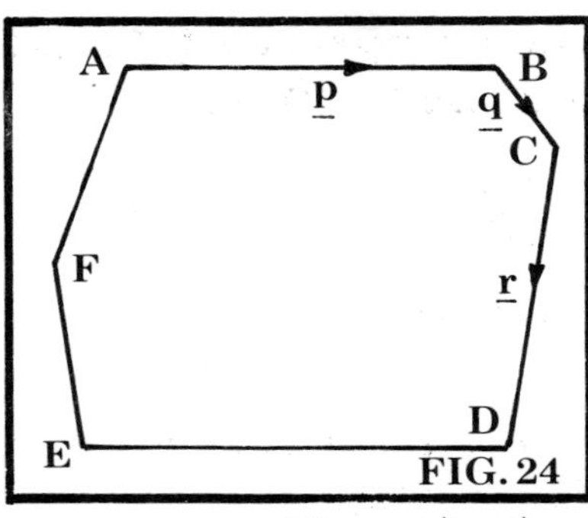

FIG. 24

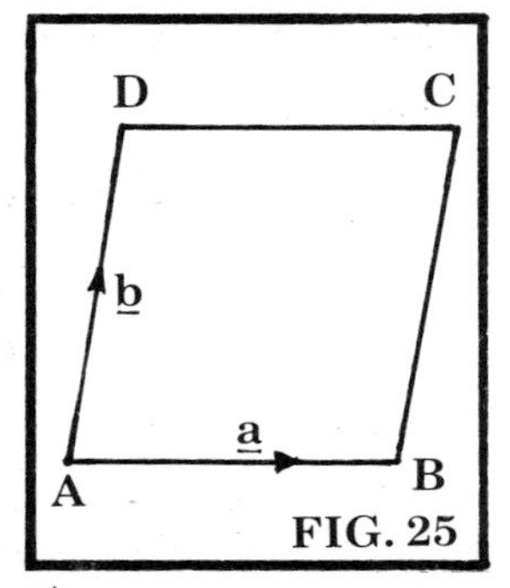

FIG. 25

2. In figure 24, $\overrightarrow{FE} = 2\overrightarrow{BC}$, $\overrightarrow{AF} = \frac{2}{3}\overrightarrow{CD}$ and AB is parallel to ED.

   Find the vector represented by $\overrightarrow{ED}$ in terms of $\boldsymbol{p}$, $\boldsymbol{q}$ and $\boldsymbol{r}$.

3. In figure 25, ABCD is a rhombus, $\overrightarrow{AB} = \boldsymbol{a}$ and $\overrightarrow{AD} = \boldsymbol{b}$.

   Express the vectors represented by $\overrightarrow{AC}$ and $\overrightarrow{DB}$ in terms of $\boldsymbol{a}$ and $\boldsymbol{b}$.

4. In figure 26, $\overrightarrow{AE}$ represents the vector $\boldsymbol{s}$ and $\overrightarrow{AB}$ the vector $\boldsymbol{t}$. $\overrightarrow{ED} = 3\overrightarrow{AB}$ and $\overrightarrow{AE} = \frac{3}{2}\overrightarrow{CD}$.

   Find the vector represented by $\overrightarrow{BC}$ in terms of $\boldsymbol{s}$ and $\boldsymbol{t}$.

5. In figure 27, $\overrightarrow{OA}$ represents $\boldsymbol{a}$ and $\overrightarrow{OC}$ represents $\boldsymbol{c}$. If $\overrightarrow{OC} = \frac{2}{5}\overrightarrow{AB}$, prove that $\overrightarrow{OD} = \frac{2}{7}\overrightarrow{OB}$ and hence express $\overrightarrow{OD}$ in terms of $\boldsymbol{a}$ and $\boldsymbol{c}$.

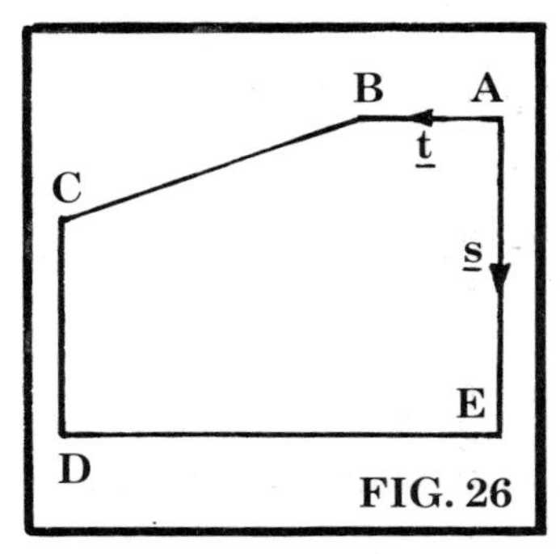

FIG. 26

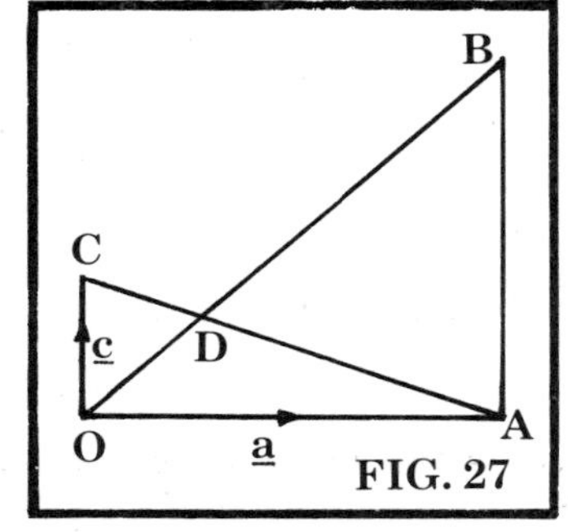

FIG. 27

6. In figure 28, G is the centroid of triangle ABC and $\overrightarrow{GA}$, $\overrightarrow{GB}$, $\overrightarrow{GC}$ represent vectors $\boldsymbol{a}$, $\boldsymbol{b}$, $\boldsymbol{c}$. If R is the mid-point of AC, show that $\overrightarrow{GR}$ represents $\frac{1}{2}(\boldsymbol{a}+\boldsymbol{c})$, and hence that $\boldsymbol{a}+\boldsymbol{b}+\boldsymbol{c} = \boldsymbol{0}$. AG is produced to D so that AG = GD.

   Find the vectors represented by $\overrightarrow{GD}$, $\overrightarrow{DC}$ and $\overrightarrow{DB}$. What kind of figure is GBDC? (Give a reason.)

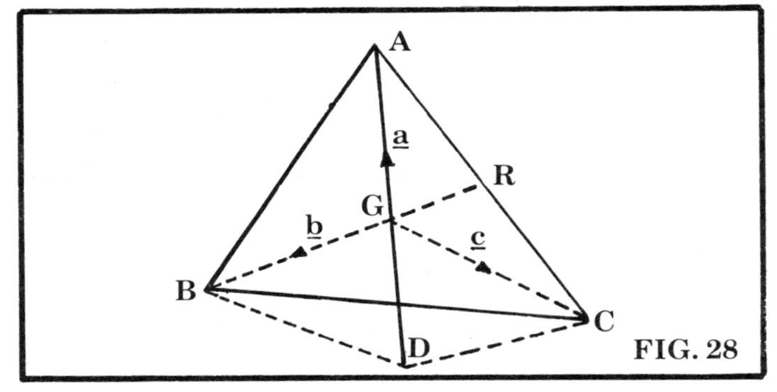

FIG. 28

7. In figure 29, PQRS is a parallelogram and PQ = QT. ST and QR intersect at M and PR and SQ intersect at N. $\overrightarrow{PS}$ and $\overrightarrow{PQ}$ represent the vectors $\boldsymbol{u}$ and $\boldsymbol{v}$.

   (i) Find the vectors represented by $\overrightarrow{QM}$, $\overrightarrow{SM}$, $\overrightarrow{NR}$.

   (ii) Find the vector represented by $\overrightarrow{TR}$ and hence deduce QTRS is a parallelogram.

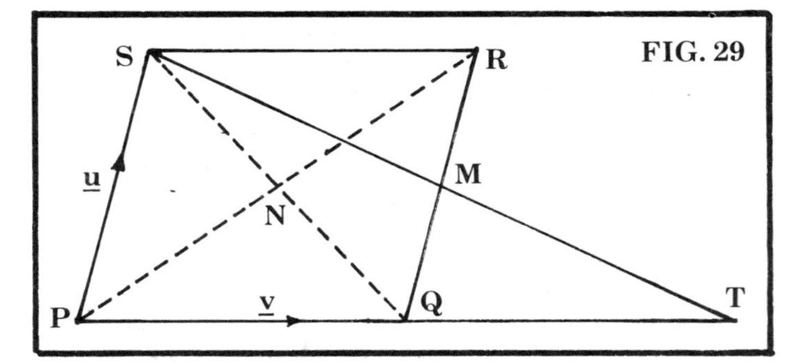

FIG. 29

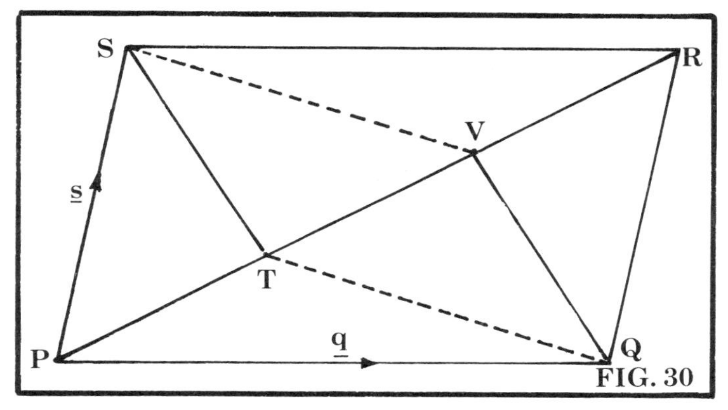

FIG. 30

8. In figure 30, PQRS is a parallelogram. T and V are points of trisection of PR. $\overrightarrow{PQ}$ represents $\boldsymbol{q}$ and $\overrightarrow{PS}$ represents $\boldsymbol{s}$.

   (a) Express the vectors represented by $\overrightarrow{PT}$, $\overrightarrow{ST}$ in terms of $\boldsymbol{q}$ and $\boldsymbol{s}$.

   (b) Express the vectors represented by $\overrightarrow{QV}$ and $\overrightarrow{RV}$ in terms of $\boldsymbol{q}$ and $\boldsymbol{s}$.

   (c) Show that STQV is a parallelogram.

POSITION VECTORS

If the point O is taken as origin and P is any other point, the vector represented by $\overrightarrow{OP}$ is called the **position vector** of the point P (with reference to the origin O). It is sometimes denoted by a small letter, namely $\boldsymbol{p}$.

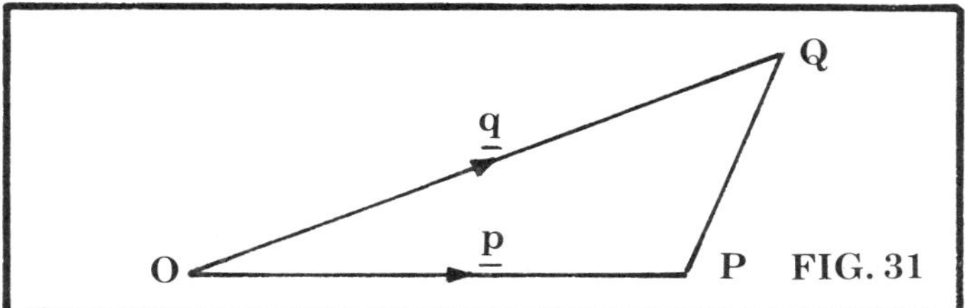

FIG. 31

Hence if P has position vector $\boldsymbol{p}$ represented by $\overrightarrow{OP}$ and Q has position vector $\boldsymbol{q}$ represented by $\overrightarrow{OQ}$ then

$$\overrightarrow{OP}+\overrightarrow{PQ}=\overrightarrow{OQ}$$

$$\Rightarrow \quad \overrightarrow{PQ}=\overrightarrow{OQ}-\overrightarrow{OP}$$

$$\Rightarrow \quad \overrightarrow{PQ} \quad \text{represents} \quad \boldsymbol{q}-\boldsymbol{p}$$

Example 1

ABCD is a parallelogram. If $\boldsymbol{a}$, $\boldsymbol{b}$, $\boldsymbol{c}$ are the position vectors of A, B, C, find the position vector of D.

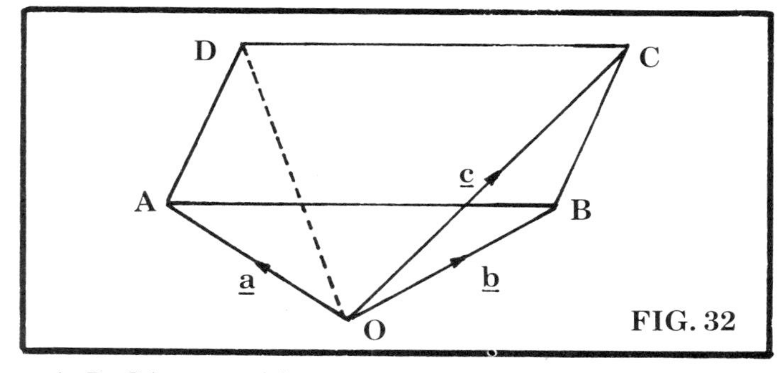

FIG. 32

A, B, C have position vectors $\boldsymbol{a}$, $\boldsymbol{b}$ and $\boldsymbol{c}$.

Hence $\overrightarrow{OA}$ represents $\boldsymbol{a}$, $\overrightarrow{OB}$ represents $\boldsymbol{b}$, $\overrightarrow{OC}$ represents $\boldsymbol{c}$, and $\overrightarrow{AB}=\overrightarrow{OB}-\overrightarrow{OA}$ represents $\boldsymbol{b}-\boldsymbol{a}$.

Since ABCD is a parallelogram

$$\overrightarrow{DC}=\overrightarrow{AB}\Rightarrow\overrightarrow{DC} \quad \text{represents} \quad \boldsymbol{b}-\boldsymbol{a}$$

Also
$$\overrightarrow{OD}=\overrightarrow{OC}+\overrightarrow{CD}$$
$$=\overrightarrow{OC}-\overrightarrow{DC}$$

Hence $\overrightarrow{OD}$ represents

$$\boldsymbol{c}-(\boldsymbol{b}-\boldsymbol{a})=\boldsymbol{a}-\boldsymbol{b}+\boldsymbol{c}$$

i.e. D has position vector $\boldsymbol{a}-\boldsymbol{b}+\boldsymbol{c}$ with reference to origin O.

3.7

Example 2

If the points A, B, C have position vectors $\boldsymbol{a}$, $\boldsymbol{b}$ and $\boldsymbol{c}$ with reference to an origin O, show that the centroid G of the triangle ABC has position vector $\frac{1}{3}(\boldsymbol{a}+\boldsymbol{b}+\boldsymbol{c})$.

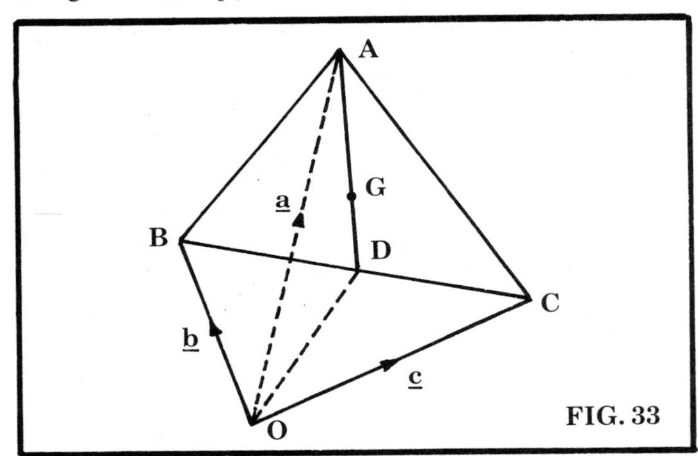

FIG. 33

3.8

A, B, C have position vectors $\boldsymbol{a}$, $\boldsymbol{b}$, $\boldsymbol{c}$ with reference to O.

Hence $\overrightarrow{OA}$ represents $\boldsymbol{a}$, $\overrightarrow{OB}$ represents $\boldsymbol{b}$, $\overrightarrow{OC}$ represents $\boldsymbol{c}$ and $\overrightarrow{BC}=\overrightarrow{OC}-\overrightarrow{OB}$ represents $\boldsymbol{c}-\boldsymbol{b}$.

If AD is the median from A to BC, then

$$\overrightarrow{BD}=\tfrac{1}{2}\overrightarrow{BC} \quad \text{represents} \quad \tfrac{1}{2}(\boldsymbol{c}-\boldsymbol{b})$$

$$\overrightarrow{OD}=\overrightarrow{OB}+\overrightarrow{BD} \quad \text{represents} \quad \boldsymbol{b}+\tfrac{1}{2}(\boldsymbol{c}-\boldsymbol{b})=\tfrac{1}{2}(\boldsymbol{c}+\boldsymbol{b})$$

G is the centroid of

$$\begin{aligned}\triangle ABC \Leftrightarrow \overrightarrow{AG} &= 2\overrightarrow{GD}\\ \Leftrightarrow \overrightarrow{OG}-\overrightarrow{OA} &= 2(\overrightarrow{OD}-\overrightarrow{OG})\\ \Leftrightarrow 3\overrightarrow{OG} &= \overrightarrow{OA}+2\overrightarrow{OD}\\ \Leftrightarrow 3\overrightarrow{OG} &\quad \text{represents} \quad \boldsymbol{a}+2.\tfrac{1}{2}(\boldsymbol{c}+\boldsymbol{b})=\boldsymbol{a}+\boldsymbol{b}+\boldsymbol{c}\\ \Leftrightarrow \overrightarrow{OG} &\quad \text{represents} \quad \tfrac{1}{3}(\boldsymbol{a}+\boldsymbol{b}+\boldsymbol{c})\end{aligned}$$

Example 3

Show that the points whose position vectors are $\boldsymbol{a}$, $\boldsymbol{b}$, and $7\boldsymbol{a}-6\boldsymbol{b}$ are collinear.

Let A, B and C be the points whose position vectors are $\boldsymbol{a}$, $\boldsymbol{b}$ and $7\boldsymbol{a}-6\boldsymbol{b}$, i.e. $\overrightarrow{OA}$ represents $\boldsymbol{a}$, $\overrightarrow{OB}$ represents $\boldsymbol{b}$ and $\overrightarrow{OC}$ represents $7\boldsymbol{a}-6\boldsymbol{b}$.

Then $\overrightarrow{AB}=\overrightarrow{OB}-\overrightarrow{OA}$ represents $\boldsymbol{b}-\boldsymbol{a}$, and

$$\overrightarrow{AC}=\overrightarrow{OC}-\overrightarrow{OA}$$

represents

$$\begin{aligned}7\boldsymbol{a}-6\boldsymbol{b}-\boldsymbol{a} &= 6\boldsymbol{a}-6\boldsymbol{b}\\ &= -6(\boldsymbol{b}-\boldsymbol{a})\end{aligned}$$

i.e.

$$\overrightarrow{AC}=-6\overrightarrow{AB}$$

Hence $\overrightarrow{AB}$ and $\overrightarrow{AC}$ are representatives of two vectors both passing through the point A and having the same direction. Thus A, B and C are collinear.

COMPONENTS OF A VECTOR IN 2-DIMENSIONS

If P in figure 34 has coordinates (x_1, y_1) then $\boldsymbol{p}$, the position vector of P, has components x_1 and y_1 and we write

$$\boldsymbol{p}=\begin{pmatrix}x_1\\y_1\end{pmatrix}$$

Similarly if Q is the point (x_2, y_2) then $\boldsymbol{q}$ has components x_2 and y_2 and

$$\boldsymbol{q}=\begin{pmatrix}x_2\\y_2\end{pmatrix}$$

Hence

$$\boldsymbol{q}-\boldsymbol{p}=\begin{pmatrix}x_2\\y_2\end{pmatrix}-\begin{pmatrix}x_1\\y_1\end{pmatrix}=\begin{pmatrix}x_2-x_1\\y_2-y_1\end{pmatrix}$$

i.e.

$$\overrightarrow{PQ} \quad \text{represents} \quad \begin{pmatrix}x_2-x_1\\y_2-y_1\end{pmatrix}$$

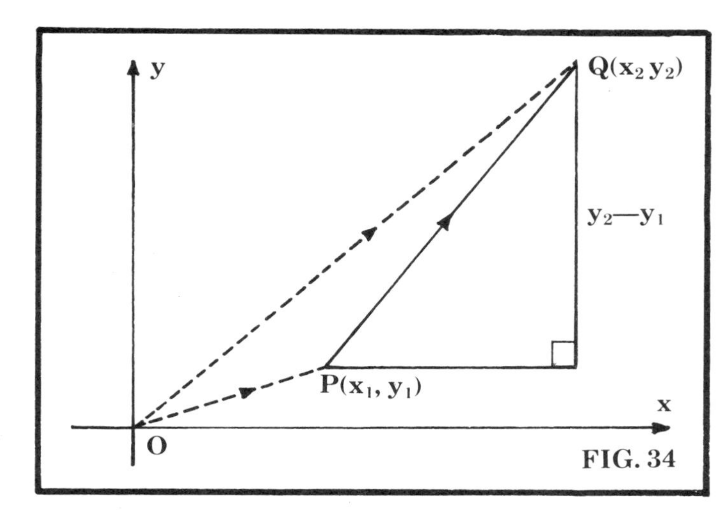

FIG. 34

Hence by Pythagoras

$$|\vec{PQ}|^2 = (x_2 - x_1)^2 + (y_2 - y_1)^2$$

where $|\vec{PQ}|$ denotes the length or magnitude of the vector represented by $\vec{PQ}$. This formula is often referred to as the **distance formula** as it gives the distance between the points $P(x_1, y_1)$ and $Q(x_2, y_2)$.

ASSIGNMENT 3.2

1. The points A, B and C have position vector $\boldsymbol{a}$, $2\boldsymbol{a}-\boldsymbol{b}$ and $-\boldsymbol{a}+2\boldsymbol{b}$ respectively. Express the vectors represented by $\vec{AB}$, $\vec{BC}$ and $\vec{CA}$ in terms of $\boldsymbol{a}$ and $\boldsymbol{b}$.

2. The position vectors of the points P, Q, R and S are respectively $\boldsymbol{p}$, $\boldsymbol{q}$, $3\boldsymbol{p}-\boldsymbol{q}$ and $-\boldsymbol{p}-2\boldsymbol{q}$. Express in terms of $\boldsymbol{p}$ and $\boldsymbol{q}$ the vectors represented by $\vec{PQ}$, $\vec{QR}$, $\vec{RS}$ and $\vec{SQ}$.

   Find also the position vectors of the mid-points of the line segments QR, RS and QS and the position vector of the centroid of triangle QRS.

3. ABCD is a rhombus. The points A, B and C have position vectors $\boldsymbol{a}$, $\boldsymbol{b}$ and $\boldsymbol{c}$. Find the position vector of the point D, and that of E the point of intersection of the diagonals.

4. The points A, B and C have position vector $\boldsymbol{a}$, $\boldsymbol{b}$ and $\boldsymbol{c}$. Show that the centroid G of triangle ABC has position vector $\frac{1}{3}(\boldsymbol{a}+\boldsymbol{b}+\boldsymbol{c})$ and hence or otherwise that

$$\vec{GA}+\vec{GB}+\vec{GC} = \boldsymbol{0}$$

5. P, Q, R are the mid-points of sides AB, BC, CA of triangle ABC. If O is any point and A, B, C have position vectors $\boldsymbol{a}$, $\boldsymbol{b}$, $\boldsymbol{c}$ with respect to O. Show that

   (i) $\vec{OP}$ represents $\frac{1}{2}(\boldsymbol{a}+\boldsymbol{b})$

   (ii) $\vec{OP}+\vec{OQ}+\vec{OR} = \vec{OA}+\vec{OB}+\vec{OC}$

6. The vectors $\boldsymbol{a}$ and $\boldsymbol{b}$ have components

$$\begin{pmatrix} 4 \\ -2 \end{pmatrix} \text{ and } \begin{pmatrix} -5 \\ 1 \end{pmatrix}.$$

   respectively.

   p and q are numbers such that

$$p\boldsymbol{a} + q\boldsymbol{b} = \begin{pmatrix} 22 \\ -8 \end{pmatrix}$$

   Find p and q.

7. The vectors $\boldsymbol{u}$, $\boldsymbol{v}$, $\boldsymbol{w}$ have components

$$\begin{pmatrix} -3 \\ 1 \end{pmatrix}, \begin{pmatrix} 1 \\ -2 \end{pmatrix} \text{ and } \begin{pmatrix} 4 \\ 4 \end{pmatrix}$$

   respectively.

   Find the components of the vectors,

   (i) $3\boldsymbol{u}-\boldsymbol{v}$ (ii) $\boldsymbol{u}-2\boldsymbol{v}+\boldsymbol{w}$

   (iii) $4\boldsymbol{v}+2\boldsymbol{w}$ (iv) $2\boldsymbol{u}-\boldsymbol{v}+3\boldsymbol{w}$

8. O, A, B, C, D are four points such that $\overrightarrow{OA}$ represents $\boldsymbol{u}$, $\overrightarrow{OC}$ represents $3\boldsymbol{u}$, $\overrightarrow{OB}$ represents $\boldsymbol{v}$ and $\overrightarrow{OD}$ represents $3\boldsymbol{v}$. Express the vectors represented by $\overrightarrow{AD}$ and $\overrightarrow{BC}$ in terms of $\boldsymbol{u}$ and $\boldsymbol{v}$.

   If the point K divides AD in the ratio 1:3 express the vectors represented by $\overrightarrow{OK}$ and $\overrightarrow{BK}$ in terms of $\boldsymbol{u}$ and $\boldsymbol{v}$.

   Show that K lies on BC.

9. Show that the points with the given position vectors are collinear.

   (i) $\boldsymbol{a}, \boldsymbol{b}, 6\boldsymbol{a}-5\boldsymbol{b}$

   (ii) $\boldsymbol{a}, \boldsymbol{b}, -2\boldsymbol{a}+3\boldsymbol{b}$

   (iii) $\boldsymbol{u}+\boldsymbol{v}, 3\boldsymbol{u}+4\boldsymbol{v}, 7\boldsymbol{u}+10\boldsymbol{v}$

   (iv) $\boldsymbol{p}+3\boldsymbol{q}, 4\boldsymbol{p}-\boldsymbol{q}, 7\boldsymbol{p}-5\boldsymbol{q}$

ASSIGNMENT 3.3

Objective items testing Sections 3.1–3.8.
Instructions for answering these items are given on page 152.

1. Points P and Q have position vectors $\boldsymbol{p}$ and $\boldsymbol{q}$ respectively with respect to some origin O. If S is the point of trisection of PQ nearer to P then S has position vector

   A. $\frac{1}{3}(\boldsymbol{p}+\boldsymbol{q})$
   B. $\frac{1}{3}(\boldsymbol{p}-\boldsymbol{q})$
   C. $\frac{1}{3}(\boldsymbol{q}-2\boldsymbol{p})$
   D. $\frac{1}{3}(\boldsymbol{p}+2\boldsymbol{q})$
   E. $\frac{1}{3}(\boldsymbol{q}+2\boldsymbol{p})$

2. PQRS is a parallelogram and T is a point on SR such that ST:TR = 1:3.

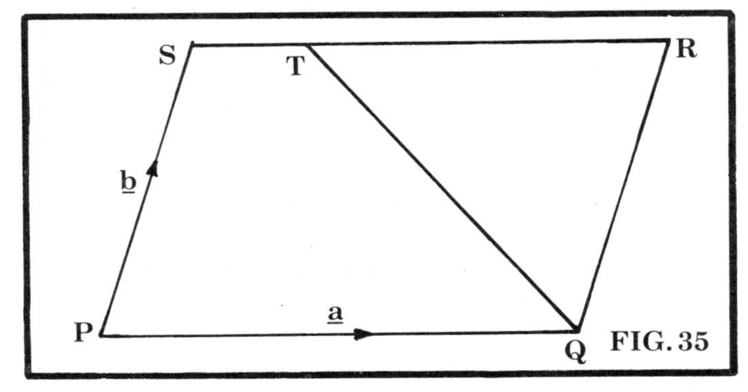

FIG. 35

   If $\overrightarrow{PQ}$ and $\overrightarrow{PS}$ represent the vectors $\boldsymbol{a}$ and $\boldsymbol{b}$ then $\overrightarrow{QT}$ represents

   A. $\boldsymbol{b}-\frac{3}{4}\boldsymbol{a}$
   B. $\boldsymbol{b}-3\boldsymbol{a}$
   C. $\boldsymbol{b}-\frac{2}{3}\boldsymbol{a}$
   D. $\boldsymbol{b}+\frac{3}{4}\boldsymbol{a}$
   E. $\boldsymbol{b}+3\boldsymbol{a}$

3. $\overrightarrow{RS}$ represents $\begin{pmatrix} 3 \\ -7 \end{pmatrix}$ and $\overrightarrow{ST}$ represents $\begin{pmatrix} -2 \\ -4 \end{pmatrix}$. If R is the point $(-1, 5)$ then T is

   A. $(2, -16)$
   B. $(-2, 16)$
   C. $(0, -6)$
   D. $(2, 6)$
   E. none of these

4. PQRS is a parallelogram with $|\overrightarrow{PQ}|=2|\overrightarrow{PS}|$. PQ, QR, RS and SP represent the vectors, $\boldsymbol{u}$, $\boldsymbol{v}$, $\boldsymbol{w}$ and $\boldsymbol{t}$ respectively. Which one of the following is false?

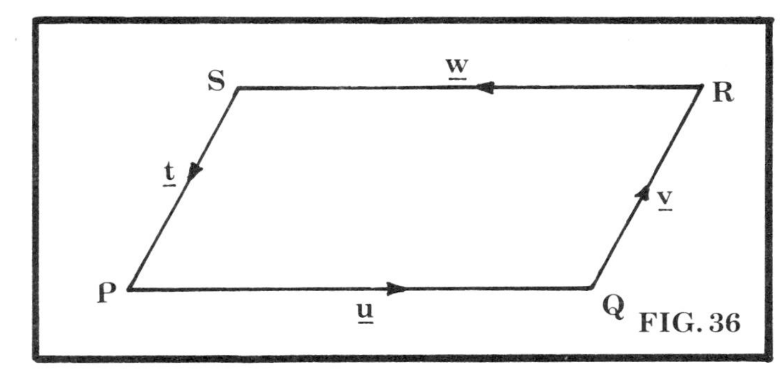

FIG. 36

A. $\mathbf{u} + \mathbf{v} = \mathbf{w} + \mathbf{t}$

B. $\mathbf{t} = -\mathbf{v}$

C. $|\mathbf{u}| = 2|\mathbf{v}|$

D. $\mathbf{t} - \mathbf{u} = \mathbf{w} - \mathbf{v}$

E. $\mathbf{u} + \mathbf{v} + \mathbf{w} + \mathbf{t} = \mathbf{0}$

5. ABCDEF is a regular hexagon (figure 37). $\vec{AB}$ and $\vec{BC}$ represent vectors $\mathbf{p}$ and $\mathbf{q}$. $\vec{BF}$ represents the vector,

A. $-\mathbf{p}-\mathbf{q}$

B. $-\mathbf{p}+\mathbf{q}$

C. $-2\mathbf{p}+\mathbf{q}$

D. $2\mathbf{p}-\mathbf{q}$

E. $-\mathbf{p}+2\mathbf{q}$

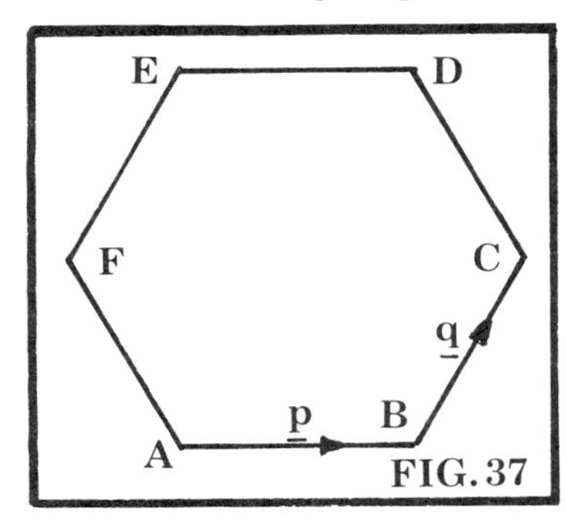

FIG. 37

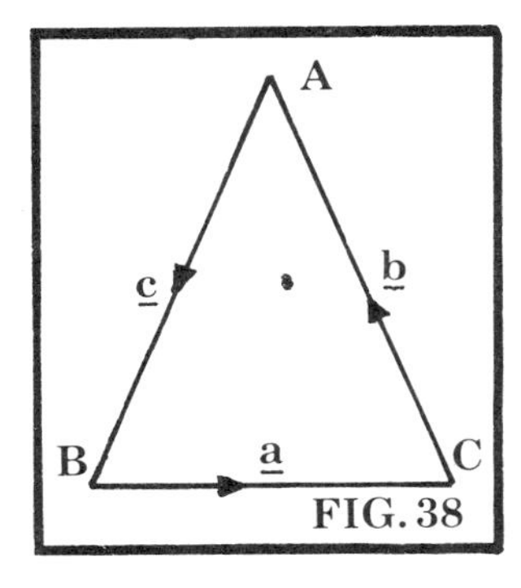

FIG. 38

6. $\vec{AB}$, $\vec{BC}$, $\vec{CD}$ and $\vec{DA}$ represent vectors with components

$$\begin{pmatrix} 3 \\ -2 \end{pmatrix}, \quad \begin{pmatrix} 2 \\ -4 \end{pmatrix}, \quad \begin{pmatrix} -3 \\ 2 \end{pmatrix} \quad \text{and} \quad \begin{pmatrix} -2 \\ 4 \end{pmatrix}$$

respectively.

Quadrilateral ABCD is

A. a parallelogram

B. a rhombus

C. a rectangle

D. a square

E. a kite

7. In triangle ABC (figure 38), AB = AC and $\vec{BC}$, $\vec{CA}$, $\vec{AB}$ represent vectors $\mathbf{a}$, $\mathbf{b}$, $\mathbf{c}$. Which one of the following statements is false?

A. $|\mathbf{a}+\mathbf{b}| = |\mathbf{c}|$

B. $|\mathbf{b}| = |\mathbf{c}|$

C. $|\mathbf{b}+\mathbf{c}| = |\mathbf{a}|$

D. $|\mathbf{c}|+|\mathbf{a}| = |\mathbf{b}|+|\mathbf{a}|$

E. $|\mathbf{b}|+|\mathbf{c}| = |\mathbf{b}+\mathbf{c}|$

8. O is the origin,

$$\vec{OA} \text{ represents } \begin{pmatrix} 3 \\ 1 \end{pmatrix}, \quad \vec{OB} \text{ represents } \begin{pmatrix} -3 \\ -1 \end{pmatrix},$$

and $\vec{OP} = k\vec{AB}$. If P lies on the line segment AB then,

A. $k \geqq \frac{1}{2}$

B. $k \leqq -\frac{1}{2}$

C. $0 \leqq k \leqq \frac{1}{2}$

D. $-\frac{1}{2} \leqq k \leqq \frac{1}{2}$

E. $-\frac{1}{2} \leqq k \leqq 0$

9. $\vec{OA}$, $\vec{OB}$, $\vec{OC}$ and $\vec{OD}$ represent the vectors $\boldsymbol{u}$, $4\boldsymbol{u}$, $\boldsymbol{v}$ and $4\boldsymbol{v}$ respectively.

(1) $\vec{AC} = \frac{1}{4}\vec{BD}$

(2) $|\vec{OB}| = |\vec{OD}|$

(3) $|\vec{AD}| = |\vec{CB}|$

10. P, Q, R have position vectors $\boldsymbol{u} - \boldsymbol{v}$, $2\boldsymbol{u} - 4\boldsymbol{v}$ and $3\boldsymbol{u} - 7\boldsymbol{v}$ respectively.

(1) $\vec{PQ}$ represents $2\boldsymbol{u} - 6\boldsymbol{v}$

(2) P, Q and R are collinear.

(3) Q is the mid-point of PR.

11. (1) Vectors $\boldsymbol{u}$ and $\boldsymbol{v}$ are such that $|\boldsymbol{u}| = |\boldsymbol{v}|$.

(2) $\boldsymbol{u} = \boldsymbol{v}$

12. (1) Vectors represented by $\vec{OA}$ and $\vec{OB}$ are such that $|\vec{OA}| = 2|\vec{OB}|$.

(2) B is the mid-point of OA.

UNIT 4: VECTORS IN THREE DIMENSIONS

VECTORS IN THREE DIMENSIONS

(a) Vectors in three dimensions operate exactly as vectors in two dimensions. If $\boldsymbol{u}$, $\boldsymbol{v}$, $\boldsymbol{w}$ are vectors in 3 dimensions, then,

(i) $\boldsymbol{u}+\boldsymbol{v} = \boldsymbol{v}+\boldsymbol{u}$ —commutative law holds

(ii) $(\boldsymbol{u}+\boldsymbol{v})+\boldsymbol{w} = \boldsymbol{u}+(\boldsymbol{v}+\boldsymbol{w})$ —associative law holds

(iii) $\boldsymbol{u}+\boldsymbol{0} = \boldsymbol{0}+\boldsymbol{u} = \boldsymbol{u}$ —$\boldsymbol{0}$ is the zero vector

(iv) $\boldsymbol{u}+(-\boldsymbol{u}) = -\boldsymbol{u}+\boldsymbol{u} = \boldsymbol{0}$ —$-\boldsymbol{u}$ is the additive inverse of $\boldsymbol{u}$

(v) $0\boldsymbol{u} = \boldsymbol{0}$ —zero times a vector gives the zero vector

(b) *Multiplication by a number*

If k and m are numbers and $\boldsymbol{u}$ and $\boldsymbol{v}$ vectors then,

(i) $k\boldsymbol{u}$ is a vector in the same direction as $\boldsymbol{u}$ but of magnitude k times the magnitude of $\boldsymbol{u}$.

If $k > 0$ the sense of $k\boldsymbol{u}$ is the same as the sense of $\boldsymbol{u}$.

If $k < 0$ the sense of $k\boldsymbol{u}$ is opposite that of $\boldsymbol{u}$.

(ii) $k(\boldsymbol{u}+\boldsymbol{v}) = k\boldsymbol{u}+k\boldsymbol{v}$

(iii) $(k+m)\boldsymbol{u} = k\boldsymbol{u}+m\boldsymbol{u}$

(iv) $k(m\boldsymbol{u}) = (km)\boldsymbol{u}$

Example 1

OABCDEFG is a cuboid. $\vec{OA}$, $\vec{OC}$ and $\vec{OD}$ represent vectors $\boldsymbol{a}$, $\boldsymbol{b}$ and $\boldsymbol{c}$ respectively.

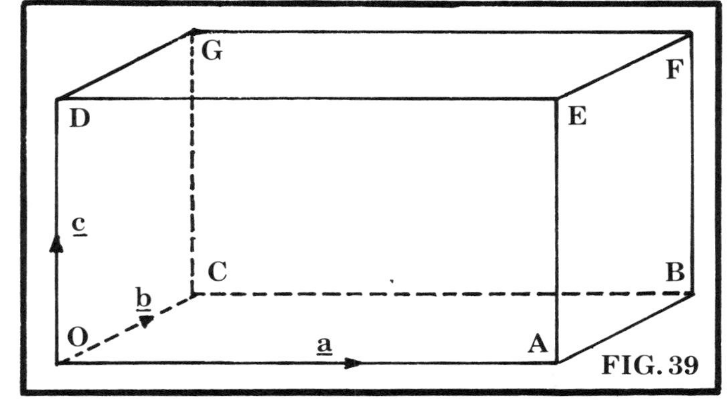

FIG. 39

(i) Express the vectors represented by $\vec{OF}$ and $\vec{GA}$ in terms of $\boldsymbol{a}$, $\boldsymbol{b}$, $\boldsymbol{c}$.

(ii) Express $|\vec{OF}|^2$ as the sum of the magnitudes of $\boldsymbol{a}$, $\boldsymbol{b}$ and $\boldsymbol{c}$.

(iii) Express the vectors represented by $\vec{EB}$ and $\vec{CF}$ in terms of $\boldsymbol{a}$, $\boldsymbol{b}$ and $\boldsymbol{c}$.

Note:

$\vec{OA} = \vec{CB} = \vec{DE} = \vec{GF}$ represents $\boldsymbol{a}$

$\vec{OC} = \vec{AB} = \vec{EF} = \vec{DG}$ represents $\boldsymbol{b}$

$\vec{OD} = \vec{AE} = \vec{BF} = \vec{CG}$ represents $\boldsymbol{c}$

Hence,

(i) $\vec{OF} = \vec{OA}+\vec{AB}+\vec{BF}$ represents $\boldsymbol{a}+\boldsymbol{b}+\boldsymbol{c}$

$\vec{GA} = \vec{GC}+\vec{CO}+\vec{OA}$ represents

$$-\boldsymbol{c}+(-\boldsymbol{b})+\boldsymbol{a} = \boldsymbol{a}-\boldsymbol{b}-\boldsymbol{c}$$

(ii) By Pythagoras,

$$OF^2 = OB^2+BF^2 = OA^2+AB^2+BF^2$$

but $OA = |\boldsymbol{a}|$, $AB = |\boldsymbol{b}|$, and $BF = |\boldsymbol{c}|$,

hence $|\vec{OF}|^2 = |\boldsymbol{a}|^2+|\boldsymbol{b}|^2+|\boldsymbol{c}|^2$

(iii) $\vec{EB} = \vec{EA}+\vec{AB}$ represents $-\boldsymbol{c}+\boldsymbol{b} = \boldsymbol{b}-\boldsymbol{c}$

$\vec{CF} = \vec{CB}+\vec{BF}$ represents $\boldsymbol{a}+\boldsymbol{c}$

ASSIGNMENT 4.1

1. In figure 40, OPQRSTUV is a cuboid and P, R, S have position vectors $\boldsymbol{p}, \boldsymbol{r}, \boldsymbol{s}$ with reference to O as origin.
   (i) Express the position vector of T in terms of $\boldsymbol{p}, \boldsymbol{r}$ and $\boldsymbol{s}$.
   (ii) What vectors are represented by $\vec{SQ}$, $\vec{RT}$ and $\vec{PV}$.
   (iii) Find $|\vec{VP}|^2$ in terms of $|\boldsymbol{p}|$, $|\boldsymbol{r}|$ and $|\boldsymbol{s}|$.

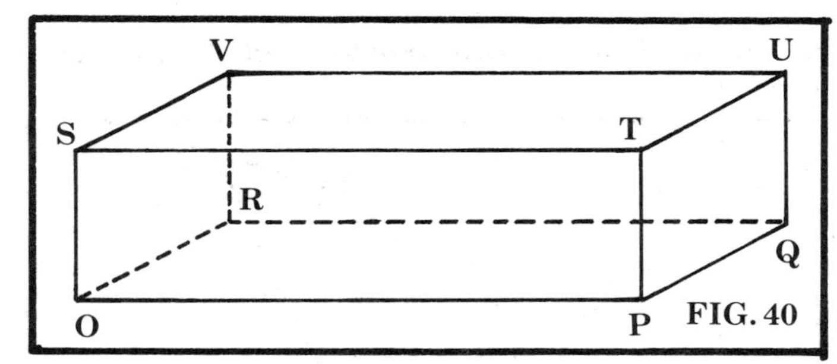

FIG. 40

2. VABCD is a regular pyramid on a square base. VP is perpendicular to the base. $\vec{AB}$, $\vec{AD}$ and $\vec{AV}$ represent vectors $\boldsymbol{u}$, $\boldsymbol{v}$ and $\boldsymbol{w}$ (see figure 41).

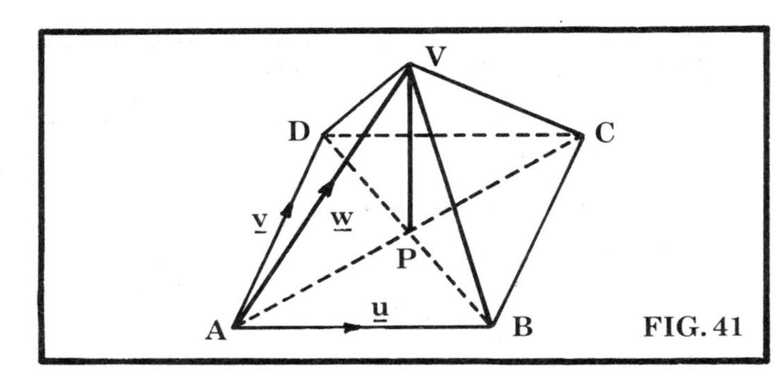

FIG. 41

   (i) Express in terms of $\boldsymbol{u}$, $\boldsymbol{v}$ and $\boldsymbol{w}$ the vectors represented by $\vec{BV}$, $\vec{CV}$ and $\vec{AP}$.
   (ii) Express the vector represented by $\vec{VP}$ in terms of $\boldsymbol{u}$, $\boldsymbol{v}$ and $\boldsymbol{w}$.
   (iii) Show that $|\vec{VP}|^2 = |\boldsymbol{w}|^2-\frac{1}{4}|(\boldsymbol{u}+\boldsymbol{v})|^2$, and hence find the length of the line segment VP if $|\boldsymbol{u}| = 6$ units, $|\boldsymbol{v}| = 8$ units and $|\boldsymbol{w}| = 13$ units.

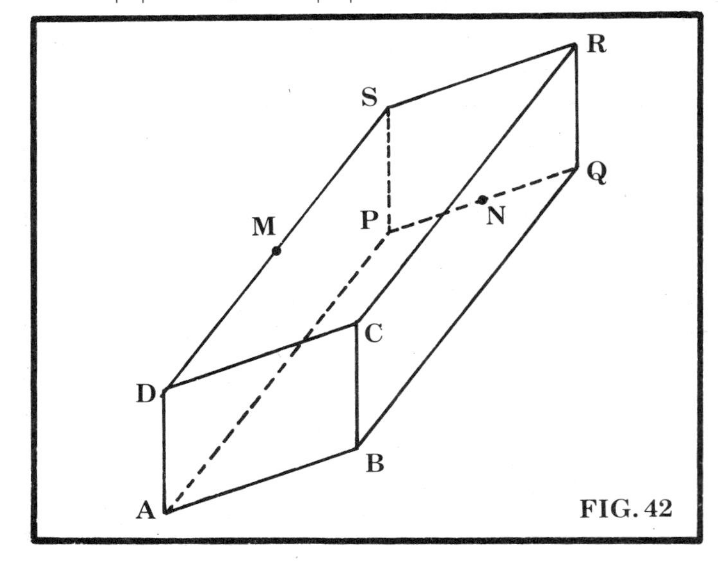

FIG. 42

3. In figure 42, ABCDPQRS is a parallelepiped (i.e. a solid with 6 faces each of which is a parallelogram).
   (i) $\overrightarrow{AR} = \overrightarrow{AB} + \overrightarrow{BQ} + \overrightarrow{QR}$. Write down 5 other similar sums for $\overrightarrow{AR}$.
   (ii) If $\overrightarrow{AB}$ represents the vector $\boldsymbol{u}$, $\overrightarrow{AP}$ the vector $\boldsymbol{v}$ and $\overrightarrow{AD}$ the vector $\boldsymbol{w}$, write down from the figure all the directed line segments that represent $\boldsymbol{u}$, $\boldsymbol{v}$ and $\boldsymbol{w}$.
   (iii) If M is the mid-point of DS and N the mid-point of PQ, express in terms of $\boldsymbol{u}$, $\boldsymbol{v}$ and $\boldsymbol{w}$ the vectors represented by $\overrightarrow{BM}$, $\overrightarrow{BN}$ and $\overrightarrow{MN}$.
4. In figure 43, PQRS is a regular tetrahedron. M, N, L are the mid-points of PQ, SR and QR. $\overrightarrow{PR}$, $\overrightarrow{PQ}$ and $\overrightarrow{PS}$ represent the vectors $\boldsymbol{a}$, $\boldsymbol{b}$ and $\boldsymbol{c}$.
   (a) Prove that
      (i) $\overrightarrow{ML}$ represents $\frac{1}{2}\boldsymbol{a}$.
      (ii) $\overrightarrow{MN}$ represents $\frac{1}{2}(\boldsymbol{a} - \boldsymbol{b} + \boldsymbol{c})$.
   (b) Express the vector represented by $\overrightarrow{SL}$ in terms of $\boldsymbol{a}$, $\boldsymbol{b}$ and $\boldsymbol{c}$.

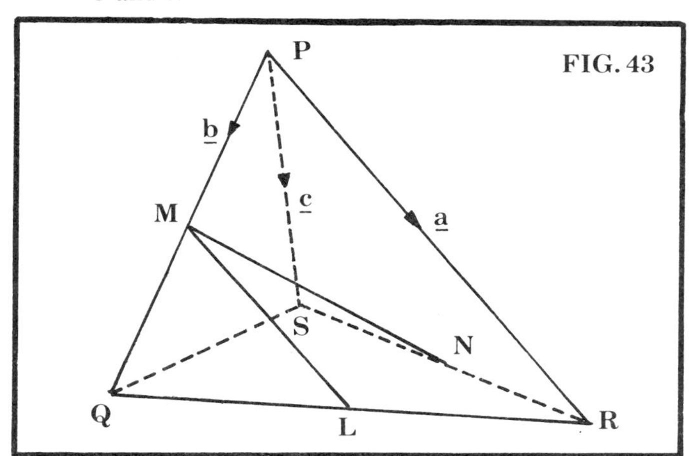

4.2 THE SECTION FORMULA (IN VECTOR FORM)

Note: A point P is said to divide the line segment AB in the ratio $m:n$ if $AP:PB = m:n$, taking sense into account. For example

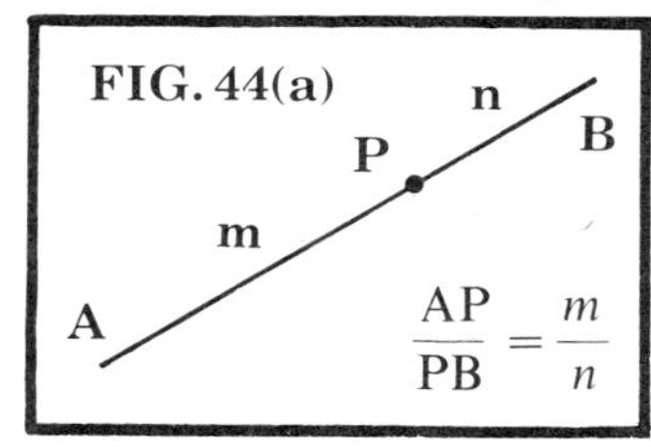

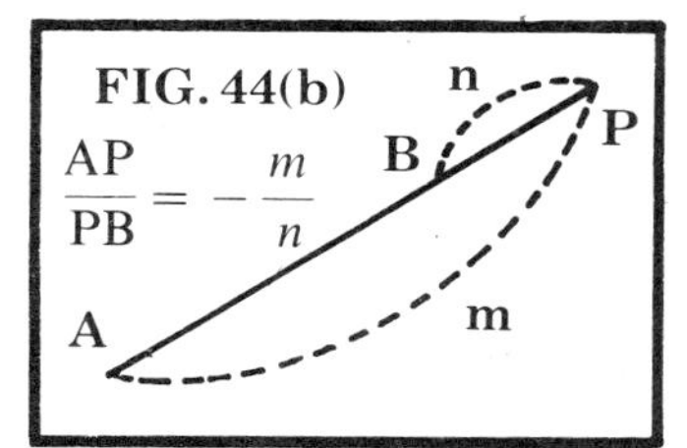

In figure 44(a), P divides AB internally, m and n are both positive.

In figure 44(b), P divides AB externally, one of m and n is positive and the other negative.

The Section Formula: If the point A has position vector $\boldsymbol{a}$, B position vector $\boldsymbol{b}$ and P divides AB in the ratio $m:n$, then the position vector of P is

$$\frac{n\boldsymbol{a} + m\boldsymbol{b}}{m+n}$$

Let A, B and P have position vectors $\boldsymbol{a}$, $\boldsymbol{b}$ and $\boldsymbol{p}$ respectively.

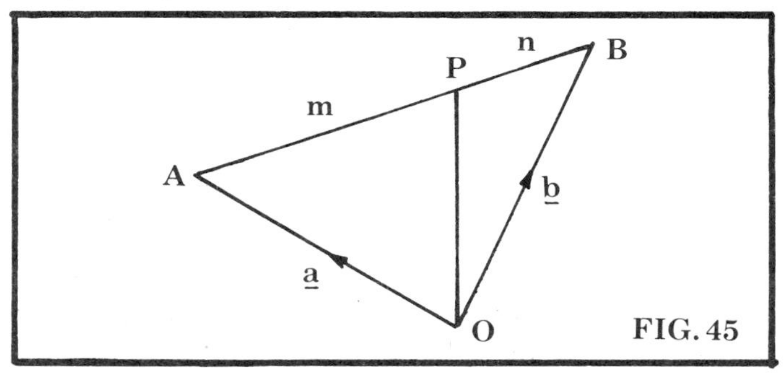

In $\triangle$OAP $\qquad \vec{OP} = \vec{OA} + \vec{AP}$ .... (1)

but $\qquad \dfrac{AP}{PB} = \dfrac{m}{n} \Rightarrow \dfrac{AP}{AB} = \dfrac{m}{m+n}$

hence $\qquad AP = \dfrac{m}{m+n} AB \Rightarrow \vec{AP} = \dfrac{m}{m+n} \vec{AB}$

and thus from (1) $\quad \vec{OP} = \vec{OA} + \dfrac{m}{m+n} \vec{AB}$

Therefore $\quad \boldsymbol{p} = \boldsymbol{a} + \dfrac{m}{m+n}(\boldsymbol{b}-\boldsymbol{a})$, since $\vec{AB}$ represents $\boldsymbol{b}-\boldsymbol{a}$

$$= \frac{(m+n)\boldsymbol{a} + m(\boldsymbol{b}-\boldsymbol{a})}{m+n}$$

$$= \frac{m\boldsymbol{a} + n\boldsymbol{a} + m\boldsymbol{b} - m\boldsymbol{a}}{m+n}$$

$$= \frac{n\boldsymbol{a} + m\boldsymbol{b}}{m+n}$$

i.e. the point P has position vector

$$\boldsymbol{p} = \frac{n\boldsymbol{a} + m\boldsymbol{b}}{m+n}$$

Note: If C is the mid-point of AB, then

$$\frac{m}{n} = \frac{1}{1}$$

Hence if $\boldsymbol{c}$ is the position vector of C then

$$\boldsymbol{c} = \frac{n\boldsymbol{a} + m\boldsymbol{b}}{m+n} = \frac{\boldsymbol{a}+\boldsymbol{b}}{2} = \tfrac{1}{2}(\boldsymbol{a}+\boldsymbol{b})$$

i.e. C has position vector $\frac{1}{2}(\boldsymbol{a}+\boldsymbol{b})$.

Example 1

C is the mid-point of the line segment AB and P is the point of trisection of AB nearer B. If A and B have position vectors $\boldsymbol{a}$ and $\boldsymbol{b}$, find

(i) the position vectors of P and C.

(ii) the ratio AC:CP.

(i) If P divides AB in the ratio $m:n$, then

$$\boldsymbol{p} = \frac{n\boldsymbol{a} + m\boldsymbol{b}}{m+n}$$

where $\boldsymbol{p}$ is the position vector of P.

But P divides AB in the ratio 2:1, hence

$$\boldsymbol{p} = \frac{\boldsymbol{a} + 2\boldsymbol{b}}{3}$$

i.e. P has position vector $\frac{1}{3}(\boldsymbol{a}+2\boldsymbol{b})$.

C is the mid-point of AB, therefore C has position vector $\frac{1}{2}(\boldsymbol{a}+\boldsymbol{b})$.

(ii) Let C divide AP in the ratio $m:n$.

Hence $\qquad \boldsymbol{c} = \dfrac{m\boldsymbol{p} + n\boldsymbol{a}}{m+n}$

where $\boldsymbol{c}$ is the position vector of C.

But $\qquad \boldsymbol{c} = \frac{1}{2}(\boldsymbol{a}+\boldsymbol{b}) \quad$ and $\quad \boldsymbol{p} = \frac{1}{3}(\boldsymbol{a}+2\boldsymbol{b})$

Hence $\frac{1}{2}(\boldsymbol{a}+\boldsymbol{b}) = \dfrac{\frac{m}{3}(\boldsymbol{a}+2\boldsymbol{b}) + n\boldsymbol{a}}{m+n}$

$$\Rightarrow 3(\boldsymbol{a}+\boldsymbol{b}) = \frac{2m(\boldsymbol{a}+2\boldsymbol{b}) + 6n\boldsymbol{a}}{m+n}$$

$$\Rightarrow 3(m+n)\boldsymbol{a} + 3(m+n)\boldsymbol{b} = (2m+6n)\boldsymbol{a} + 4m\boldsymbol{b}$$

$$\Rightarrow (3m+3n-2m-6n)\boldsymbol{a} + (3m+3n-4n)\boldsymbol{b} = \boldsymbol{0}$$

$$\Rightarrow (m-3n)\boldsymbol{a} + (3n-m)\boldsymbol{b} = \boldsymbol{0}$$

Since $\boldsymbol{a}$ and $\boldsymbol{b}$ are both not zero, nor parallel, it follows that $m-3n = 0$ and $3n-m = 0$

$$\Rightarrow \frac{m}{n} = \frac{3}{1}$$

i.e. C divides AP in the ratio 3:1.

Example 2

ABCD is a tetrahedron and P, Q, R, S, T, U are the mid-points of AB, BC, CD, DA, AC, BD respectively. Find the position vector of M the mid-point of PR and hence show that PR, QS, TU are concurrent at M.

Let A, B, C, D have position vectors $\boldsymbol{a}$, $\boldsymbol{b}$, $\boldsymbol{c}$, and $\boldsymbol{d}$ respectively, with reference to some origin O.

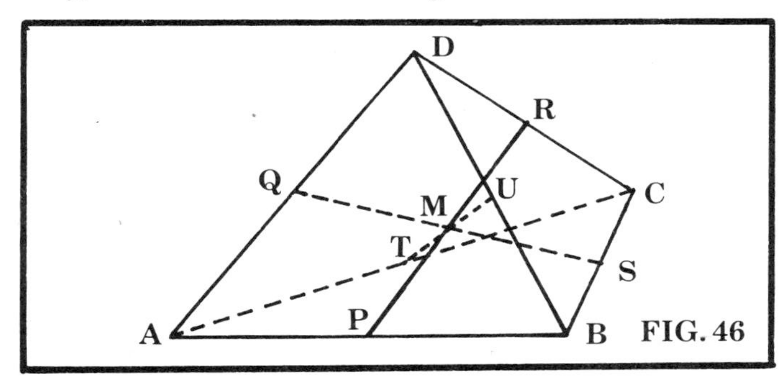

FIG. 46

P is the mid-point of AB $\Rightarrow$ $\boldsymbol{p} = \frac{1}{2}(\boldsymbol{a}+\boldsymbol{b})$

R is the mid-point of DC $\Rightarrow$ $\boldsymbol{r} = \frac{1}{2}(\boldsymbol{c}+\boldsymbol{d})$

M is the mid-point of PR $\Rightarrow$ $\boldsymbol{m} = \frac{1}{2}(\boldsymbol{p}+\boldsymbol{r})$

$$= \tfrac{1}{2}[\tfrac{1}{2}(\boldsymbol{a}+\boldsymbol{b})+\tfrac{1}{2}(\boldsymbol{c}+\boldsymbol{d})]$$
$$= \tfrac{1}{4}(\boldsymbol{a}+\boldsymbol{b}+\boldsymbol{c}+\boldsymbol{d})$$

i.e. M the mid-point of PR has position vector

$$\tfrac{1}{4}(\boldsymbol{a}+\boldsymbol{b}+\boldsymbol{c}+\boldsymbol{d})$$

Similarly it can be shown that N the mid-point of QS has position vector $\boldsymbol{n} = \frac{1}{4}(\boldsymbol{a}+\boldsymbol{b}+\boldsymbol{c}+\boldsymbol{d})$, and that L the mid-point of TU has position vector $\boldsymbol{l} = \frac{1}{4}(\boldsymbol{a}+\boldsymbol{b}+\boldsymbol{c}+\boldsymbol{d})$,

i.e. $\boldsymbol{m} = \boldsymbol{n} = \boldsymbol{l} = \frac{1}{4}(\boldsymbol{a}+\boldsymbol{b}+\boldsymbol{c}+\boldsymbol{d}) \Rightarrow$ M, N, L

are the same point.

i.e. PR, QS, TU are concurrent at M

Example 3

In figure 47, PQRS is a parallelogram with T the mid-point of PQ. K divides ST in the ratio 2:1. If $\overrightarrow{PQ}$ represents the vector $\boldsymbol{q}$ and $\overrightarrow{PS}$ the vector $\boldsymbol{s}$, show that $\overrightarrow{PK}$ represents $\frac{1}{3}(\boldsymbol{q}+\boldsymbol{s})$.

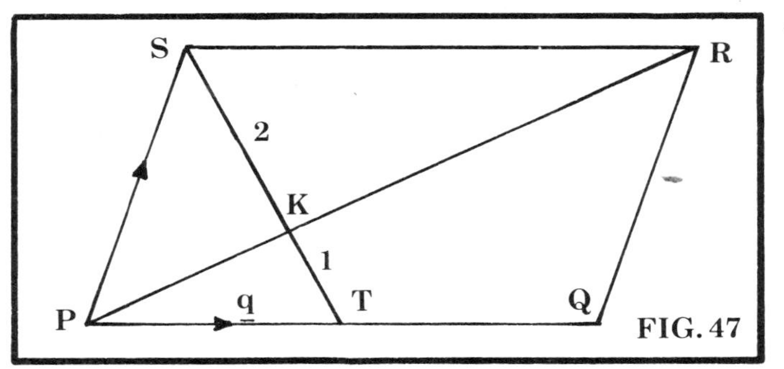

FIG. 47

Prove that P, K and R are collinear and find the ratio PK:KR.

$\overrightarrow{PQ}$ represents $\boldsymbol{q} \Rightarrow \overrightarrow{PT}$ represents $\frac{1}{2}\boldsymbol{q}$, $\overrightarrow{PS}$ represents $\boldsymbol{s}$. Using P as origin, then

$$\overrightarrow{PK} = \frac{2.\overrightarrow{PT}+1.\overrightarrow{PS}}{3} \quad \text{represents} \quad \frac{2.\frac{1}{2}\boldsymbol{q}+\boldsymbol{s}}{3}$$

i.e. $\overrightarrow{PK}$ represents $\frac{1}{3}(\boldsymbol{q}+\boldsymbol{s})$.

$\overrightarrow{RS}$ represents $-\boldsymbol{q}$ and $\overrightarrow{RT} = \overrightarrow{RQ}+\overrightarrow{QT}$ represents

$$-\boldsymbol{s}+(-\tfrac{1}{2}\boldsymbol{q}) = -(\boldsymbol{s}+\tfrac{1}{2}\boldsymbol{q})$$

But

$$\overrightarrow{RK} = \frac{2.\overrightarrow{RT}+1.\overrightarrow{RS}}{3} \quad \text{represents} \quad \frac{-2\boldsymbol{s}-\boldsymbol{q}-\boldsymbol{q}}{3} = -\tfrac{2}{3}(\boldsymbol{s}+\boldsymbol{q})$$

Thus $\overrightarrow{PK}$ represents $\frac{1}{3}(\boldsymbol{q}+\boldsymbol{s})$

and $\overrightarrow{KR}$ represents $\frac{2}{3}(\boldsymbol{q}+\boldsymbol{s})$

$\Rightarrow$ P, K, R collinear.

Also $\overrightarrow{PK}$ represents $\frac{1}{3}(q+s)$

and $\overrightarrow{KR}$ represents $\frac{2}{3}(q+s)$

$\Rightarrow \overrightarrow{PK} = \frac{1}{2}\overrightarrow{KR} \Rightarrow PK:KR = 1:2$

ASSIGNMENT 4.2

1. P has position vector $\boldsymbol{p}$ and Q has position vector $\boldsymbol{q}$.
   (i) Find $\boldsymbol{r}$ the position vector of R the point which divides PQ in the ratio 3:5.
   (ii) Find $\boldsymbol{s}$ the position vector of the point S which divides PQ in the ratio $-5:2$.
   (iii) Express the vector represented by $\overrightarrow{RS}$ in terms of $\boldsymbol{p}$ and $\boldsymbol{q}$ and hence deduce what fraction RS is of PQ.

2. The vertices A, B, C and D of a parallelogram ABCD have position vectors $\boldsymbol{a}$, $\boldsymbol{b}$, $\boldsymbol{c}$ and $\boldsymbol{d}$ respectively. Find
   (i) $\boldsymbol{m}$, the position vector of M the mid-point of AB in terms of $\boldsymbol{a}$ and $\boldsymbol{b}$.
   (ii) $\boldsymbol{p}$, the position vector of P which divides DM in the ratio 2:1 in terms of $\boldsymbol{a}$, $\boldsymbol{b}$ and $\boldsymbol{d}$.
   (iii) $\boldsymbol{c}$, the position vector of C in terms of $\boldsymbol{a}$, $\boldsymbol{b}$ and $\boldsymbol{d}$, and hence prove that A, P and C are collinear.

3. PQRS is a parallelogram. M is the mid-point of QR and T is the point of intersection of PM and QS. If P, Q, R and S have position vectors $\boldsymbol{p}$, $\boldsymbol{q}$, $\boldsymbol{r}$ and $\boldsymbol{s}$, express
   (i) the position vector of M in terms of $\boldsymbol{q}$ and $\boldsymbol{r}$.
   (ii) the position vector of S in terms of $\boldsymbol{p}$, $\boldsymbol{q}$ and $\boldsymbol{r}$.

   If T divides PM in the ratio 2:1 find the position vector of T and deduce that T divides SQ in the ratio 2:1.

4. In triangle ABC, P divides AB in the ratio 3:2, Q divides BC in the ratio 3:2 and R divides CA in the ratio 3:2. The points A, B, C, P, Q, R have position vectors $\boldsymbol{a}$, $\boldsymbol{b}$, $\boldsymbol{c}$, $\boldsymbol{p}$, $\boldsymbol{q}$ and $\boldsymbol{r}$.
   (a) Prove that,
      (i) $\boldsymbol{p}+\boldsymbol{q}+\boldsymbol{r} = \boldsymbol{a}+\boldsymbol{b}+\boldsymbol{c}$
      (ii) $\overrightarrow{AQ}+\overrightarrow{BR}+\overrightarrow{CP} = \boldsymbol{0}$
   (b) If M is the mid-point of AB, write down the position vector of M in terms of $\boldsymbol{a}$ and $\boldsymbol{b}$ and hence deduce that G the centroid of triangle ABC has position vector $\frac{1}{3}(\boldsymbol{a}+\boldsymbol{b}+\boldsymbol{c})$.
      [*Note:* G divides CM in the ratio 2:1.]
   (c) Is G the centroid of triangle PQR? If yes give a reason. If not find the position vector of the centroid of triangle PQR.

5. D, E, F are the mid-points of the sides BC, CA and AB of triangle ABC. If $\boldsymbol{a}$, $\boldsymbol{b}$, $\boldsymbol{c}$ are the position vectors of A, B, C, find
   (i) the position vector of D and the position vector of G the point which divides the median AD in the ratio 2:1.
   (ii) the position vector of E and of H the point which divides the median BE in the ratio 2:1.
   (iii) the position vector of F and of K the point which divides the median CF in the ratio 2:1.

   What can you say about the points G, H and K? Hence what can you say about the centroid of the triangle ABC?

6. In figure 48, A, B, C and D the vertices of a tetrahedron have position vectors $\boldsymbol{a}$, $\boldsymbol{b}$, $\boldsymbol{c}$ and $\boldsymbol{d}$. P and Q are the mid-points of the opposite edges AD and BC.

   Find the position vectors of P and Q and of M the mid-point of PQ.

   If R and S are the mid-points of the edges AB and DC, show that M is the mid-point of RS.

   Show also that M is the mid-point of TU where T and U are the mid-points of AC and BD.

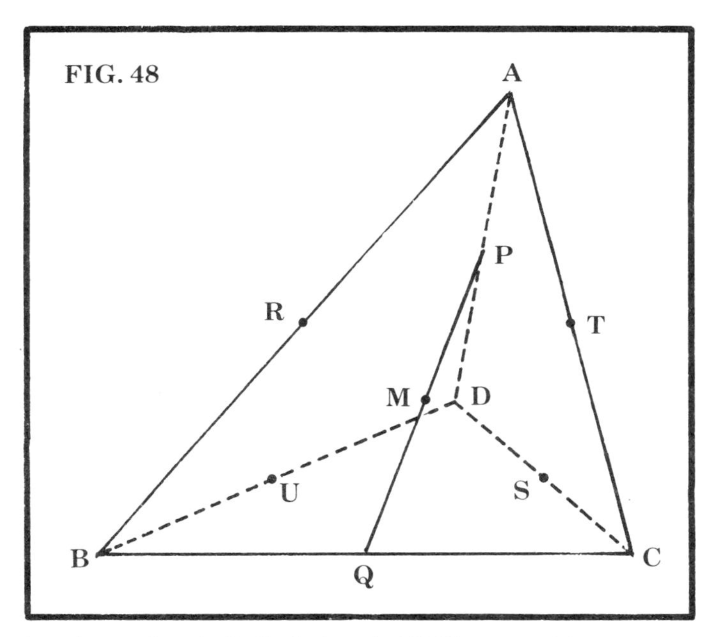

7. The vertices A, B, C of triangle ABC have position vectors $\boldsymbol{a}$, $\boldsymbol{b}$ and $\boldsymbol{c}$ respectively. U divides BC in the ratio $-2:1$, V divides CA in the ratio $3:4$ and W divides AB in the ratio $2:3$. If U, V and W have position vectors $\boldsymbol{u}$, $\boldsymbol{v}$ and $\boldsymbol{w}$,
   (i) Find $\boldsymbol{u}$, $\boldsymbol{v}$ and $\boldsymbol{w}$ in terms of $\boldsymbol{a}$, $\boldsymbol{b}$ and $\boldsymbol{c}$.
   (ii) Show that U, V and W are collinear.

8. In triangle ABC the vertices A, B and C have position vectors $\boldsymbol{a}$, $\boldsymbol{b}$ and $\boldsymbol{c}$. M divides AB in the ratio $3:4$ and N divides AC in the ratio $3:2$.
   (i) Find $\boldsymbol{m}$ and $\boldsymbol{n}$ the position vectors of M and N.
   (ii) BC is produced to L such that $BL:LC = 2:1$, find $\boldsymbol{l}$ the position vector of L.
   (iii) Show that M, N and L are collinear and find the ratio $MN:NL$.

COMPONENTS OF A VECTOR IN THREE DIMENSIONS

If P is a point in 3 dimensions with coordinates (x_1, y_1, z_1) (see figure 49) then the vector represented by $\overrightarrow{OP}$ has components x_1, y_1, z_1.

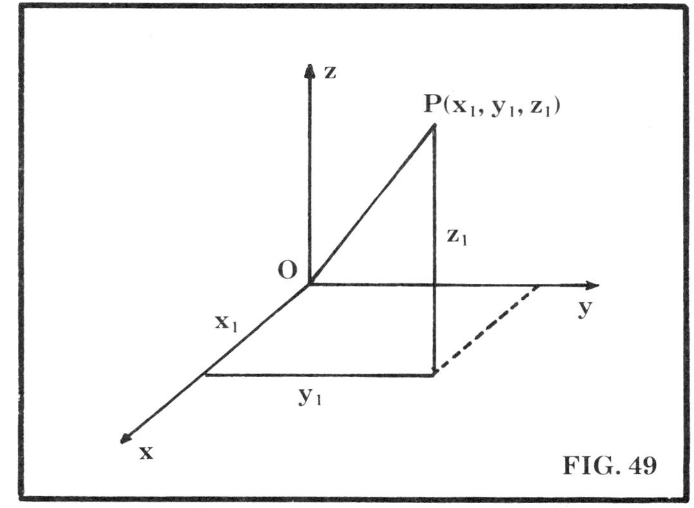

FIG. 49

We write that $\overrightarrow{OP}$ represents

$$\begin{pmatrix} x_1 \\ y_1 \\ z_1 \end{pmatrix}$$

and relative to the origin O, P has position vector

$$\boldsymbol{p} = \begin{pmatrix} x_1 \\ y_1 \\ z_1 \end{pmatrix}$$

also $|\overrightarrow{OP}|^2 = x_1^2 + y_1^2 + z_1^2$, the square of the line segment OP.

If Q is the point (x_2, y_2, z_2) then $\overrightarrow{OQ}$ represents

$$\begin{pmatrix} x_2 \\ y_2 \\ z_2 \end{pmatrix}$$

and has position vector

$$\boldsymbol{q} = \begin{pmatrix} x_2 \\ y_2 \\ z_2 \end{pmatrix}$$

Also $\overrightarrow{PQ} = \overrightarrow{OQ} - \overrightarrow{OP}$ represents

$$\begin{pmatrix} x_2 \\ y_2 \\ z_2 \end{pmatrix} - \begin{pmatrix} x_1 \\ y_1 \\ z_1 \end{pmatrix} = \begin{pmatrix} x_2 - x_1 \\ y_2 - y_1 \\ z_2 - z_1 \end{pmatrix}$$

giving the components of the vector $\overrightarrow{PQ}$ with position vector

$$\boldsymbol{q} - \boldsymbol{p} = \begin{pmatrix} x_2 - x_1 \\ y_2 - y_1 \\ z_2 - z_1 \end{pmatrix}$$

Hence $|\overrightarrow{PQ}|^2 = (x_2 - x_1)^2 + (y_2 - y_1)^2 + (z_2 - z_1)^2$

the square of the distance between P and Q, and

$$|\overrightarrow{PQ}| = \sqrt{[(x_2 - x_1)^2 + (y_2 - y_1)^2 + (z_2 - z_1)^2]}$$

the distance formula in 3 dimensions.

Again if $\overrightarrow{PQ}$ represents the vector $\boldsymbol{a}$, i.e. $\overrightarrow{PQ}$ represents $\boldsymbol{q} - \boldsymbol{p} = \boldsymbol{a}$, we write

$$\boldsymbol{a} = \begin{pmatrix} x_2 - x_1 \\ y_2 - y_1 \\ z_2 - z_1 \end{pmatrix} = \begin{pmatrix} a_1 \\ a_2 \\ a_3 \end{pmatrix}$$

and say that the vector $\boldsymbol{a}$ has components (a_1, a_2, a_3).

Thus $\boldsymbol{a} = (a_1, a_2, a_3)$ means the vector $\boldsymbol{a}$ has components a_1 in the x-direction, a_2 in the y-direction and a_3 in the z-direction and if $\boldsymbol{a}$ is represented by the directed line segment PQ, to go from P to Q we move,

a_1 units parallel to the x-axis
a_2 units parallel to the y-axis
a_3 units parallel to the z-axis, as in figure 50.

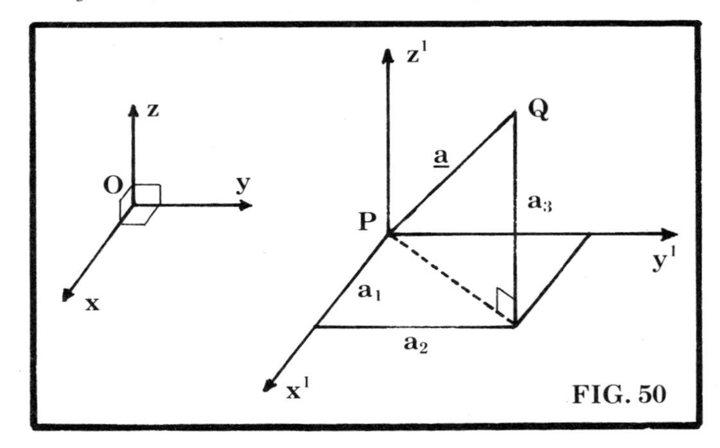

FIG. 50

PX′, PY′ and PZ′ are parallel to OX, OY and OZ, and hence if $\boldsymbol{a}$ has components (a_1, a_2, a_3) then

$$|\boldsymbol{a}|^2 = a_1{}^2 + a_2{}^2 + a_3{}^2$$

Example 1

P, Q and R are the points (4, 7, 1), (3, 2, − 1) and (5, 12, 3) respectively.

(i) Find an expression in component form for the vector represented by $\overrightarrow{PQ}$ and hence the length of the line segment PQ.
(ii) Show that the points P, Q and R are collinear.
(iii) Find the coordinates of the point S such that $\overrightarrow{PS} = 3\overrightarrow{PQ}$.

(i) P has coordinates $(4, 7, 1) \Rightarrow \overrightarrow{OP}$ represents

$$\begin{pmatrix} 4 \\ 7 \\ 1 \end{pmatrix}$$

Q has coordinates $(3, 2, -1) \Rightarrow \overrightarrow{OQ}$ represents

$$\begin{pmatrix} 3 \\ 2 \\ -1 \end{pmatrix}$$

hence $\overrightarrow{PQ} = \overrightarrow{OQ} - \overrightarrow{OP}$ represents

$$\begin{pmatrix} 3 \\ 2 \\ -1 \end{pmatrix} - \begin{pmatrix} 4 \\ 7 \\ 1 \end{pmatrix} = \begin{pmatrix} -1 \\ -5 \\ -2 \end{pmatrix}$$

i.e. $\overrightarrow{PQ}$ represents a vector with components

$$(-1, -5, -2)$$

Hence

$$|\overrightarrow{PQ}|^2 = (-1)^2 + (-5)^2 + (-2)^2 = 1 + 25 + 4 = 30$$

i.e. $|\overrightarrow{PQ}| = \sqrt{30}$ units.

Note: The writing in this example may be simplified thus

P is $(4. 7, 1)$, Q is $(3, 2, -1) \Rightarrow \overrightarrow{PQ}$ represents $\begin{pmatrix} -1 \\ -5 \\ -2 \end{pmatrix}$

i.e. $\overrightarrow{PQ}$ has components $(-1, -5, -2)$ and

$$|\overrightarrow{PQ}| = \sqrt{(1 + 25 + 4)} = \sqrt{30} \text{ units}$$

(ii) Q is $(3, 2, -1)$, R is $(5, 12, 3) \Rightarrow \overrightarrow{QR}$ represents

$$\begin{pmatrix} 2 \\ 10 \\ 4 \end{pmatrix} = 2\begin{pmatrix} 1 \\ 5 \\ 2 \end{pmatrix} = -2\begin{pmatrix} -1 \\ -5 \\ -2 \end{pmatrix}$$

i.e. $\qquad \overrightarrow{QR} = -2\overrightarrow{PQ}$

Hence $\qquad \overrightarrow{PQ} = -\tfrac{1}{2}\overrightarrow{QR}$

i.e. points P, Q and R are collinear.

(iii) Let S have coordinates (x, y, z), then $\overrightarrow{PS}$ represents

$$\begin{pmatrix} x-4 \\ y-7 \\ z-1 \end{pmatrix}$$

and $\overrightarrow{PQ}$ represents $\begin{pmatrix} -1 \\ -5 \\ -2 \end{pmatrix} \Rightarrow 3\overrightarrow{PQ}$ represents $\begin{pmatrix} -3 \\ -15 \\ -6 \end{pmatrix}$.

But $\overrightarrow{PS} = 3\overrightarrow{PQ}$

$$\Rightarrow x - 4 = -3, \quad y - 7 = -15, \quad z - 1 = -6$$

$$\Rightarrow \quad x = 1, \qquad y = -8, \qquad z = -5$$

i.e. S has coordinates $(1, -8, -5)$.

Example 2

The points A and B have coordinates $(3, -1, -4)$ and $(6, -1, -1)$ respectively and P divides AB in the ratio 2:1. Find the coordinates of P.

2 1 B
A P FIG. A

P divides AB in the ratio $2:1 \Rightarrow \dfrac{AP}{PB} = \dfrac{2}{1} \Rightarrow \dfrac{m}{n} = \dfrac{2}{1}$.

Let A have position vector $\boldsymbol{a} \Rightarrow \boldsymbol{a} = \begin{pmatrix} 3 \\ -1 \\ -4 \end{pmatrix}$

and B have position vector $\boldsymbol{b} \Rightarrow \boldsymbol{b} = \begin{pmatrix} 6 \\ -1 \\ -1 \end{pmatrix}$

Also $\overrightarrow{OP}$ represents $\dfrac{n\boldsymbol{a}+m\boldsymbol{b}}{m+n} = \dfrac{\boldsymbol{a}+2\boldsymbol{b}}{3}$

But $$\boldsymbol{a}+2\boldsymbol{b} = \begin{pmatrix} 3 \\ -1 \\ -4 \end{pmatrix} + \begin{pmatrix} 12 \\ -2 \\ -2 \end{pmatrix} = \begin{pmatrix} 15 \\ -3 \\ -6 \end{pmatrix}$$

Hence $$\overrightarrow{OP} \text{ represents } \begin{pmatrix} 5 \\ -1 \\ -2 \end{pmatrix}$$

i.e. P has position vector $\boldsymbol{p} = \begin{pmatrix} 5 \\ -1 \\ -2 \end{pmatrix}$

$\Rightarrow$ P has coordinates $(5, -1, -2)$

4.4 THE SECTION FORMULA IN COMPONENT FORM

If A and B are the points (x_1, y_1, z_1) and (x_2, y_2, z_2) and P with coordinates (x_p, y_p, z_p) divides AB in the ratio $m:n$, find the coordinates of P in terms of the coordinates of A and B.

Let O the origin of coordinates be taken as the origin of position vectors, then if A, B and P have position vectors $\boldsymbol{a}$, $\boldsymbol{b}$ and $\boldsymbol{p}$,

$$\boldsymbol{a} = \begin{pmatrix} x_1 \\ y_1 \\ z_1 \end{pmatrix} \quad \boldsymbol{b} = \begin{pmatrix} x_2 \\ y_2 \\ z_2 \end{pmatrix} \quad \boldsymbol{p} = \begin{pmatrix} x_p \\ y_p \\ z_p \end{pmatrix}$$

By the section formula in vector form,

$$\boldsymbol{p} = \frac{m\boldsymbol{b}+n\boldsymbol{a}}{m+n}$$

$$\Rightarrow \begin{pmatrix} x_p \\ y_p \\ z_p \end{pmatrix} = \frac{1}{m+n}\left[m\begin{pmatrix} x_2 \\ y_2 \\ z_2 \end{pmatrix} + n\begin{pmatrix} x_1 \\ y_1 \\ z_1 \end{pmatrix}\right] = \frac{1}{m+n}\begin{bmatrix} mx_2+nx_1 \\ my_2+ny_1 \\ mz_2+ny_1 \end{bmatrix}$$

Hence

$$x_p = \frac{mx_2+nx_1}{m+n}, \quad y_p = \frac{my_2+ny_1}{m+n}, \quad z_p = \frac{mz_2+nz_1}{m+n}$$

Note: For the coordinates in a plane (i.e. in two dimensions) put $z_1 = z_2 = 0$.

Thus if a point $P(x_p, y_p)$ in the x-y plane divides the line joining the points $A(x_1, y_1)$ and $B(x_2, y_2)$ in the ratio $m:n$, then

$$x_p = \frac{mx_2+nx_1}{m+n}, \quad y_p = \frac{my_2+ny_1}{m+n}$$

Example 1

With reference to three mutually perpendicular axes and origin O, A and B are the points $(4, 6, -2)$ and $(1, 3, 1)$ respectively.

(i) If P divides AB internally in the ratio $1:2$ and Q divides AB externally in the ratio $5:2$, find the coordinates of P and Q.

(ii) A point R on PQ has coordinates $(x_R, y_R, 2)$. Find the ratio in which R divides PQ and state the coordinates of R.

(i) P divides AB internally in the ratio

$$1:2 \Rightarrow \frac{m}{n} = \frac{1}{2}$$

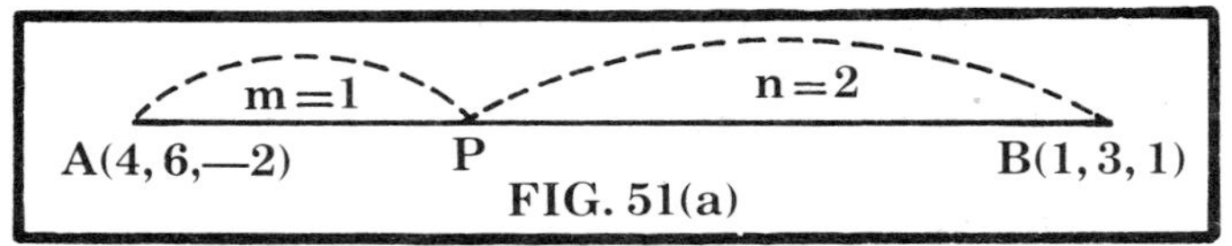

FIG. 51(a)

$$\begin{pmatrix} x_p \\ y_p \\ z_p \end{pmatrix} = \frac{1}{m+n}\begin{pmatrix} mx_2+nx_1 \\ my_2+ny_1 \\ mz_2+ny_1 \end{pmatrix} = \frac{1}{3}\begin{pmatrix} 1+8 \\ 3+12 \\ 1-4 \end{pmatrix} = \frac{1}{3}\begin{pmatrix} 9 \\ 15 \\ -3 \end{pmatrix}$$

$\Rightarrow$ P is the point $(3, 5, -1)$

Q divides AB externally in the ratio

$$5:2 \Rightarrow \frac{m}{n} = \frac{5}{-2}$$

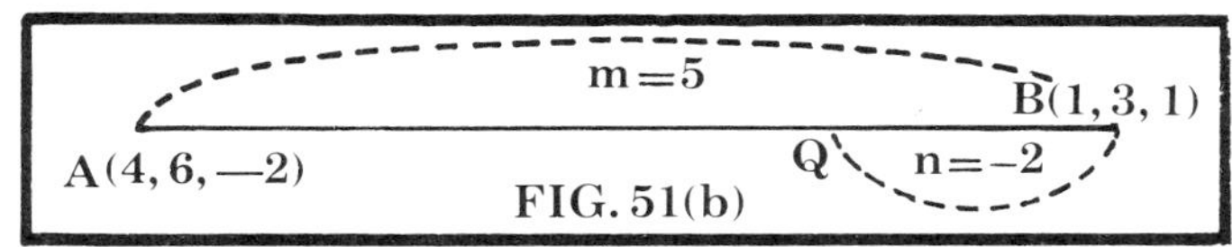

FIG. 51(b)

$$\begin{pmatrix} x_Q \\ y_Q \\ z_Q \end{pmatrix} = \frac{1}{m+n}\begin{pmatrix} mx_2+nx_1 \\ my_2+ny_1 \\ mz_2+nz_1 \end{pmatrix} = \frac{1}{3}\begin{pmatrix} 5-8 \\ 15-12 \\ 5+4 \end{pmatrix} = \frac{1}{3}\begin{pmatrix} -3 \\ 3 \\ 9 \end{pmatrix}$$

$\Rightarrow$ Q is the point $(-1, 1, 3)$

(iii) Let R divide PQ in the ratio $m:n$ where R is the point $(x_R, y_R, 2)$,

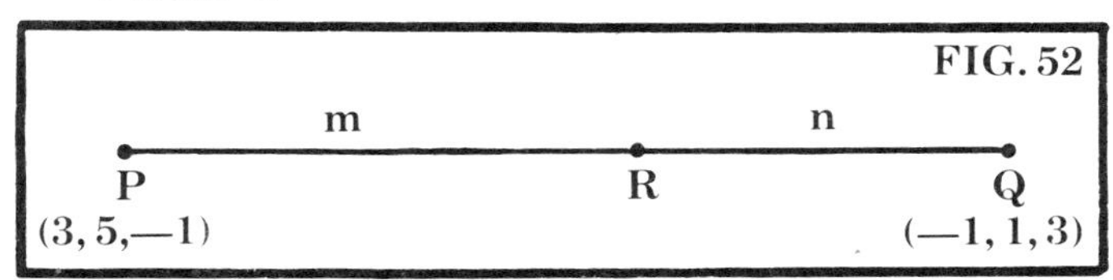

FIG. 52

and $\quad z_R = \dfrac{mz_2+nz_1}{m+n} \Rightarrow 2 = \dfrac{3m-n}{m+n}$

$$\Rightarrow 2m+2n = 3m-n$$

$$\Rightarrow 3n = m$$

$$\Rightarrow m:n = 3:1$$

i.e. R divides PQ internally in the ratio 3 : 1,

hence $\quad y_R = \dfrac{3+5}{4} = 2 \quad$ and $\quad x_R = \dfrac{-3+3}{4} = 0$

i.e. R is the point $(0, 2, 2)$.

ASSIGNMENT 4.3

1. A and B are the points $(1, -2, 4)$ and $(3, 6, -4)$. Write down the components of $\overrightarrow{OA}$, $\overrightarrow{OB}$ and $\overrightarrow{AB}$. What are the components of $\overrightarrow{BA}$? Find the length of the line segment AB.

2. Calculate the distance between the following sets of points:
   (i) $A(6, 3, -3)$ and $B(5, 1, -1)$
   (ii) $P(0, -2, -6)$ and $Q(7, -6, -2)$
   (iii) $R(3, -1, -2)$ and $S(1, 1, -2)$

3. Show that the points $A(5, 1, -6)$, $B(-7, -5, -2)$ and $C(11, 5, 6)$ are the vertices of an isosceles triangle.

4. P is a point on the line AB. Find the coordinates of P if,
   (i) A is $(4, -1, 2)$, B is $(4, -4, -1)$ and $AP:PB = 2:1$
   (ii) A is $(3, -7, 6)$, B is $(7, -1, 2)$ and $AP:PB = 1:3$
   (iii) A is $(4, -3, 5)$, B is $(6, 3, -1)$ and $AP:PB = -3:2$
   (iv) A is $(-1, -6, 1)$, B is $(2, -9, 4)$ and $AP:PB = -4:3$

5. Find the coordinates of the point P on the line AB, if
   (i) A is $(7, 1)$, B is $(2, -4)$ and $AP:PB = 2:3$
   (ii) A is $(4, -4)$, B is $(-2, 2)$ and $AP:PB = 2:1$
   (iii) A is $(-2, 4)$, B is $(0, 10)$ and $AP:PB = -3:2$
   (iv) A is $(5, 1)$, B is $(7, -1)$ and $AP:PB = -3:5$

6. Find the coordinates of A if
   (i) B is the point $(6, -3)$, P the point $(2, 1)$ and $AP:PB = 1:4$
   (ii) B is the point $(2, -3)$, P the point $(4, -5)$ and $AP:PB = -5:2$

7. A and B are the point with coordinates $(3, -4)$ and $(5, 4)$ respectively. A point $P(2, -8)$ lies on AB. Find the ratio in which P divides AB.

8. A is the point $(1, 4)$ and $B(-2, 1)$. Find the ratio in which $P(-5, t)$ divides AB, and hence write down the coordinates of P.

9. The point $Q(3, p, q)$ divides PR in the ratio $m:n$. Find the ratio $m:n$ if P is $(4, 3, 3)$ and $R(1, -3, -6)$. State the coordinates of Q.

10. The point $S(r, t, -2)$ divides RT in the ratio $m:n$. Find the ratio $m:n$ if R is $(2, 1, -4)$ and T is $(-1, -2, -1)$. State the coordinates of S.

11. Points A, B and C have coordinates $(3, 2, -4)$, $(6, 2, -1)$ and $(0, 2, -7)$ respectively.

    (a) Show that the points A, B and C are collinear.

    (b) Find the coordinates of the point T such that $\overrightarrow{AT} = \frac{1}{3}\overrightarrow{AB}$.

12. If the vectors $\boldsymbol{p}$, $\boldsymbol{q}$ and $\boldsymbol{r}$ have components

$$\begin{pmatrix} 1 \\ 4 \\ -3 \end{pmatrix}, \quad \begin{pmatrix} -4 \\ 4 \\ -2 \end{pmatrix} \quad \text{and} \begin{pmatrix} -1 \\ 4 \\ -4 \end{pmatrix}$$

respectively, find the components of the vectors,

(i) $2\boldsymbol{p} - \boldsymbol{q} + \boldsymbol{r}$ (ii) $3\boldsymbol{p} - 5\boldsymbol{q} - 2\boldsymbol{r}$

(iii) $5\boldsymbol{p} - 2\boldsymbol{q}$

13. Which of the following sets of 3 points are collinear?

    (i) $(3, 0, 2)$, $(4, -2, 1)$, $(1, 4, 4)$

    (ii) $(1, 5, 5)$, $(-1, 3, -1)$, $(2, 6, 9)$

    (iii) $(-1, 3, 0)$, $(5, 6, 1)$, $(-3, 2, -\frac{1}{3})$

14. The points A and B have coordinates $(2, 7, 6)$ and $(-1, 1, -6)$ respectively. P divides AB such that $\overrightarrow{AP} = 2\overrightarrow{PB}$, find the coordinates of P.

15. If in question 14 the point Q divides AB externally in the ratio $2:1$ (i.e. $\overrightarrow{AQ} = 2\overrightarrow{BQ}$), find the position vector of Q.

16. The points A and B have position vectors $\boldsymbol{a}$ and $\boldsymbol{b}$ given by

$$\boldsymbol{a} = \begin{pmatrix} 2 \\ 0 \\ -6 \end{pmatrix} \quad \text{and} \quad \boldsymbol{b} = \begin{pmatrix} 7 \\ 4 \\ -1 \end{pmatrix}$$

Find the position vectors $\boldsymbol{p}$ and $\boldsymbol{q}$ of the points P and Q such that

(i) P divides AB internally in the ratio $2:3$.

(ii) Q divides AB externally in the ratio $2:3$ (i.e. $3\overrightarrow{AQ} = -2\overrightarrow{QB}$).

[*Note:* part (ii) may be done by letting Q have co-ordinates (x, y, z) and therefore position vector

$$\boldsymbol{q} = \begin{pmatrix} x \\ y \\ z \end{pmatrix}$$

and using

$$3\overrightarrow{AQ} = -2\overrightarrow{QB} \Rightarrow 3(\boldsymbol{q} - \boldsymbol{a}) = -2(\boldsymbol{b} - \boldsymbol{q})$$

Or Q divides AB externally in the ratio

$$2:3 \Rightarrow \frac{AQ}{QB} = \frac{2}{-3}$$

and therefore

$$\boldsymbol{q} = \frac{n\boldsymbol{a} + m\boldsymbol{b}}{m+n} = \frac{-3\boldsymbol{a} + 2\boldsymbol{b}}{2-3}$$]

17. The components of a vector represented by $\overrightarrow{AB}$ are

$$\begin{pmatrix} 5 \\ 10 \\ -5 \end{pmatrix}$$

A point P on AB divides AB in the ratio 3:2. If P has coordinates (3, 4, −4) find the coordinates of A and B.

18. If

$$\boldsymbol{a} = \begin{pmatrix} 3 \\ 0 \\ -2 \end{pmatrix}, \quad \boldsymbol{b} = \begin{pmatrix} 4 \\ -1 \\ -3 \end{pmatrix} \quad \text{and} \quad \boldsymbol{c} = \begin{pmatrix} -2 \\ 6 \\ 3 \end{pmatrix}$$

find numbers k, m and n such that

$$k\boldsymbol{a} + m\boldsymbol{b} + n\boldsymbol{c} = \begin{pmatrix} -1 \\ 3 \\ 1 \end{pmatrix}$$

19. P, Q, R are the points (5, 2, −2), (7, −2, 3) and (9, −6, 8) respectively.
    (i) Find the length of the line segment PQ.
    (ii) Show that P, Q and R are collinear and that Q is the mid-point of PR.
    (iii) Find the coordinates of the point S such that $\overrightarrow{PS} = 4\overrightarrow{PQ}$.

UNIT VECTORS IN THREE MUTUALLY PERPENDICULAR DIRECTIONS

A **unit vector** is one which has magnitude 1, for example, if $\boldsymbol{a}$ is a unit vector then $|\boldsymbol{a}| = 1$.

We usually take $\boldsymbol{i}, \boldsymbol{j}$ and $\boldsymbol{k}$ as three mutually perpendicular unit vectors along OX, OY and OZ.

Since $\boldsymbol{i}, \boldsymbol{j}$ and $\boldsymbol{k}$ are unit vectors then

$$|\boldsymbol{i}| = |\boldsymbol{j}| = |\boldsymbol{k}| = 1$$

Thus the components of $\boldsymbol{i}, \boldsymbol{j}$ and $\boldsymbol{k}$ are,

$$\boldsymbol{i} = \begin{pmatrix} 1 \\ 0 \\ 0 \end{pmatrix}, \quad \boldsymbol{j} = \begin{pmatrix} 0 \\ 1 \\ 0 \end{pmatrix} \quad \text{and} \quad \boldsymbol{k} = \begin{pmatrix} 0 \\ 0 \\ 1 \end{pmatrix}$$

If A has coordinates (a_1, a_2, a_3) and position vector $\boldsymbol{a}$, then

$$\boldsymbol{a} = \begin{pmatrix} a_1 \\ a_2 \\ a_3 \end{pmatrix}$$

or

$$\boldsymbol{a} = a_1\boldsymbol{i} + a_2\boldsymbol{j} + a_3\boldsymbol{k}$$

i.e. a has components a_1 along OX,
a_2 parallel to OY, and
a_3 parallel to OZ.

Figure 53(a) and (b) show two different ways of drawing the vector $\boldsymbol{a}$ in three dimensions.

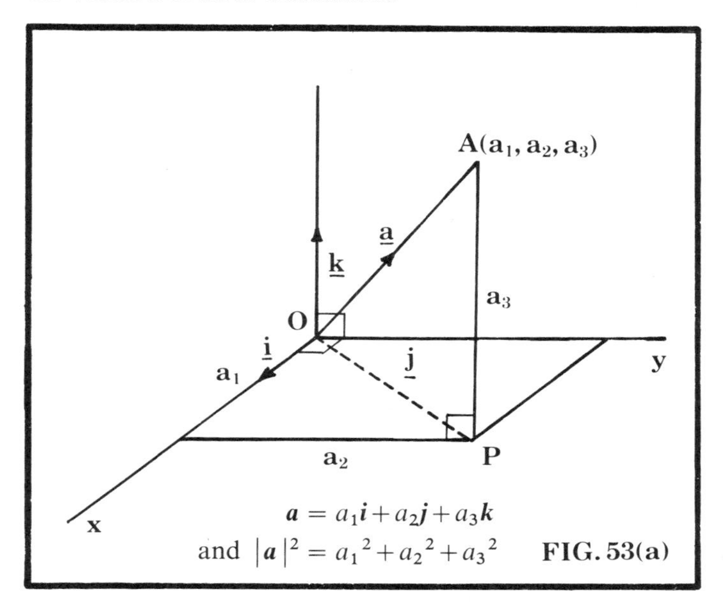

$\boldsymbol{a} = a_1\boldsymbol{i} + a_2\boldsymbol{j} + a_3\boldsymbol{k}$

and $|\boldsymbol{a}|^2 = a_1^2 + a_2^2 + a_3^2$ **FIG. 53(a)**

4.5

Note: In the diagrams $\boldsymbol{p} = a_1\boldsymbol{i}+a_2\boldsymbol{j}+0\boldsymbol{k} = a_1\boldsymbol{i}+a_2\boldsymbol{j}$, where $\boldsymbol{p}$ is the position vector of P.

Example 1

4.6

If $\boldsymbol{a} = \begin{pmatrix} 2 \\ 1 \\ 4 \end{pmatrix}$, $\boldsymbol{b} = \begin{pmatrix} -2 \\ 0 \\ 3 \end{pmatrix}$, $\boldsymbol{c} = \begin{pmatrix} 1 \\ 4 \\ -2 \end{pmatrix}$

express the vectors $\boldsymbol{a}$, $\boldsymbol{b}$, $\boldsymbol{c}$, $2\boldsymbol{a}-3\boldsymbol{b}$, $\boldsymbol{a}+4\boldsymbol{b}-3\boldsymbol{c}$ in terms of $\boldsymbol{i}, \boldsymbol{j}$ and $\boldsymbol{k}$ and find the value of $|2\boldsymbol{a}-3\boldsymbol{b}|$.

$$\boldsymbol{a} = 2\boldsymbol{i}+\boldsymbol{j}+4\boldsymbol{k}, \quad \boldsymbol{b} = -2\boldsymbol{i}+3\boldsymbol{k}, \quad \boldsymbol{c} = \boldsymbol{i}+4\boldsymbol{j}-2\boldsymbol{k}$$

$$2\boldsymbol{a}-3\boldsymbol{b} = 4\boldsymbol{i}+2\boldsymbol{j}+8\boldsymbol{k}-(-6\boldsymbol{i}+9\boldsymbol{k}) = 10\boldsymbol{i}+2\boldsymbol{j}-\boldsymbol{k}$$

hence $$|2\boldsymbol{a}-3\boldsymbol{b}|^2 = 10^2+2^2+(-1)^2 = 105$$

$$|2\boldsymbol{a}-3\boldsymbol{b}| = \sqrt{105} \text{ units}$$

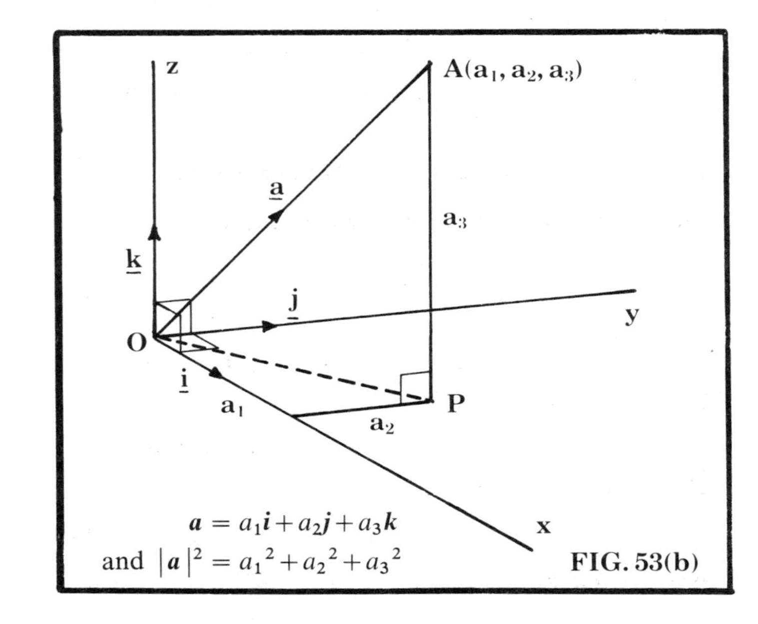

$\boldsymbol{a} = a_1\boldsymbol{i}+a_2\boldsymbol{j}+a_3\boldsymbol{k}$

and $|\boldsymbol{a}|^2 = a_1{}^2+a_2{}^2+a_3{}^2$

FIG. 53(b)

Again
$$\begin{aligned} \boldsymbol{a} &= \ \ 2\boldsymbol{i}+\ \ \boldsymbol{j}+\ \ 4\boldsymbol{k} \\ 4\boldsymbol{b} &= -8\boldsymbol{i} \qquad\ +12\boldsymbol{k} \\ -3\boldsymbol{c} &= -3\boldsymbol{i}-12\boldsymbol{j}+\ \ 6\boldsymbol{k} \end{aligned}$$

hence $$\boldsymbol{a}+4\boldsymbol{b}-3\boldsymbol{c} = -9\boldsymbol{i}-11\boldsymbol{j}+22\boldsymbol{k}$$

THE SCALAR PRODUCT OF TWO VECTORS

If $\boldsymbol{a}$ and $\boldsymbol{b}$ are two non-zero vectors such that the angle between them is θ then the **scalar product** of $\boldsymbol{a}$ and $\boldsymbol{b}$ (which is denoted by $\boldsymbol{a}.\boldsymbol{b}$) is defined to be the number $|\boldsymbol{a}||\boldsymbol{b}|\cos\theta$. θ is taken to be that value which lies in the interval $0 \leqq \theta \leqq \pi$.

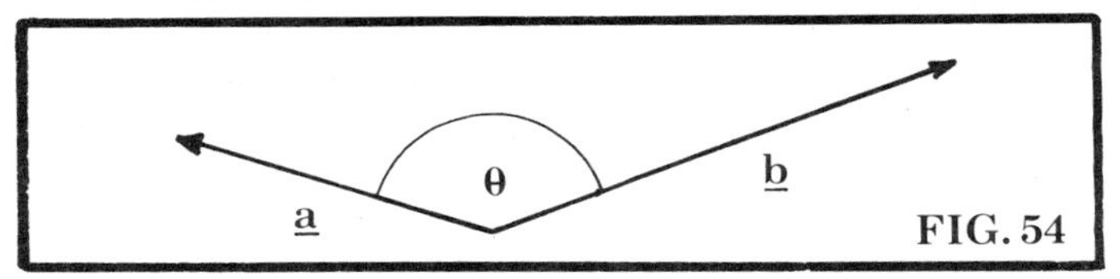

FIG. 54

Hence the scalar product of two vectors $\boldsymbol{a}$ and $\boldsymbol{b}$ is

$$\boldsymbol{a}.\boldsymbol{b} = |\boldsymbol{a}||\boldsymbol{b}|\cos\theta \quad \text{where} \quad 0 \leqq \theta \leqq \pi$$

Note: The following properties follow from this definition of the scalar product:

(1) If either $\boldsymbol{a}$ or $\boldsymbol{b}$ is the zero vector $\boldsymbol{0}$, then $\boldsymbol{a}.\boldsymbol{b}$ is defined as the zero number 0.

i.e. $$\boldsymbol{0}.\boldsymbol{a} = \boldsymbol{a}.\boldsymbol{0} = 0$$

(2) *The commutative property for scalar products*

Since $\boldsymbol{a}.\boldsymbol{b} = |\boldsymbol{a}||\boldsymbol{b}|\cos\theta$

and $\boldsymbol{b}.\boldsymbol{a} = |\boldsymbol{b}||\boldsymbol{a}|\cos\theta = |\boldsymbol{a}||\boldsymbol{b}|\cos\theta$

then $\boldsymbol{a}.\boldsymbol{b} = \boldsymbol{b}.\boldsymbol{a}$

(3) If k is a number then,

$$\begin{aligned} \boldsymbol{a}.(k\boldsymbol{b}) &= (k\boldsymbol{b}).\boldsymbol{a} \quad \text{—by the commutative property} \\ &= k(\boldsymbol{b}.\boldsymbol{a}) \\ &= k(\boldsymbol{a}.\boldsymbol{b}) \end{aligned}$$

(4) *The distributive property for scalar products*

$$\boldsymbol{a}.(\boldsymbol{b}+\boldsymbol{c}) = \boldsymbol{a}.\boldsymbol{b}+\boldsymbol{a}.\boldsymbol{c}$$

(5) $\boldsymbol{a}.\boldsymbol{a} = |\boldsymbol{a}||\boldsymbol{a}|\cos 0 = |\boldsymbol{a}|^2$ since $\cos 0 = 1$

The scalar product $\boldsymbol{a}.\boldsymbol{a}$ is often denoted simply by a^2.

(6) *Perpendicular vectors*

(i) If two non-zero vectors $\boldsymbol{a}$ and $\boldsymbol{b}$ are perpendicular, then $\boldsymbol{a}.\boldsymbol{b} = |\boldsymbol{a}|\,|\boldsymbol{b}|\cos 90° = 0$, since $\cos 90° = 0$.

(ii) If two non-zero vectors $\boldsymbol{a}$ and $\boldsymbol{b}$ are such that $\boldsymbol{a}.\boldsymbol{b} = 0$, then the vectors $\boldsymbol{a}$ and $\boldsymbol{b}$ are perpendicular.

Hence $\boldsymbol{a}.\boldsymbol{b} = 0 \Leftrightarrow \boldsymbol{a}$ and $\boldsymbol{b}$ are perpendicular ($\boldsymbol{a} \neq \boldsymbol{0}$, $\boldsymbol{b} \neq \boldsymbol{0}$).

These six consequences of the definition of scalar products are very important in vector work and should be studied carefully and known.

Example 1

If the vectors $\boldsymbol{a}$ and $\boldsymbol{b}$ have equal magnitude (i.e. $|\boldsymbol{a}| = |\boldsymbol{b}|$), prove that the vectors $(\boldsymbol{a}+\boldsymbol{b})$ and $(\boldsymbol{a}-\boldsymbol{b})$ are perpendicular.

Consider the scalar property of $\boldsymbol{a}+\boldsymbol{b}$ and $\boldsymbol{a}-\boldsymbol{b}$.

$(\boldsymbol{a}+\boldsymbol{b}).(\boldsymbol{a}-\boldsymbol{b}) = (\boldsymbol{a}+\boldsymbol{b}).\boldsymbol{a}-(\boldsymbol{a}+\boldsymbol{b}).\boldsymbol{b}$

—using the distributive property

$= \boldsymbol{a}.(\boldsymbol{a}+\boldsymbol{b})-\boldsymbol{b}.(\boldsymbol{a}+\boldsymbol{b})$

—using the commutative property

$= \boldsymbol{a}.\boldsymbol{a}+\boldsymbol{a}.\boldsymbol{b}-\boldsymbol{b}.\boldsymbol{a}-\boldsymbol{b}.\boldsymbol{b}$

—using the distributive property

$= \boldsymbol{a}.\boldsymbol{a}-\boldsymbol{b}.\boldsymbol{b}$ (since $\boldsymbol{a}.\boldsymbol{b} = \boldsymbol{b}.\boldsymbol{a}$)

$= |\boldsymbol{a}|^2-|\boldsymbol{b}|^2$

—using property (5) above

$= 0$ since $|\boldsymbol{a}| = |\boldsymbol{b}|$

And $(\boldsymbol{a}+\boldsymbol{b}).(\boldsymbol{a}-\boldsymbol{b}) = 0 \Rightarrow \boldsymbol{a}+\boldsymbol{b}$ is perpendicular to $\boldsymbol{a}-\boldsymbol{b}$, from property (6) above.

Example 2

The altitudes AD and BE of triangle ABC intersect at H, prove that the third altitude CF passes through H.

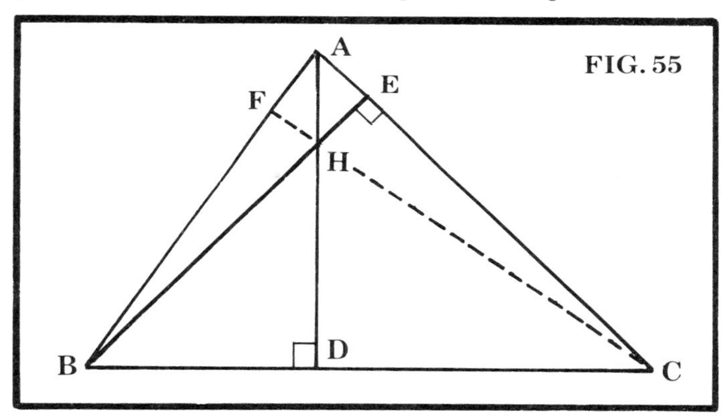

FIG. 55

AD and BE are the two altitudes intersecting at H. Let the line CH meet AB at F.

We have to show that CHF is perpendicular to AB.

Take H as the origin and let A, B, C have position vectors $\boldsymbol{a}$, $\boldsymbol{b}$ and $\boldsymbol{c}$ with reference to H.

i.e. $\overrightarrow{HA}$ represents $\boldsymbol{a}$, $\overrightarrow{HB}$ represents $\boldsymbol{b}$, $\overrightarrow{HC}$ represents $\boldsymbol{c}$

Hence $\overrightarrow{AB} = \overrightarrow{HB}-\overrightarrow{HA}$ represents $\boldsymbol{b}-\boldsymbol{a}$ and similarly

$\overrightarrow{BC}$ represents $\boldsymbol{c}-\boldsymbol{b}$, $\overrightarrow{CA}$ represents $\boldsymbol{a}-\boldsymbol{c}$

$\overrightarrow{HA}$ is perpendicular to $\overrightarrow{BC}$

$$\Rightarrow \boldsymbol{a}.(\boldsymbol{b}-\boldsymbol{c}) = 0 \Rightarrow \boldsymbol{a}.\boldsymbol{b} = \boldsymbol{a}.\boldsymbol{c} \qquad (1)$$

$\overrightarrow{HB}$ is perpendicular to $\overrightarrow{AC}$

$$\Rightarrow \boldsymbol{b}.(\boldsymbol{c}-\boldsymbol{a}) = 0 \Rightarrow \boldsymbol{b}.\boldsymbol{c} = \boldsymbol{b}.\boldsymbol{a} \qquad (2)$$

Hence from (1) and (2)

$$\boldsymbol{b}.\boldsymbol{c} = \boldsymbol{a}.\boldsymbol{c} \quad \ldots\ldots\ldots\ldots\ldots (3)$$

(since $\boldsymbol{a}.\boldsymbol{b} = \boldsymbol{b}.\boldsymbol{a}$).

Now $\vec{HC}.\vec{AB}$ represents $c.(b-a) = c.b - c.a$

$$= b.c - a.c$$
$$= 0 \quad \text{from (3)}$$
$$\Rightarrow \vec{HC} \text{ is perpendicular to } \vec{AB}$$

i.e. CHF is perpendicular to AB and hence the three altitudes of a triangle are concurrent.

4.7 THE SCALAR PRODUCT IN COMPONENT FORM

In figure 56 let A have coordinates (x_1, y_1, z_1) and B (x_2, y_2, z_2) with reference to three mutually perpendicular axes OX, OY and OZ.

Let $\vec{OA}$ represent $\boldsymbol{a}$ and $\vec{OB}$ represent $\boldsymbol{b}$.

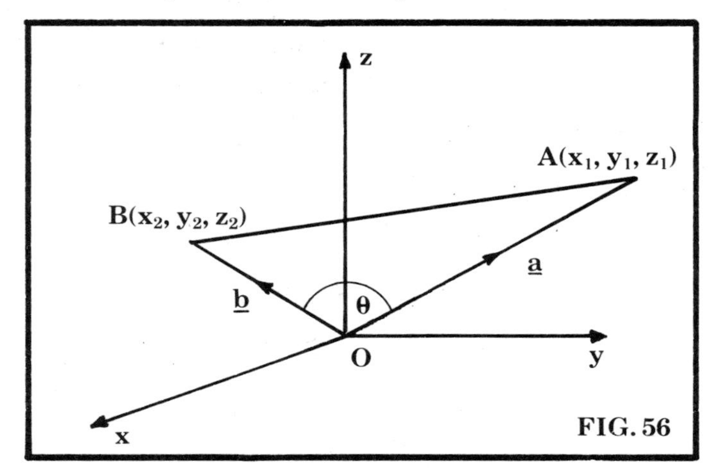

FIG. 56

i.e. $\boldsymbol{a} = \begin{pmatrix} x_1 \\ y_1 \\ z_1 \end{pmatrix}$ and $\boldsymbol{b} = \begin{pmatrix} x_2 \\ y_2 \\ z_2 \end{pmatrix}$

$$|\vec{OA}|^2 = |\boldsymbol{a}|^2 = x_1{}^2 + y_1{}^2 + z_1{}^2$$
$$|\vec{OB}|^2 = |\boldsymbol{b}|^2 = x_2{}^2 + y_2{}^2 + z_2{}^2$$
$$|\vec{AB}|^2 = |\boldsymbol{b}-\boldsymbol{a}|^2 = (x_2-x_1)^2 + (y_2-y_1)^2 + (z_2-z_1)^2$$
$$= x_2{}^2 - 2x_1x_2 + x_1{}^2 + y_2{}^2 - 2y_1y_2$$
$$+ y_2{}^2 + z_2{}^2 - 2z_1z_2 + z_2{}^2$$
$$= (x_1{}^2 + y_1{}^2 + z_1{}^2) + (x_2{}^2 + y_2{}^2 + z_2{}^2)$$
$$- 2(x_1x_2 + y_1y_2 + z_1z_2)$$

hence

$$|\vec{AB}|^2 = |\boldsymbol{a}|^2 + |\boldsymbol{b}|^2 - 2(x_1x_2 + y_1y_2 + z_1z_2) \qquad (1)$$

Applying the cosine rule to triangle AOB

$$|\vec{AB}|^2 = |\boldsymbol{a}|^2 + |\boldsymbol{b}|^2 - 2|\boldsymbol{a}||\boldsymbol{b}|\cos\theta$$

and from (1) $\quad |\vec{AB}|^2 = |\boldsymbol{a}|^2 + |\boldsymbol{b}|^2 - 2(x_1x_2 + y_1y_2 + z_1z_2)$

Hence $\quad |\boldsymbol{a}||\boldsymbol{b}|\cos\theta = x_1x_2 + y_1y_2 + z_1z_2$

and hence the scalar product of $\boldsymbol{a}$ and $\boldsymbol{b}$ may be defined as

$$\boldsymbol{a}.\boldsymbol{b} = |\boldsymbol{a}||\boldsymbol{b}|\cos\theta$$

Or $\quad \boldsymbol{a}.\boldsymbol{b} = x_1x_2 + y_1y_2 + z_1z_2$

—this is known as the **scalar product in component form.**

It follows from these definitions that if θ is the angle between the vectors $\boldsymbol{a}$ and $\boldsymbol{b}$, then

$$\cos\theta = \frac{\boldsymbol{a}.\boldsymbol{b}}{|\boldsymbol{a}||\boldsymbol{b}|} = \frac{x_1x_2 + y_1y_2 + z_1z_2}{|\boldsymbol{a}||\boldsymbol{b}|}$$

where $\quad \boldsymbol{a} = \begin{pmatrix} x_1 \\ y_1 \\ z_1 \end{pmatrix}$ and $\boldsymbol{b} = \begin{pmatrix} x_2 \\ y_2 \\ z_2 \end{pmatrix}$

Example 1

A, B, C are the points $(4, -1, 3)$, $(3, -1, 4)$ and $(4, 2, -5)$ respectively. If $\overrightarrow{BA}$ represents the vector $\boldsymbol{a}$ and $\overrightarrow{BC}$ represents the vector $\boldsymbol{b}$, find the value of $\boldsymbol{a}.\boldsymbol{b}$ and the cosine of angle ABC.

$$\left.\begin{array}{l}\text{A is the point } (4, -1, 3)\\ \text{B is the point } (3, -1, 4)\end{array}\right\} \Rightarrow \boldsymbol{a} = \begin{pmatrix}-1\\0\\1\end{pmatrix}$$

$$\left.\begin{array}{l}\text{B is the point } (3, -1, 4)\\ \text{C is the point } (4, 2, -5)\end{array}\right\} \Rightarrow \boldsymbol{b} = \begin{pmatrix}1\\3\\-9\end{pmatrix}$$

Hence $\quad \boldsymbol{a}.\boldsymbol{b} = x_1x_2 + y_1y_2 + z_1z_2 = -1+0-9 = -10$

also $\quad |\boldsymbol{a}|^2 = (-1)^2 + 0^2 + 1^2 = 2 \Rightarrow |\boldsymbol{a}| = \sqrt{2}$

and $\quad |\boldsymbol{b}|^2 = 1^2 + 3^2 + (-9)^2 = 91 \Rightarrow |\boldsymbol{b}| = \sqrt{91}$

But $\quad \boldsymbol{a}.\boldsymbol{b} = |\boldsymbol{a}||\boldsymbol{b}| \cos \text{ABC} = \sqrt{2}.\sqrt{91}.\cos \text{ABC}$

Hence

$$\sqrt{2}.\sqrt{91} \cos \text{ABC} = -10 \Rightarrow \cos \text{ABC} = \frac{-10}{\sqrt{2}.\sqrt{91}} = \frac{-10}{\sqrt{182}}$$

Example 2

With reference to three mutually perpendicular axes OX, OY and OZ, triangle ABC has vertices $A(-7, -2, -3)$, $B(6, 2, 2)$ and $C(-1, -5, 9)$. P is the point $(2, -2, 6)$.

(i) Prove that AP is perpendicular to BC.

(ii) Show that the cosine of angle ABC is

$$\frac{4}{\sqrt{70}}$$

(i) A is the point $(-7, -2, -3)$ and P the point $(2, -2, 6)$.

If $\overrightarrow{AP}$ represents $\boldsymbol{l}$ then $\boldsymbol{l} = \begin{pmatrix}9\\0\\9\end{pmatrix}$

B is the point $(6, 2, 2)$ and C the point $(-1, -5, 9)$.

If $\overrightarrow{BC}$ represents $\boldsymbol{m}$ then $\boldsymbol{m} = \begin{pmatrix}-7\\-7\\7\end{pmatrix}$

$\boldsymbol{l}.\boldsymbol{m} = x_1x_2 + y_1y_2 + z_1z_2 = -63+0+63 = 0$

$\Rightarrow$ AP is perpendicular to BC

(ii) $\overrightarrow{BA} = \overrightarrow{OA} - \overrightarrow{OB}$ represents $\begin{pmatrix}-13\\-4\\-5\end{pmatrix} = \boldsymbol{n}$ say

$\overrightarrow{BC} = \overrightarrow{OC} - \overrightarrow{OB}$ represents $\begin{pmatrix}-7\\-7\\7\end{pmatrix} = \boldsymbol{m}$

$$\begin{aligned}|\boldsymbol{n}|^2 = |\overrightarrow{BA}|^2 &= (13)^2 + (-4)^2 + (-5)^2\\ &= 169+16+25 = 210\\ &\Rightarrow |\boldsymbol{n}| = \sqrt{210}\\ |\boldsymbol{m}|^2 = |\overrightarrow{BC}|^2 &= (-7)^2 + (-7)^2 + 7^2\\ &= 49+49+49 = 3\times 49\\ &\Rightarrow |\boldsymbol{m}| = 7\sqrt{3}\end{aligned}$$

But

$$\begin{aligned}\cos \text{ABC} &= \frac{\boldsymbol{n}.\boldsymbol{m}}{|\boldsymbol{n}||\boldsymbol{m}|} = \frac{84}{\sqrt{210}.7\sqrt{3}}\\ &= \frac{12}{\sqrt{(3.7.10)}.\sqrt{3}}\\ &= \frac{12}{3\sqrt{70}}\\ &= \frac{4}{\sqrt{70}}\end{aligned}$$

ASSIGNMENT 4.4

1. Express the vectors

$$a = \begin{pmatrix} 3 \\ -1 \\ 2 \end{pmatrix}, \quad b = \begin{pmatrix} -1 \\ -2 \\ 0 \end{pmatrix} \quad \text{and} \quad c = \begin{pmatrix} 2 \\ 2 \\ -1 \end{pmatrix}$$

in terms of the unit vectors i, j and k.

2. $$a = \begin{pmatrix} 2 \\ -4 \\ 0 \end{pmatrix}, \quad b = \begin{pmatrix} -3 \\ -2 \\ 1 \end{pmatrix} \quad \text{and} \quad c = \begin{pmatrix} -2 \\ -1 \\ 3 \end{pmatrix}$$

   (i) Express a, b and c in terms of i, j and k.

   (ii) Express the vectors $2a - b, a + 2b - c$ and $4c$ in terms of i, j and k.

3. Given

$$a = 2i - 3j + k, \; b = -2i - 3j + k \text{ and } c = i - j + 2k$$

where i, j and k are three mutually perpendicular unit vectors.

   (i) Calculate $|a|$.

   (ii) Express $2a - b$ in terms of i, j and k and calculate $|2a - b|$.

   (iii) Express $a + b - c$ and $3b - c$ in terms of i, j and k.

4. In figure 57, $\overrightarrow{OA}$ represents the vector a, and

$$a = \begin{pmatrix} 3 \\ -2 \\ 5 \end{pmatrix}$$

   (i) State the coordinates of A.

   (ii) Write a in terms of i, j and k.

   (iii) Write the components of the vectors represented by $\overrightarrow{OM}$, $\overrightarrow{MP}$, $\overrightarrow{PA}$ and $\overrightarrow{OP}$ in column form and in terms of i, j and k.

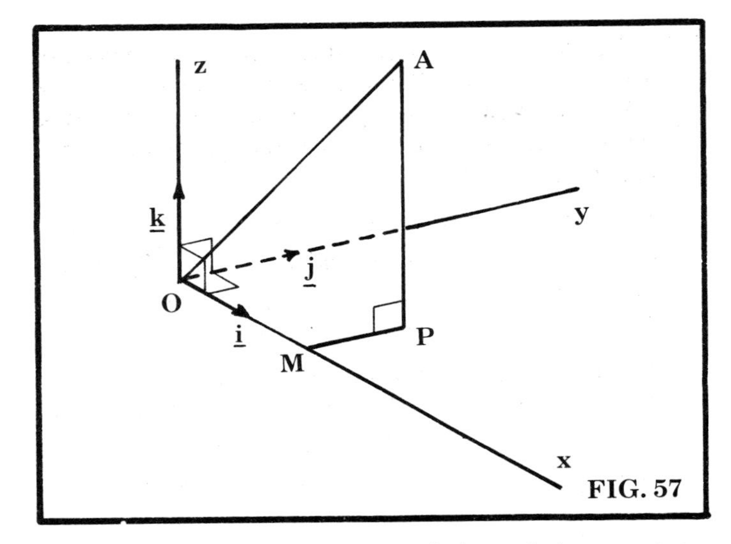

FIG. 57

5. a and b are vectors such that $|a| = 4$, $|b| = 7$ and the angle between a and b is 45°. Find the value of the scalar product $a \cdot b$.

6. If $a \neq 0$ and $a \cdot b = a \cdot c$,

   (i) Is $b = c$? Give a reason.

   (ii) What can you conclude about a and $b - c$?

7. Expand the following scalar products:

   (i) $p \cdot (q + r)$ (ii) $(p + q) \cdot (r + s)$

   (iii) $(p + q) \cdot (r - s)$ (iv) $(p + q) \cdot (p - q)$

8. Show that $(a + b) \cdot (a + b) = |a|^2 + |b|^2 + 2a \cdot b$.
   If $|a| = |b|$ simplify this expression.

9. If $|a| = |b|$ show that $(a + b) \cdot (a - b) = 0$.

10. In triangle POQ, $\overrightarrow{OP}$ represents the vector p, $\overrightarrow{OQ}$ the vector q and A denotes the area of the triangle POQ. Show that the area A is given by

$$4A^2 = |p|^2 |q|^2 - (p \cdot q)^2$$

11. OABC is a rhombus. $\vec{OA}$ and $\vec{OC}$ represent the vectors $\boldsymbol{a}$ and $\boldsymbol{c}$ respectively. Express the vectors represented by OB and CA in terms of $\boldsymbol{a}$ and $\boldsymbol{c}$ and hence using scalar products prove that the diagonals of a rhombus are at right angles.

12. P, Q, R and S are the points $(3, -3, -1)$, $(4, -1, 2)$, $(3, -1, 3)$ and $(4, 4, -6)$ respectively. If $\vec{PQ}$, $\vec{RS}$, $\vec{PR}$ and $\vec{QS}$ represent vectors $\boldsymbol{a}$, $\boldsymbol{b}$, $\boldsymbol{c}$ and $\boldsymbol{d}$ respectively, find the scalar products $\boldsymbol{a}.\boldsymbol{b}$ and $\boldsymbol{c}.\boldsymbol{d}$.

13. P is the point $(2, -3, 1)$ and $Q(-1, 5, -4)$. Calculate the size of the angle POQ where O is the origin of three mutually perpendicular axes.

14. With reference to three mutually perpendicular unit vectors,

$$\boldsymbol{a} = \begin{pmatrix} a_1 \\ a_2 \\ a_3 \end{pmatrix} \quad \text{and} \quad \boldsymbol{b} = \begin{pmatrix} b_1 \\ b_2 \\ b_3 \end{pmatrix}$$

Prove that $|\boldsymbol{a}||\boldsymbol{b}|\cos\theta = a_1b_1 + a_2b_2 + a_3b_3$, where θ is the angle between $\boldsymbol{a}$ and $\boldsymbol{b}$.

[*Hint:* Let $\vec{PA} = \boldsymbol{a}$ and $\vec{PB} = \boldsymbol{b}$ and apply the cosine rule to triangle PAB.]

15. With reference to three mutually perpendicular axes, A, B, C and D are the points $(-4, 5, 8)$, $(3, 7, 7)$, $(0, 1, 1)$ and $(-3, -5, -5)$ respectively.

(i) Prove that the angle DAB is right.

(ii) Show that C is the mid-point of DB and hence that D, A, B lie on a circle with centre C.

(iii) State the length of the radius of this circle.

16. Give a reason to show whether the points $A(2, 4, 1)$. $B(-2, 2, -1)$ and $C(1, 5, -1)$ are collinear or not.
If collinear find the ratio AB:BC.
If non-collinear calculate the size of the angle ABC and the area of triangle ABC.

17. $\boldsymbol{i}, \boldsymbol{j}$ and $\boldsymbol{k}$ are three mutually perpendicular unit vectors. Show that $\boldsymbol{i}.\boldsymbol{i} = 1$, $\boldsymbol{i}.\boldsymbol{j} = 0$ and hence complete the table of scalar products

.	$\boldsymbol{i}$	$\boldsymbol{j}$	$\boldsymbol{k}$
$\boldsymbol{i}$	1	0	
$\boldsymbol{j}$	0		
$\boldsymbol{k}$			

With reference to three mutually perpendicular unit vectors $\boldsymbol{i}, \boldsymbol{j}$ and $\boldsymbol{k}$,

$$\boldsymbol{a} = \begin{pmatrix} a_1 \\ a_2 \\ a_3 \end{pmatrix} \quad \text{and} \quad \boldsymbol{b} = \begin{pmatrix} b_1 \\ b_2 \\ b_3 \end{pmatrix}$$

Express $\boldsymbol{a}$ and $\boldsymbol{b}$ in terms of $\boldsymbol{i}, \boldsymbol{j}$ and $\boldsymbol{k}$ and hence show that

$$\boldsymbol{a}.\boldsymbol{b} = a_1b_1 + a_2b_2 + a_3b_3$$

18. OABCDEFG is a cube of side 1 unit. Prove that the acute angle between any two space diagonals has cosine $\frac{1}{3}$.

19. A triangle PQR has vertices $P(3, -2, 6)$, $Q(0, -2, 3)$ and $R(4, 0, -1)$. Show that angle Q is a right angle and that $\cos R = \sqrt{2}\cos P$.

20. Relative to an origin O the points P and Q of triangle OPQ have position vectors $\boldsymbol{p}$ and $\boldsymbol{q}$.

(i) Show that the cosine of angle OPQ may be written as

$$\frac{|\boldsymbol{p}|^2 + |\vec{PQ}|^2 - |\boldsymbol{q}|^2}{2|\boldsymbol{p}||\vec{PQ}|}$$

and write down a similar expression for cos OQP.

(ii) Show that angle OPQ = angle OQP if and only if $|\boldsymbol{p}| = |\boldsymbol{q}|$.

21. $\boldsymbol{p}$ and $\boldsymbol{q}$ are vectors with components

$$\begin{pmatrix} p_1 \\ p_2 \\ p_3 \end{pmatrix} \text{ and } \begin{pmatrix} q_1 \\ q_2 \\ q_3 \end{pmatrix}$$

respectively and θ is the angle between $\boldsymbol{p}$ and $\boldsymbol{q}$.

(i) Write down two expressions for $\boldsymbol{p}.\boldsymbol{q}$.

(ii) If

$$\overrightarrow{OP} \text{ represents } \begin{pmatrix} 1 \\ -4 \\ -8 \end{pmatrix} \text{ and } \overrightarrow{OQ} \text{ represents } \begin{pmatrix} -4 \\ -2 \\ -4 \end{pmatrix}$$

find the size of the angle between $\overrightarrow{OP}$ and $\overrightarrow{OQ}$.

(iii) Find the value of the scalar product of the vectors represented by $\overrightarrow{PQ}$ and $\overrightarrow{OQ}$ and deduce the size of the angle OQP.

22. In figure 58, VOABC is a pyramid on a square base OABC of side 4 units. VM = 12 units is perpendicular to the base OABC. P, Q and R are the mid-points of the edges AB, VB and OV respectively.

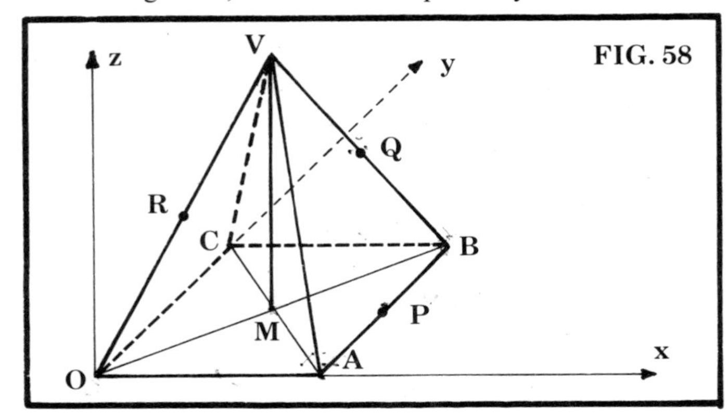

(a) With reference to three mutually perpendicular axes OX, OY and OZ, write down the coordinates of P, Q and R.

(b) Calculate the components of the vectors represented by $\overrightarrow{QR}$, $\overrightarrow{RP}$ and $\overrightarrow{PQ}$ and hence deduce that triangle PQR is right angled,

(i) by using scalar products,

(ii) by using the Converse of Pythagoras.

23. AB is a diameter of a circle centre O and C is any point on the circumference. A, B and C have position vectors $\boldsymbol{a}$, $\boldsymbol{b}$ and $\boldsymbol{c}$ respectively with reference to O.

Find the position vectors of the vectors represented by $\overrightarrow{CA}$ and $\overrightarrow{CB}$ and hence by using scalar products prove that angle ACB is a right angle.

24. P, Q and R are the points $(5, 1, 6)$, $(-1, -2, 0)$ and $(1, -4, 2)$ respectively. The point S divides QR externally in the ratio $3:1$.

(i) Find the coordinates of the point S.

(ii) Show that triangle PQR is obtuse-angled at R.

(iii) Show that PS is perpendicular to QR.

25. P, Q and R are the points $(6, -1, 1)$, $(3, 1, -1)$ and $(4, 0, 5)$ respectively.

(i) Prove that PQ and PR are perpendicular.

(ii) If S is the point $(6, 0, 2)$ show that PQ and PS are perpendicular.

(iii) Show that if θ is the acute angle between the planes SPQ and PQR then

$$\cos\theta = \frac{5}{\sqrt{41}}$$

ASSIGNMENT 4.5

Objective items testing Sections 4.1–4.7.
Instructions for answering these items are given on page 152.

1. PQRSTUVW is a parallelepiped (i.e. a solid of six faces each of which is a parallelogram).

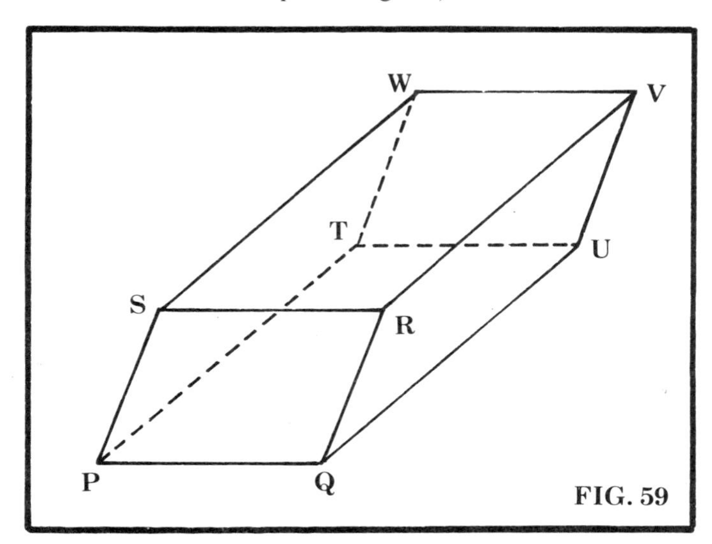

FIG. 59

$\vec{PQ}+\vec{QT}+\vec{TW}$ is equal to

A. $\vec{QW}$
B. $\vec{WP}$
C. $\vec{QT}$
D. $\vec{RW}$
E. none of these

2. A and B have position vectors $\boldsymbol{a}$ and $\boldsymbol{b}$ respectively. The point P divides AB externally in the ratio 5:2. The position vector of P is given by

A. $\dfrac{2\boldsymbol{a}-5\boldsymbol{b}}{3}$
B. $\dfrac{5\boldsymbol{b}+2\boldsymbol{a}}{3}$
C. $\dfrac{5\boldsymbol{a}-2\boldsymbol{b}}{3}$
D. $\dfrac{5\boldsymbol{a}+2\boldsymbol{b}}{3}$
E. $\dfrac{5\boldsymbol{b}-2\boldsymbol{a}}{3}$

3. $\vec{AB}$ represents $\begin{pmatrix}-1\\1\\1\end{pmatrix}$ and $\vec{BC}$ represents $\begin{pmatrix}-1\\0\\2\end{pmatrix}$. If A is the point $(-2,0,3)$, then C has coordinates

A. $(-2,0,2)$
B. $(-4,1,6)$
C. $(2,0,-4)$
D. $(4,1,6)$
E. $(2,0,-2)$

4. P has position vector $\begin{pmatrix}1\\0\\-2\end{pmatrix}$ and Q $\begin{pmatrix}-1\\-2\\1\end{pmatrix}$. The length of PQ equals

A. $\sqrt{5}$
B. $\sqrt{6}$
C. $\sqrt{11}$
D. $\sqrt{13}$
E. $\sqrt{17}$

5. P is the point $(10, -2, 4)$, $Q(4, -3, -1)$ and R divides PQ externally in the ratio $3:1$. The x-coordinate of R is

A. 1

B. $\frac{11}{2}$

C. $\frac{17}{2}$

D. 10

E. 13

6. Which one of the following vectors is not perpendicular to the vector $\begin{pmatrix} 1 \\ 2 \\ -3 \end{pmatrix}$?

A. $\begin{pmatrix} 3 \\ 3 \\ 3 \end{pmatrix}$ B. $\begin{pmatrix} 0 \\ 3 \\ 2 \end{pmatrix}$ C. $\begin{pmatrix} -3 \\ -3 \\ -3 \end{pmatrix}$

D. $\begin{pmatrix} 1 \\ 1 \\ -1 \end{pmatrix}$ E. $\begin{pmatrix} -2 \\ 4 \\ 2 \end{pmatrix}$

7. $\boldsymbol{a}$ and $\boldsymbol{b}$ are two vectors. Which one of the following products is a vector?

A. $(\boldsymbol{a}+\boldsymbol{b}).(\boldsymbol{a}-\boldsymbol{b})$

B. $(\boldsymbol{a}+\boldsymbol{b}).(\boldsymbol{a}+\boldsymbol{b})$

C. $\boldsymbol{a}.(\boldsymbol{a}+\boldsymbol{b})$

D. $(\boldsymbol{a}.\boldsymbol{b})\boldsymbol{a}$

E. $(\boldsymbol{a}-\boldsymbol{b}).\boldsymbol{b}$

8. KLMN is a parallelogram such that the vectors represented by $\overrightarrow{LM}$ and $\overrightarrow{NM}$ have components

$$\begin{pmatrix} 5 \\ 1 \\ 2 \end{pmatrix} \text{ and } \begin{pmatrix} 5 \\ -3 \\ -2 \end{pmatrix}$$

respectively. The vector represented by $\overrightarrow{KM}$ has components:

A. $\begin{pmatrix} 10 \\ -2 \\ 0 \end{pmatrix}$ B. $\begin{pmatrix} 0 \\ 4 \\ 4 \end{pmatrix}$ C. $\begin{pmatrix} 0 \\ -4 \\ -4 \end{pmatrix}$

D. $\begin{pmatrix} 5 \\ -2 \\ 0 \end{pmatrix}$ E. $\begin{pmatrix} -10 \\ 2 \\ 0 \end{pmatrix}$

9. A is the point $(6, 0, 3)$ and B is $(2, 5, -4)$. The position vectors of A and B are representatives of vectors $\boldsymbol{a}$ and $\boldsymbol{b}$.

(1) $\boldsymbol{a}$ and $\boldsymbol{b}$ are perpendicular to each other.

(2) $\boldsymbol{a}+\boldsymbol{b}$ and $\boldsymbol{a}-\boldsymbol{b}$ are perpendicular to each other.

(3) $|\boldsymbol{a}+\boldsymbol{b}| = |\boldsymbol{a}-\boldsymbol{b}|$.

10. $\boldsymbol{p}$ is a vector such that $|\boldsymbol{p}| = \sqrt{(a^2+b^2+c^2)}$.

(1) $\boldsymbol{p} = \begin{pmatrix} a \\ b \\ c \end{pmatrix}$

(2) If $\overrightarrow{OP}$ represents $\boldsymbol{p}$, where O is the origin, then P is the point (a, b, c).

(3) If $\boldsymbol{p}$ is a unit vector then $a^2+b^2+c^2 = 1$.

11. (1) $\boldsymbol{u}$, $\boldsymbol{v}$ and $\boldsymbol{w}$ are vectors such that $\boldsymbol{u}.\boldsymbol{v} = \boldsymbol{u}.\boldsymbol{w}$

(2) $\boldsymbol{v} = \boldsymbol{w}$

12. (1) Vectors

$$\boldsymbol{a} = \begin{pmatrix} a_1 \\ a_2 \\ a_3 \end{pmatrix} \text{ and } \boldsymbol{b} = \begin{pmatrix} b_1 \\ b_2 \\ b_3 \end{pmatrix}$$

are perpendicular.

(2) $a_1b_1+a_2b_2+a_3b_3 = 0$.

ASSIGNMENT 4.6

SUPPLEMENTARY EXAMPLES

1. Two vectors $\boldsymbol{a}$ and $\boldsymbol{b}$ have components

$$\begin{pmatrix} -1 \\ 2 \\ -4 \end{pmatrix} \text{ and } \begin{pmatrix} 4 \\ 6 \\ 2 \end{pmatrix}$$

respectively.

(i) Show that $\boldsymbol{a}$ is perpendicular to $\boldsymbol{b}$.

(ii) If a third vector $\boldsymbol{c}$ has components

$$\begin{pmatrix} p \\ q \\ r \end{pmatrix}$$

and is perpendicular to both the vectors $\boldsymbol{a}$ and $\boldsymbol{b}$, find p and q in terms of r.

2. In triangle PQR, $\overrightarrow{QR}$ and $\overrightarrow{RP}$ represent the vectors $\boldsymbol{p}$ and $\boldsymbol{q}$ respectively. Find in terms of $\boldsymbol{p}$ and $\boldsymbol{q}$ the vector represented by $\overrightarrow{PQ}$. Find also in terms of $\boldsymbol{p}$ and $\boldsymbol{q}$ the vectors represented by $\overrightarrow{KL}$, $\overrightarrow{LM}$ and $\overrightarrow{MK}$, where K, L and M are the mid-points of QR, RP and PQ respectively.

3. $\boldsymbol{p}$, $\boldsymbol{q}$ and $\boldsymbol{r}$ are three vectors such that $\boldsymbol{p}.\boldsymbol{r} = 1$ and $\boldsymbol{q}.\boldsymbol{r} = -1$ with $\boldsymbol{r} \neq \boldsymbol{0}$. Show that either $\boldsymbol{p} = -\boldsymbol{q}$ or $\boldsymbol{r}$ is perpendicular to $\boldsymbol{p}+\boldsymbol{q}$.

4. Prove that

$$(\boldsymbol{a}+\boldsymbol{b}).(\boldsymbol{a}+\boldsymbol{b}) = \boldsymbol{a}.\boldsymbol{a}+2\boldsymbol{a}.\boldsymbol{b}+\boldsymbol{b}.\boldsymbol{b}$$
$$= |\boldsymbol{a}|^2+2\boldsymbol{a}.\boldsymbol{b}+|\boldsymbol{b}|^2$$

If the vectors $\boldsymbol{a}$ and $\boldsymbol{b}$ have components

$$\begin{pmatrix} p \\ q \\ r \end{pmatrix} \text{ and } \begin{pmatrix} s \\ t \\ u \end{pmatrix}$$

respectively, express the vectors $\boldsymbol{a}$ and $\boldsymbol{b}$ in terms of $\boldsymbol{i}$, $\boldsymbol{j}$ and $\boldsymbol{k}$, where $\boldsymbol{i}, \boldsymbol{j}$ and $\boldsymbol{k}$ are three mutually perpendicular unit vectors and hence or otherwise prove that

(i) $|\boldsymbol{a}| = (p^2+q^2+r^2)^{\frac{1}{2}}$

(ii) $\boldsymbol{a}.\boldsymbol{b} = ps+qt+ru$

5. If vector $\boldsymbol{a} = \alpha\boldsymbol{i}+\beta\boldsymbol{j}+\gamma\boldsymbol{k}$ where $\boldsymbol{i}$, $\boldsymbol{j}$ and $\boldsymbol{k}$ are three mutually perpendicular vectors, find in terms of α, β and γ the value of $\boldsymbol{a}.\boldsymbol{a}$.

6. The vectors $\boldsymbol{a}$ and $\boldsymbol{b}$ have components

$$\begin{pmatrix} 4 \\ -6 \\ 3 \end{pmatrix} \text{ and } \begin{pmatrix} -3 \\ -4 \\ 2 \end{pmatrix}$$

respectively. A third vector $\boldsymbol{c}$ with components

$$\begin{pmatrix} p \\ q \\ 1 \end{pmatrix}$$

is perpendicular to both $\boldsymbol{a}$ and $\boldsymbol{b}$. Calculate the values of p and q.

7. P, Q and R are the points $(-2, -2, 1)$, $(4, -5, 7)$ and $(2, -4, 5)$ respectively.

(i) Show that P, Q and R are collinear and find the ratio in which R divides $\overrightarrow{PQ}$.

(ii) If S is the point $(k, 1, 4)$ and PS is perpendicular to QS, calculate the value of k.

(iii) Show that

$$\cos \text{SPR} = \frac{1}{\sqrt{3}}$$

8. $\boldsymbol{a}$ and $\boldsymbol{b}$ are two vectors such that $|\boldsymbol{a}| = |\boldsymbol{b}|$, show that $(\boldsymbol{a}+\boldsymbol{b}).(\boldsymbol{a}-\boldsymbol{b}) = 0$. If $\overrightarrow{OA}$ represents $\boldsymbol{a}$ and $\overrightarrow{OB}$ represents $\boldsymbol{b}$ interpret this result geometrically.

9. P, Q, R are the mid-points of sides BC, CA, AB of triangle ABC respectively. If A, B and C have position vectors $\boldsymbol{a}$, $\boldsymbol{b}$ and $\boldsymbol{c}$ respectively, show that,
   (i) $\boldsymbol{p} = \frac{1}{2}(\boldsymbol{b}+\boldsymbol{c})$, where $\boldsymbol{p}$ is the position vector of P.
   (ii) $\overrightarrow{AP}+\overrightarrow{BQ}+\overrightarrow{CR} = 0$.
   (iii) the point of intersection of AP, BQ and CR has position vector $\frac{1}{3}(\boldsymbol{a}+\boldsymbol{b}+\boldsymbol{c})$.

10. P and Q are the mid-points of the sides AB and BC of the parallelogram ABCD. Show (by using position vectors) that PQ is parallel to AC. Show also that DP divides AC in the ratio 1:2.

11. Show that $\boldsymbol{a}.\boldsymbol{a} = |\boldsymbol{a}|^2$ and hence or otherwise that $|\boldsymbol{a}+\boldsymbol{b}|^2 = |\boldsymbol{a}|^2+|\boldsymbol{b}|^2+2\boldsymbol{a}.\boldsymbol{b}$. If $\boldsymbol{a}$ and $\boldsymbol{b}$ are two vectors such that $|\boldsymbol{a}+\boldsymbol{b}| = |\boldsymbol{a}-\boldsymbol{b}|$, prove that $\boldsymbol{a}.\boldsymbol{b} = 0$.

12. Figure 60 shows a tetrahedron PQRS in which PQ is perpendicular to RS and PS is perpendicular to QR. Prove that PR is perpendicular to QS.

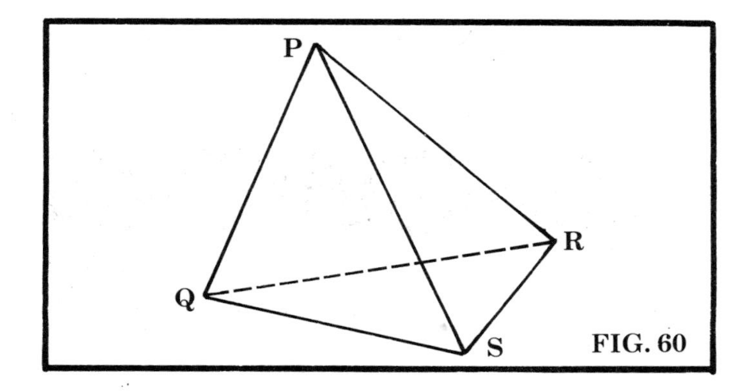

FIG. 60

[*Hint:* Take P as origin and let Q, R and S have position vectors $\boldsymbol{q}$, $\boldsymbol{r}$ and $\boldsymbol{s}$ respectively relative to P.]

13. PQRS is a rectangle. $\overrightarrow{PQ}$ represents the vector $\boldsymbol{a}$ and $\overrightarrow{PS}$ represents the vector $\boldsymbol{b}$. Write in terms of $\boldsymbol{a}$ and $\boldsymbol{b}$ the vectors represented by $\overrightarrow{PR}$ and $\overrightarrow{QS}$ and hence or otherwise show that the diagonals of a rectangle are equal.
    If $|\boldsymbol{a}| > |\boldsymbol{b}|$ show that angle POQ must be obtuse, where O is the point of intersection of the diagonals.

14. If A is the point $(2, 2, 5)$, $B(-1, -2, 6)$ and $C(-4, 0, 5)$ show that triangle ABC is right-angled at B. If P is the point $(\alpha, \beta, 5)$ show that
    (i) $3\alpha+\beta = -5$ if BP is perpendicular to AC.
    (ii) $\alpha-3\beta = -4$ if P lies on AC.
    Hence find the coordinates of P, if P lies on AC and BP is perpendicular to AC.

15. ABCD is a parallelogram. If $\overrightarrow{AB}$ represents the vector $\boldsymbol{p}$ and $\overrightarrow{AD}$ the vector $\boldsymbol{q}$,
    (i) write in terms of $\boldsymbol{p}$ and $\boldsymbol{q}$ the vectors represented by $\overrightarrow{AC}$ and $\overrightarrow{BD}$.
    (ii) show that $|\overrightarrow{AC}|^2 = |\boldsymbol{p}|^2+|\boldsymbol{q}|^2+2\boldsymbol{p}.\boldsymbol{q}$. Find a similar expression for $|\overrightarrow{BD}|^2$ and hence show that for any parallelogram ABCD,
    $$AC^2+BD^2 = AB^2+BC^2+CD^2+DA^2$$

16. P, Q, R and S are the points $(1, 4, 2)$, $(0, -1, 0)$, $(2, 0, -1)$ and $(-1, -1, 3)$ respectively. Show that PQ is perpendicular to RS. If T is the point $(2, 9, 4)$, show that P, Q and T are collinear and that P is the mid-point of QT.

ANSWERS

ALGEBRA

Page 10, Assignment 1.1

1. (i) $2, 5, 8, 11; 29$. (ii) $4, 16, 36, 64; 400$.
   (iii) $6, 1, -4, -9; -39$. (iv) $2, 4, 8, 16; 1024$.
   (v) $0, 2, 6, 12; 90$. (vi) $\frac{1}{2}, \frac{2}{3}, \frac{3}{4}, \frac{4}{5}; \frac{10}{11}$.
   (vii) $1, -1, 1, -1; -1$. (viii) $-1, 1, -1, 1; 1$.
2. 4 terms.
3. $0, x, 2x^2$.
4. $9, 108, 3(2k-1), 3(4k-1), 3(2n-3)$.
5. 4; 5th.
6. (i) n. (ii) $2n$. (iii) $5n$. (iv) n^2. (v) 2^n. (vi) $(\frac{1}{3})^n$.
   (vii) $3n+1$. (viii) $15-4n$. (ix) $(-1)^n$. (x) $(-\frac{1}{2})^{n-1}$.
7. $-6, -10, -12$; 9th.
8. 7th.
9. $-\frac{9}{4}$.
10. (i) 3, (ii) 15th.
11. (i) $u_n = \dfrac{1}{2^{n-1}}, u_{n-1} = \dfrac{1}{2^{n-2}}, u_{n+1} = \dfrac{1}{2^n}$.
    (ii) $\dfrac{u_{n+1}}{u_n} = \dfrac{1}{2}, \dfrac{u_{n-1}}{u_n} = 2$.
12. $2, 3; u_n = 3n-1; 3$.
13. $4, 4{\cdot}4, 4{\cdot}44, 4{\cdot}444, 4{\cdot}4444$.
14. (i) $1, -\frac{1}{4}, \frac{1}{16}, -\frac{1}{64}$. (ii) $1, \dfrac{x}{2}, \dfrac{x^2}{3}, \dfrac{x^3}{4}$.
15. $0, 2, 0, 2, 0, 2, 0, \ldots$
16. $1, -1, -5, -13, -29, -61$.

Page 13, Assignment 1.2

1. (i) 2. (ii) -2. (iii) 5. (iv) k.
2. (i) $3; 1$. (ii) $3; 4$. (iii) $a+3; 3$. (iv) $1; -2$.
   (v) $b+2; b$.
3. (i) 900. (ii) 165. (iii) -63. (iv) 117. (v) -99.
   (vi) $\dfrac{n}{2}(n+1)$. (vii) $k(k+1)$. (viii) $4n^2$.
4. 2500.
5. 1050.
6. 55,944
7. 5.
8. 15 terms; 59.
9. $24, 22, 20; 4$ or 21.
10. $-5; 7; -5, 2, 9, 16, 23$.
11. 8; 30p, 42p, 54p, 66p, 78p, 90p, 102p, 114p.
12. -5.
13. $-5, -4, 3, 16; -5, 1, 7, 13; 6n-11$
14. 265; 765.
16. $1, 4, 7, 10, \ldots$ or $7, 4, 1, -2, \ldots$
17. $5, 11, 17, 23; u_n = 6n-1$.
19. yes, $p-5, k$.
20. No (d and kd).
21. 15.
22. -125.

Page 16, Assignment 1.3

1. (i) $3^n, 3^{n-2}$. (ii) $5.2^n, 5.2^{n-2}$. (iii) $\dfrac{1}{4^n}, \dfrac{1}{4^{n-2}}$.
   (iv) $\dfrac{3^{n+1}}{2^n}, \dfrac{3^{n-1}}{2^{n-2}}$.
2. (i) $1, \frac{1}{3}, \frac{1}{9}, \frac{1}{27}; \frac{1}{3}$. (ii) $-6, 18, -54, 162; -3$
   (iii) $8, -2, \frac{1}{2}, -\frac{1}{8}; -\frac{1}{4}$. (iv) $\frac{3}{4}, \frac{1}{2}, \frac{1}{3}, \frac{2}{9}; \frac{2}{3}$.
3. (i) 2^{n-1}. (ii) $(-3)^{n-1}$. (iii) $\dfrac{1}{2^{n-1}}$. (iv) 5^{3-n}.
   (v) $(-b)^{3-n}$. (vi) ar^{n-1}.
4. 8th.
5. 5th.
6. (i) $2 - \dfrac{1}{2^7}$. (ii) $\frac{1}{4}(1-3^8)$. (iii) $3^8 - 3$. (iv) $\frac{2}{3}(1-2^{10})$
   (v) $\dfrac{3}{4}\left(1 - \dfrac{1}{3^8}\right)$. (vi) $3^3 - \dfrac{1}{3^7}$. (vii) $\dfrac{a^n - 1}{a-1}$.
   (viii) $\dfrac{3\left(1 - \left(\dfrac{x}{3}\right)^{12}\right)}{3+x}$
7. 6.
8. 7.
10. 6.
11. 4.

12. 8, −12, 18, −27, ... 13. 5th.

14. 3, 12, 48.

15. 12, 6, 3, $\frac{3}{2}$...; 12, −18, 27, $-40\frac{1}{2}$, ...
8, 4, 2, 1...; 8, −12, 18, −27, ...

Page 18, Assignment 1.4

1. (i) 1. (ii) 2. (iii) 0. (iv) 1. (v) $\frac{1}{2}$. (vi) 1.
2. $\frac{1}{2}, \frac{2}{3}, \frac{3}{4}, \frac{4}{5}, \frac{5}{6}, \frac{6}{7}$; 1 ; 99th.
3. 3.
4. (i) $\frac{4}{3}$. (ii) $\frac{1}{2}$. (iii) no limiting sum. (iv) 9.
(v) no limiting sum. (vi) $\frac{1}{3}$. (vii) no limiting sum.
(viii) $\frac{100}{9}$. (ix) $\frac{16}{3}$. (x) no limiting sum.
5. $2, \frac{4}{3}, \frac{8}{9}, \frac{16}{27}, \ldots$ 6. $\frac{256}{2187}$.
7. $a > 1$ or $a < -1$; $\frac{5}{4}$. 8. $-1 < a < 1$; $\pm\frac{1}{2}$.
9. $0{\cdot}\dot{4} = \frac{4}{10} + \frac{4}{100} + \frac{4}{1000} + \ldots; \frac{4}{9}$.
10. $0{\cdot}\dot{6}\dot{3} = \frac{63}{10^2} + \frac{63}{10^4} + \frac{63}{10^6} + \ldots; \frac{7}{11}$
11. 15 m. 12. 23 m.
13. 48 cm. 14. $3, 2, \frac{4}{3}, \frac{8}{9}, \ldots$; 9.
15. $2, \frac{5}{4}, 1, \frac{7}{8}; \frac{1}{2}; \frac{3}{400}$ 16. 6, −6, 6, ...; no sum
36, 24, 16, ...; 108
17. 1.

Page 21, Assignment 1.5

1. (i) 1·05. (ii) 0·8. (iii) 0·88. (iv) 1·03.
2. £137. 3. 0·85 ; £1275 ; £780.
4. 3·57 kg ; 0·63 kg. 5. 4 years.
6. 1·08 ; £3080. 7. 1·35 m, 1·57 m, 1·82 m.
8. (i) 0·98. (ii) £96,000. (iii) £66,000.
9. (i) 1·05, 0·965. (ii) 237,000

Page 22, Assignment 1.6

1. B. 2. D. 3. E. 4. D.
5. A. 6. D. 7. B. 8. C.
9. D. 10. E. 11. A. 12. D.

Page 27, Assignment 2.1

1. (i) $\begin{pmatrix} -6 & 3 \\ 12 & -9 \end{pmatrix}_{2\times 2}$ (ii) $\begin{pmatrix} 3 & 2 & 1 \\ 7 & 6 & 5 \end{pmatrix}_{2\times 3}$
2. (i) $\begin{pmatrix} 1 & 5 \\ 8 & 5 \end{pmatrix}$. (ii) not possible.
(iii) $\begin{pmatrix} -3 & 1 \\ 2 & -1 \end{pmatrix}$. (iv) $\begin{pmatrix} a-p & b-q \\ c-r & d-s \end{pmatrix}$
3. (i) $\begin{pmatrix} 1 & 3 \\ -2 & 6 \end{pmatrix}$. (ii) $\begin{pmatrix} 3 & 3 \\ 3 & 3 \end{pmatrix}$; The commutative and associative laws for addition.
4. (i) $\begin{pmatrix} 1 & 5 \\ 3 & 1 \end{pmatrix}$. (ii) $\begin{pmatrix} 6 & 1 \\ -2 & -4 \end{pmatrix}$. (iii) $\begin{pmatrix} 1 & 0 \\ -2 & 3 \end{pmatrix}$.
5. $\begin{pmatrix} -5 & 4 \\ 2 & -1 \end{pmatrix}$. 6. $\begin{pmatrix} -3 & -1 \\ 1 & 4 \end{pmatrix}$.
7. A and D. 8. $x = -2, y = -1, a = 3$.
9. $A' = \begin{pmatrix} 2 & 0 \\ -4 & 1 \end{pmatrix}_{2\times 2}$ $B' = \begin{pmatrix} 7 & -2 & 0 \\ 1 & 3 & 2 \end{pmatrix}_{2\times 3}$
$C' = \begin{pmatrix} 2 & 1 \\ 0 & -1 \\ 1 & 0 \end{pmatrix}_{3\times 2}$
10. (i) $x = 2, y = -2, t = 1$. (ii) $x = 0, y = 1, t = 0$.
11. (a) The amount of pocket money each received in four weeks.

(b) Mary gets 60p, John 70p, Paul 72p, $\begin{pmatrix} M \\ J \\ P \end{pmatrix} = \begin{pmatrix} 60 \\ 70 \\ 72 \end{pmatrix}$.

(c) $\begin{pmatrix} M \\ J \\ P \end{pmatrix} = \begin{pmatrix} 45 \\ 55 \\ 57 \end{pmatrix}$

12. $y = z$; (i), (ii). 13. 0 *or* 1.

14. $b = d, c = g, f = h; b = -1, a = 2, c = -4$.

16. $p = -3, q = 1, r = -4, s = -7$.

17. $x = 5, y = -7, z = 6, t = -4$.

Page 30, Assignment 2.2

1. (i) $\begin{pmatrix} 3 & -9 \\ 6 & 0 \end{pmatrix}, \begin{pmatrix} -12 & -8 \\ 16 & 4 \end{pmatrix}$. (ii) $\begin{pmatrix} 9 & -5 \\ -2 & -2 \end{pmatrix}$.

   (iii) $\begin{pmatrix} -4 & -10 \\ 12 & 2 \end{pmatrix}$.

3. $p = 2, q = -1$. 4. $\begin{pmatrix} 3 & \frac{3}{2} \\ -\frac{3}{2} & -1 \end{pmatrix}$.

5. (i) $(31)_{1\times 1}$. (ii) $\begin{pmatrix} 1 & -3 \\ 20 & 24 \end{pmatrix}_{2\times 2}$. (iii) $\begin{pmatrix} 9 & 12 & 15 \\ 19 & 26 & 33 \end{pmatrix}_{2\times 3}$.

(iv) not possible. (v) $\begin{pmatrix} -1 \\ 7 \end{pmatrix}_{2\times 1}$. (vi) not possible.

   (vii) $\begin{pmatrix} -1 \\ -1 \end{pmatrix}_{2\times 1}$. (viii) not possible. (ix) $\begin{pmatrix} 6 \\ -2 \\ 15 \end{pmatrix}_{3\times 1}$.

   (x) not possible.

6. (i) $\begin{pmatrix} 6 & -6 \\ 14 & -12 \end{pmatrix}, \begin{pmatrix} 2 & -2 \\ 7 & -8 \end{pmatrix}$.

   (ii) $\begin{pmatrix} -36 & 6 \\ -74 & 16 \end{pmatrix}, \begin{pmatrix} -36 & 6 \\ -74 & 16 \end{pmatrix}$.

   The associative law holds
   The commutative law does not hold.

7. (i) $\begin{pmatrix} 5 & -\frac{5}{2} \\ 8 & \frac{13}{2} \end{pmatrix}$. (ii) $\begin{pmatrix} a^2-b^2 & -a \\ a & -1 \end{pmatrix}$. (iii) $\begin{pmatrix} 4 & -1 \\ 8 & 0 \\ -4 & -1 \end{pmatrix}$.

   (iv) $\begin{pmatrix} ak & b \\ ck & d \end{pmatrix}$.

8. (i) $\begin{pmatrix} -6 & 5 \\ -5 & 25 \end{pmatrix}$. (ii) $\begin{pmatrix} -10 & 3 \\ -15 & 5 \end{pmatrix}$. (iii) $\begin{pmatrix} 23 & 1 \\ 33 & -4 \end{pmatrix}$.

   The distributive law holds.

9. (i) $\begin{pmatrix} -4 & -2 \\ 8 & 0 \end{pmatrix}, \begin{pmatrix} -8 & 0 \\ 0 & -8 \end{pmatrix}$.

   (ii) $\begin{pmatrix} -3 & -10 \\ 10 & 12 \end{pmatrix}, \begin{pmatrix} 3 & -9 \\ 2 & 6 \end{pmatrix}$;

   $(A+B)^2 = A^2 + AB + BA + B^2$.

   (iii) $(A-B)^2 = A^2 - AB - BA + B^2$. (iv) No.

12. $k = 2, m = 2; BA = \begin{pmatrix} 4 & 0 \\ -4 & 0 \end{pmatrix}$.

13. $\begin{pmatrix} 17 & 1 \\ 5 & 3 \end{pmatrix}, \begin{pmatrix} 17 & 5 \\ 1 & 3 \end{pmatrix}; \begin{pmatrix} 7 & -2 \\ 1 & 3 \end{pmatrix}, \begin{pmatrix} 2 & 3 \\ 0 & 1 \end{pmatrix}, \begin{pmatrix} 17 & 5 \\ 1 & 3 \end{pmatrix}$;

   $(AB)' = B'A'; (ABC)' = C'B'A'$.

14. $x = -2, y = -4$.

17. $x' = x, y' = 3x + y$; A′(2, 6), B′(2, 7), C′(4, 15);

   $x = x', y = y' - 3x', \begin{pmatrix} 1 & 0 \\ -3 & 1 \end{pmatrix}$.

18. Parallelogram.

19. $\begin{pmatrix} 1 & 0 \\ 0 & -1 \end{pmatrix}\begin{pmatrix} x' \\ y' \end{pmatrix} = \begin{pmatrix} x'' \\ y'' \end{pmatrix}; \begin{pmatrix} \cos\theta & -\sin\theta \\ -\sin\theta & -\cos\theta \end{pmatrix}$; No.

20. $A = \begin{pmatrix} 20 & 3 & 4 & 0 \\ 20 & 8 & 4 & 0 \\ 15 & 5 & 3 & 2 \end{pmatrix}$, $B = \begin{pmatrix} 8 \\ 15 \\ 6 \\ 20 \end{pmatrix}$; $\begin{pmatrix} 229 \\ 304 \\ 253 \end{pmatrix}$.

The cost of manufacturing one batch of one kind of biscuit; Total cost £7·86.

22. (i) $\begin{pmatrix} 2 & 0 \\ -4 & -3 \end{pmatrix}, \begin{pmatrix} 0 & 3 \\ 2 & -1 \end{pmatrix}, \begin{pmatrix} -5 & 1 \\ -4 & -2 \end{pmatrix}$.

(ii) $\begin{pmatrix} 0 & \frac{3}{2} \\ -\frac{2}{3} & -\frac{5}{3} \end{pmatrix}$.

23. (i), (ii).

26. $\begin{pmatrix} 3 & -2 \\ 2 & -1 \end{pmatrix}, \begin{pmatrix} 4 & -3 \\ 3 & -2 \end{pmatrix}, \begin{pmatrix} 5 & -4 \\ 4 & -3 \end{pmatrix}$.

Page 35, Assignment 2.3

1. (i) $\begin{pmatrix} \frac{3}{2} & -2 \\ -2 & 3 \end{pmatrix}$. (ii) $\begin{pmatrix} 1 & -4 \\ -\frac{1}{2} & \frac{5}{2} \end{pmatrix}$

(iii) No inverse. (iv) $\begin{pmatrix} -\frac{1}{6} & \frac{5}{6} \\ \frac{1}{3} & -\frac{2}{3} \end{pmatrix}$

(v) No inverse. (vi) $\begin{pmatrix} \frac{1}{2} & \frac{1}{2} \\ 0 & 1 \end{pmatrix}$. (vii) $\begin{pmatrix} -2 & -1 \\ 3 & 2 \end{pmatrix}$.

(viii) No inverse. (ix) $\begin{pmatrix} -\frac{1}{p} & \frac{2}{p} \\ \frac{1}{q} & \frac{1}{q} \end{pmatrix}$.

(x) $\frac{1}{a^2+b^2}\begin{pmatrix} a & -b \\ b & a \end{pmatrix}$.

2. (i) $\begin{pmatrix} \frac{1}{3} & -\frac{2}{3} \\ 0 & 1 \end{pmatrix}$. (ii) $\begin{pmatrix} -3 & 1 \\ -5 & 2 \end{pmatrix}$. (iii) $\begin{pmatrix} \frac{7}{3} & -1 \\ -5 & 2 \end{pmatrix}$.

(iv) $\begin{pmatrix} -1 & 3 \\ -\frac{5}{3} & \frac{16}{3} \end{pmatrix}$. (v) $\begin{pmatrix} -1 & 3 \\ -\frac{5}{3} & \frac{16}{3} \end{pmatrix}$.

3. (i) yes. (ii) $\begin{pmatrix} 4 & 5 \\ 2 & 3 \end{pmatrix}$.

(iii) $(A^{-1})' = \begin{pmatrix} \frac{3}{2} & -1 \\ -\frac{5}{2} & 2 \end{pmatrix}$, $(A')^{-1} = \begin{pmatrix} \frac{3}{2} & -1 \\ -\frac{5}{2} & 2 \end{pmatrix}$.

4. $\begin{pmatrix} 1 & -2 \\ -2 & 3 \end{pmatrix}$; $\begin{pmatrix} -3 & -2 \\ -2 & -1 \end{pmatrix}$.

5. $a = \pm 1$. 6. Rotate it through an angle $-\theta$.

7.

x	P	Q	R	S
P	Q	P	S	R
Q	P	Q	R	S
R	S	R	Q	P
S	R	S	P	Q

(i) Q. (iii) Yes.

8. $P^{-1} = \frac{1}{6}\begin{pmatrix} 4 & -1 \\ -2 & 2 \end{pmatrix}$; $X = \begin{pmatrix} \frac{1}{3} & \frac{4}{3} \\ \frac{1}{3} & -\frac{2}{3} \end{pmatrix}$.

9. (i), (ii), (iii). 10. (i), (ii), (iii).

11. $\begin{pmatrix} c & d \\ a & b \end{pmatrix}, \begin{pmatrix} b & a \\ d & c \end{pmatrix}, \begin{pmatrix} d & -b \\ -c & a \end{pmatrix}$.

Page 37, Assignment 2.4

1. (i) $X = \begin{pmatrix} 4 & 3 \\ -2 & -2 \end{pmatrix}$. (ii) $X = \begin{pmatrix} -3 & \frac{19}{2} \\ -2 & \frac{9}{2} \end{pmatrix}$.

(iii) $X = \begin{pmatrix} 3 & -1 \\ 1 & -1 \end{pmatrix}$.

2. $X = A^{-1}B$. 3. (i), (ii), (iii).

4. (i) $X = \begin{pmatrix} \frac{7}{2} & -\frac{1}{2} \\ 4 & -2 \end{pmatrix}$. (ii) $X = \begin{pmatrix} \frac{20}{11} & \frac{5}{11} \\ \frac{10}{11} & -\frac{3}{11} \end{pmatrix}$.

5. $2x+y = 2$, $3x+4y = -1$.

6. $\begin{pmatrix} 3 & -2 \\ -4 & 3 \end{pmatrix}$, $x = -2$, $y = -4$.

7. $a = -5$, $b = -3$.

8. (i) $\{(1,2)\}$, (ii) $\{(2,3)\}$, (iii) $\{(-1,1)\}$, (iv) $\{(4,11)\}$, (v) $\{(0,2)\}$, (vi) $\{(1,1)\}$, (vii) $\{\frac{4}{7}, -\frac{20}{7}\}$, (viii) $\left\{\left(\frac{dp-bq}{ad-bc}, \frac{aq-pc}{ad-bc}\right)\right\}$.

9. (i) $\begin{pmatrix} 1 & 1 & 1 \\ 2 & 1 & 1 \\ 1 & 1 & -1 \end{pmatrix}\begin{pmatrix} x \\ y \\ z \end{pmatrix} = \begin{pmatrix} 8 \\ 7 \\ 6 \end{pmatrix}$

(ii) $\begin{pmatrix} 1 & 3 & -2 \\ 4 & 1 & -1 \\ 8 & -2 & 1 \end{pmatrix}\begin{pmatrix} x \\ y \\ z \end{pmatrix} = \begin{pmatrix} 6 \\ 7 \\ 4 \end{pmatrix}$.

10. $\{(-1, -1, 1)\}$. 11. $\{(1, 1, 1)\}$.

Page 39, Assignment 2.5

1. D. 2. C. 3. E. 4. B. 5. A. 6. D.
7. E. 8. D. 9. C. 10. A. 11. A. 12. C.

Page 44, Assignment 3.1

1. (ii) $\{(2,1), (3,1), (3,2), (4,1), (4,2), (4,3), (5,1), (5,2), (5,3), (5,4)\}$.
(iv) $\{(x, y): x > y, x \in S, y \in S\}$.

2. $D = \{2, 4, 6, 8, 10\}$, $R = \{1, 2, 3, 4, 5\}$.
(i) $\{(2,1), (4,2), (6,3), (8,4), (10,5)\}$.
(ii) $\{(x, y): x = 2y, x \in S, y \in S\}$.

3. (i) True, (ii) True, (iii) True, (iv) False.

4. (i) 1–1 correspondence, (ii) mapping, (iii) not a mapping, (iv) 1–1 correspondence, (v) 1–1 correspondence.

5. (i) 1–1 correspondence, (ii) 1–1 correspondence, (iii) mapping, (iv) mapping, (v) mapping.

6. 1, 2, 10, 5.

7. (i) $(1,1), (-1,1), (0,0), (-2,2), (2,2)$.
(ii) $(-4,-9), (-3,-7), (-2,-5), (-1,-3), (0,1), (1,2), (2,3), (3,4), (4,5)$.

8. (i) $f: x \to 2x$, $f: x \to x+1$. (ii) $f: x \to 6$.

9. $\{y: y = 3\}$.

10. $\{y: -4 \leqq y \leqq 12, y \in R\}$; -4; 4; $\{y: -12 \leqq y \leqq 4, y \in R\}$.

12. (a) $\{x: x > 0, x \in R\}$; (b) $\{x: x \geqq 0, x \in R\}$; (c) $\{x: x \in R, x \neq 0\}$.

13. (i) 5, (ii) 6, (iii) $\{x: 4 \leqq x \leqq 5, x \in R\}$, (iv) $\{x: x \geqq 4, x \in Z\}$.

Page 46, Assignment 3.2

1. 5; 1.

2. (i) $1, -1$, (ii) $-2, -2$, (iii) 4, (iv) $D = \{-2, -1, 0, 2\}$, $R = \{-2, 0, 2\}$.

3. (i) $(2-x)^2, R$. (ii) $x^2 - 1, R$. (iii) $x^2 + 3x + 1, R$. (iv) $3x + 1, \{x: x \geqq 0, x \in R\}$. (v) $\sqrt{(x^2+1)}, R$.
In (i) $\{y: y \geqq 0, y \in R\}$. (ii) $\{y: y \geqq -1, y \in R\}$. (v) $\{y: y \geqq 1, y \in R\}$.

4. (i) $\frac{-(x^2+2x)}{(x+1)^2}$, $(k \circ h)(x)$ does not exist.
(ii) $-2x^3, -8x^3$. (iii) 9, 4.

6. $(-3, -1), (2, -4)$; mappings are the same.

7. (i) 49. (ii) 4; $(g \circ f)(x) = 2x^2 + 1$.

8. (i) 8. (ii) 48, 36; $(k \circ h)(x) = x^2 - 1$.

9. $2(y^3 + 1)$, -729.

11. $(g \circ f)(x) = x - 1$, $D = \{x: x \in R, x \neq 1\}$, $R = \{x: x \in R, x \neq 0\}$.

12. (a) (i) x, (ii) $-x$, (iii) $-x$, (iv) $\frac{1}{x}$,

(v) Does not exist.

(b) $-3, 3, 3$. (c) (i), (ii), (iii), (iv).

(d) $\{n: n = 2k, k \in N\}$; no.

13. $(x+1)^4, R$; $(x^2+1)^2, R$.

14. $g \circ f$ not possible since image set of f is not a subset of the domain of g.

15. $\left(\frac{1}{x}+1\right)^2$; $g \circ f$ not possible since image set of f is not a subset of the domain of g.

Page 49, Assignment 3.3

1. (i) $\frac{x}{2}$. (ii) $\frac{x-1}{3}$. (iii) $\frac{x-4}{2}$. (iv) $-y$.

(v) $-\frac{t}{2}$. (vi) $\frac{1-s}{2}$.

3. (i) p, (ii) does not exist since f^{-1} does not exist, (iii) t, (iv) t, (v) s, (vi) s, (vii) does not exist, (viii) does not exist.

4. (i) $x+1$, domain and range R.

(ii) $1-x$, domain and range R.

(iii) x, domain and range R. (iv) no inverse.

(v) $\frac{1}{x}$, $x \neq 0$, domain $= \{x: x \neq 0, x \in R\}$,

range $= \{x: x \neq 0, x \in R\}$.

(vi) $\frac{1-x}{3}$, domain and range R.

(vii) $2(x+3)$, domain and range R.

(viii) $\frac{x-1}{x}$, domain $= \{x: x \neq 0, x \in R\}$,

range $= \{x: x \neq 1, x \in R\}$.

(ix) $\sqrt[3]{(x+1)}$, domain and range R.

5. $g^{-1}(x) = \frac{x-3}{2}$, $h^{-1}(x) = \frac{x-1}{x}$. $x \neq 0$.

6. (i) $2(y^2+2)$, (ii) $\frac{y-2}{2}$; 22, 1.

7. $\frac{x-1}{2}$, $x+2$; $2x-1$.

8. $f^{-1}(x) = \sqrt[3]{\left(\frac{x-1}{x}\right)}$, $g^{-1}(x) = 1+x$;

$f^{-1}(x)$ does not exist for $x = 0$; -1.

9. (ii) $\frac{1}{2}(x-1)$, R, R

(iii) $\frac{x+1}{x-1}$, $\{x: x \in R, x \neq 1\}$, $\{x: x \in R, x \neq 1\}$.

11. $\frac{1+x}{1-x}$, $x \neq 1$; 1.

Page 50, Assignment 3.4

1. D.	2. A.	3. D.	4. D.	5. B.	6. C.
7. E.	8. B.	9. C.	10. E.	11. B.	12. D.

Page 54, Assignment 4.1

1. (i) 3, (ii) 3, (iii) 4, (iv) 2, (v) 3.

2. (i) -2, (ii) 4, (iii) -2, (iv) $0, -1$.

3. (i) 3, (ii) $p = -2, q = 3$, (iii) $a = 1, b = -2, c = 1$, (iv) $b = -2, c = 4$.

4. 1.
5. 0.
6. $1\frac{1}{4}$.
7. 18.
8. 233.
9. 69.
10. $8k^3 - 12k^2 + 16k - 4$.
11. -5.
12. $p = -4, q = -9$.
13. $12, \frac{53}{4}, 10$.
15. $-\frac{3}{2}$
16. $0, \frac{3}{10}$.

Page 56, Assignment 4.2

1. $8, -15$.
2. $u-1, 3$.
3. $2x, -3$.
4. $4x, 5$.
5. $3y-1, 4$.
6. $x^2-4x+11, -21$.
7. $2x^2+3x-\frac{1}{2}, -2\frac{1}{2}$.
8. $f(x)=(x-2)(x^2+6)+11$.
9. $f(x)=(x+3)(2x^2-7x+23)-66$.
10. $f(x)=(x-k)(x^2+2kx+k^2)+2k^3$.
11. $g(t)=(t+2)(t^2-2t+1)-4$.
12. $p=-4, q=-3$.
13. $a=9\frac{1}{2}$.
16. $7, -38, 3\frac{5}{8}$.
17. $u^3-3u^2+4u-6, 3$; $2y^3+3y^2+4y+9, 6$.
18. -3.

Page 58, Assignment 4.3

3. $(x+1), (x+2), (x-4)$.
6. $u+5$.
7. $(y-2), (y+4), (2y+1)$; $f(-2)=0$.
8. x^3-2x^2-5x+6.
9. $(x-1)$.
10. $p=11$.
11. $p=-\frac{1}{4}, q=-4$.
12. $(x-3), (x-5)$.
13. $(x-5), (x-3), (x-1)$.
14. $(y-1), (y+1), (y-3)$.
15. $(z+1), (z+3), (z-2)$.
16. $(t-2), (2t-1), (2t+1)$.
17. $(x+2), (x-1)$.
18. $(x-1)(x+1)(x-2)(x+2)$.
19. $(x+1)(x+2)(x+3)$; $\{-1, -2, -3\}$.
20. $\{-3, -\frac{1}{2}, 2\}$.
21. $\{-2, -1, 3\}$.
22. $\{-3, -1, 2\}$.
23. $\{-1, \frac{1}{3}, 2\}$.
24. $k=-6$; $\{\frac{1}{2}, 2, 3\}$.
25. $\{a, 2a, 3a\}$.
26. 9.
27. $8, 2; 2, -22$.

Page 59, Assignment 4.4

1. C. 2. D. 3. C. 4. C. 5. D. 6. C.
7. E. 8. E. 9. D. 10. D. 11. E. 12. C.

Page 63, Assignment 4.5

1. $f(-2)=0$; $f(0)=2$, $f(1)=-3$; $f(2)=-4$, $f(3)=5$.
2. $f(-1)=0$.
3. -4.
4. $-0{\cdot}84$.
5. $2\frac{1}{8}$.
6. $0{\cdot}62$.
7. $0{\cdot}56$.
8. $1{\cdot}41$.
9. $1{\cdot}34$.
10. $-1{\cdot}25, 1{\cdot}96$.
11. $1{\cdot}89, -1{\cdot}89$.
12. $0{\cdot}125$.

Page 63, Assignment 4.6

1. $2, 8, 20; 2, 6, 12; n(n+1)$.
2. $-1, -5, -13$.
3. $\frac{1}{9}, \frac{1}{3}, 6$.
4. $1-(-1)^n, 9(1-3^{-n}); 9$.
5. $-\frac{1}{2}$.
6. (i) $\begin{pmatrix}-2 & 1\\ -5 & 3\end{pmatrix}, \begin{pmatrix}-3 & -5\\ 1 & 2\end{pmatrix}, \begin{pmatrix}-6 & -15\\ 3 & 9\end{pmatrix}$

   (ii) $\begin{pmatrix}2 & 0\\ -1 & -2\end{pmatrix}, \begin{pmatrix}\frac{1}{2} & -\frac{1}{4}\\ 0 & -\frac{1}{2}\end{pmatrix}, \begin{pmatrix}-2 & 1\\ 0 & 2\end{pmatrix}$

   (iii) $\begin{pmatrix}0 & -6\\ 1 & 1\end{pmatrix}, \begin{pmatrix}-4 & -4\\ 1 & 5\end{pmatrix}$
8. $\begin{pmatrix}\frac{1}{\sqrt{2}} & -\frac{1}{\sqrt{2}}\end{pmatrix}$; rotation of 90° about the origin; 8.
9. $\begin{pmatrix}4 & -2\\ 3 & -1\end{pmatrix}\begin{pmatrix}a\\ b\end{pmatrix}=\begin{pmatrix}-2\\ 4\end{pmatrix}$; $(5, 11)$
10. (i) $f^{-1}=\frac{1}{2}(x-3)$, g^{-1} does not exist, $2x^2+3$, $(2x+3)^2$.
    (ii) 0, 3, 1.
11. $2-2k$; or -1.
12. (i) $f \circ g$ does exist and $(f \circ g)(x)=\dfrac{2-x}{x-1}, x \neq 1$.

    $g \circ f$ does not exist since the image set of f is not a subset of the domain of g.
13. (i) $2x^2-2x-1, 4x^2+2x-1; 0, -2$, (ii) $4x+3, 8x+7$.
14. $p=-1, q=-4; 0, \frac{1}{3}, -\frac{1}{3}$.

Page 68, Assignment 5.1

1. (i) $x = -2 \pm \sqrt{13}$. (ii) $x = 2$ or $-\frac{1}{2}$. (iii) $x = \frac{3}{2}$.
(iv) $x = \dfrac{3 \pm \sqrt{17}}{4}$.

2. (i) $x = \dfrac{3 \pm \sqrt{29}}{2}$ (ii) $x = \dfrac{7 \pm \sqrt{5}}{2}$. (iii) $x = \frac{7}{2}$ or 1.
(iv) no roots. (v) $x = \frac{2}{3}$. (vi) $x = -2 \pm \sqrt{11}$.
(vii) $x = \frac{5}{2}$. (viii) $x = 4$ or -7.

3. (i) real, unequal, irrational.
(ii) no real roots.
(iii) real, unequal, rational.
(iv) no real roots.
(v) real, equal, rational.
(vi) real, unequal, irrational.
(vii) real, unequal, irrational.
(viii) real, unequal, rational.
(ix) real, equal, rational.
(x) real, unequal, rational.
(xi) no real roots.
(xii) no real roots.

4. $p = 16$, $q = \frac{16}{3}$, $r = \pm 10$, $s = 5$. 5. $k = -1$ or -5.

6. $p \geqq 2$ or $p \leqq -2$. 7. 9. 8. $k \leqq -1$ or $k \geqq -\frac{1}{9}$.

9. (i) $k = -2$ or 6, (ii) $k \geqq 6$ or $k \leqq -2$.

10. root is $x = 1$.

11. $q^2 - 4pr < 0$; real and equal. 12. $k \geqq 8$ or $k \leqq -8$.

13. $x^2 + (2n-4)x + 10 - 5n = 0$. 14. $-5 < p < -1$.

15. (i) no rational factors. (ii) $(2x+1)$, $(x-1)$.
(iii) $(x-4)$, $(x+3)$. (iv) $(3x+1)$, $(x+1)$.
(v) no rational factors. (vi) no rational factors.

16. $y^2 + 6y + 9 = 0$. 17. $m = \pm 1$. 18. $k > -\frac{4}{3}$; $k = -\frac{4}{3}$.

19. $2y + x = 6$ and $2y - x = 6$.

20. (i) $x^2 - 3x - 10 = 0$, (ii) $x^2 + 18x + 77 = 0$,
(iii) $x^2 - 2ax + a^2 - b^2 = 0$.

21. $-\frac{1}{2}$.

23. (iii) $(3x+2)(x-4)$, (iv) $(2x-3)(2x-3)$,
(v) $(2x+1)(x+2)$.

24. 3, 4.

Page 71, Assignment 5.2

1. (i) $(x+1)^2 + 2$, (ii) $(x+2)^2 - 9$, (iii) $\frac{9}{4} - (x+\frac{1}{2})^2$,
(iv) $2 - (x-1)^2$, (v) $2(x-1)^2 + 3$, (vi) $2(x-\frac{3}{4})^2 - \frac{1}{8}$.

2. (i) Minimum value $-2\frac{1}{4}$; $x = \frac{1}{2}$; $(\frac{1}{2}, -2\frac{1}{4})$.
(ii) Minimum value $-6\frac{1}{8}$; $x = -\frac{5}{4}$; $(-\frac{5}{4}, -6\frac{1}{8})$.
(iii) Maximum value 9; $x = 2$; (2, 9).
(iv) Minimum value $\frac{7}{4}$; $x = \frac{3}{2}$; $(\frac{3}{2}, \frac{7}{4})$.
(v) Maximum value $-1\frac{3}{4}$; $x = \frac{3}{2}$; $(\frac{3}{2}, -1\frac{3}{4})$.
(vi) Minimum value -1; $x = 1$; $(1, -1)$.
(vii) Maximum value -4; $x = 0$; $(0, -4)$.
(viii) Minimum value -8; $x = 2$; $(2, -8)$.

3. (i) E, (ii) A, (iii) F, (iv) B, (v) C, (vi) D.

Page 74, Assignment 5.3

1. (i) 3, 4. (ii) -2, $-\frac{5}{2}$. (iii) $-q/p$, r/p.
(iv) $-\dfrac{b+1}{a}$, $-\dfrac{1}{a}$. (v) $(a+b)$, ab. (vi) 3, -2.

2. $a = \frac{1}{6}$, $b = \frac{1}{3}$.

3. (i) $x^2 - 5x + 6 = 0$. (ii) $x^2 - 2x - 3 = 0$.
(iii) $2x^2 + 3x - 2 = 0$. (iv) $6x^2 + 5x + 1 = 0$.
(v) $x^2 - (p+q)x + pq = 0$.

4. (i) $\frac{4}{3}$, 3. (ii) $\frac{4}{9}$. (iii) $\frac{16}{9}$. (iv) $-\frac{92}{9}$.

5. (i) 2, $\frac{1}{2}$. (ii) $3\frac{1}{2}$. (iii) $2\frac{1}{2}$.

6. (i) p, q. (ii) p/q. (iii) $p^2 - 2q$. (iv) $p^2 - q$.

(v) $\dfrac{(q+1)^2}{q}$. (vi) $q-p+1$. (vii) $\dfrac{p^2-2q}{q}$.

7. $a = \pm 8$. 8. $p = 1$. 9. $4q^2 = 25pr$.

10. (i) $6x^2+3x+2=0$. (ii) $4x^2+15x+36=0$.

11. $2x^2-2x-5=0$. 12. $2x^2+3x-2=0$.

13. $4x^2-x+4=0$. 14. $\frac{2}{3}, -\frac{4}{3}, \frac{28}{9}$.

15. 27. 16. ± 33.

17. $\pm 2\frac{1}{2}$. 18. $x^2-3x-5=0$.

Page 76, Assignment 5.4

1. $-\sqrt{2}$, $2\sqrt{3}$, $3-\sqrt{2}$, $-3-\sqrt{2}$, $-\sqrt{3}+1$, $\sqrt{3}+1$, $a-\sqrt{b}$, $-\sqrt{b}-1$, $3+2\sqrt{3}$, $\dfrac{1+4\sqrt{2}}{2}$, $\dfrac{4-2\sqrt{2}}{5}$.

2. $-\sqrt{3}$; $x^2-3=0$.

3. $1+\sqrt{2}$; $x^2-2x-1=0$.

4. (i) $14-6\sqrt{5}$, (ii) $17+12\sqrt{2}$, (iii) $23-4\sqrt{15}$, (iv) $6\sqrt{2}+\sqrt{6}-2\sqrt{3}-6$, (v) $5+\sqrt{6}$, (vi) 2, (vii) 7.

5. $k=1$. 6. $a=1, b=-6, c=7$.

7. (i) $\sqrt{2}+1$, (ii) $\dfrac{\sqrt{5}-1}{4}$, (iii) $\dfrac{2\sqrt{3}+3}{3}$, (iv) $\dfrac{7(4-\sqrt{3})}{13}$, (v) $-\dfrac{\sqrt{5}+5}{8}$.

8. (i) $7-4\sqrt{3}$, (ii) $\dfrac{\sqrt{3}+1}{2}$, (iii) $6-\sqrt{35}$, (iv) $\sqrt{3}$.

9. 6, 28. 10. $\dfrac{7\sqrt{2}}{6}$. 11. $\dfrac{2\sqrt{6}}{3}$.

12. $\sqrt{2}$. 13. $6\sqrt{5}$.

14. (i) $x^2-6x+4=0$, (ii) $x^2+4x+1=0$. 16. $\dfrac{-5\sqrt{3}}{3}$.

17. $\dfrac{5-\sqrt{5}}{5}$, $\dfrac{5+\sqrt{5}}{2}$; $\pm\sqrt{2}$.

Page 77, Assignment 5.5

1. E. 2. A. 3. D. 4. D. 5. B. 6. A.
7. D. 8. D. 9. D. 10. D. 11. D. 12. A.

Page 80, Assignment 6.1

1. $\{1, 2, -1\}$. 2. $\{-2, 1, -2\}$.

3. $\{-1, 3, -2\}$. 4. $\{4, 3, 2\}$.

5. $\{-\frac{4}{3}, -\frac{11}{6}, -\frac{1}{6}\}$. 6. $\{0, \frac{5}{2}, -1\}$.

7. $\{-1, 0, -2\}$. 8. $\{-7, -2, 1\}$.

9. $a=-1,\ b=4,\ c=-2;\ y=-x^2+4x-2$.

10. $g=-\frac{7}{2},\ f=\frac{1}{2},\ c=10;\ x^2+y^2-7x+y+10=0$.

11. $a=14d,\ b=21d,\ c=-2d;\ 14x+21y-2z=1$.

12. $\{-2, -\frac{1}{2}, 6\}$.

13. $\{3, -1, 2\}$.

14. $a=1,\ b=2,\ c=3$.

Page 82, Assignment 6.2

1. $\{(2,4), (-1,1)\}$. 2. $\{(4, 0, (-1, -5)\}$.

3. $\{(8, -1), (-1, 8)\}$. 4. $\{(1, 3)\}$.

5. $\{(-1, 1), (-\frac{1}{5}, \frac{7}{5})\}$. 6. $\{(-1, -1), (\frac{1}{5}, \frac{7}{5})\}$.

7. $\{(1, 2), (-\frac{2}{3}, \frac{8}{9})\}$. 8. $\{(3, 3), (-\frac{3}{5}, -\frac{21}{5})\}$.

9. $\{(1, 3)\}$. 10. $\{(3, 1), (\frac{9}{5}, \frac{7}{5})\}$.

11. $\{(-1, -1), (-\frac{1}{2}, 0)\}$. 12. $\{(4, 3), (\frac{1}{11}, -4\frac{9}{11})\}$.

13. $\{(-2, 2), (\frac{17}{8}. -\frac{3}{4})\}$. 14. $\{(-4, 2), (-\frac{5}{2}, \frac{7}{2})\}$.

15. $\{(-\frac{5}{2}, \frac{1}{2}), (-1, 1)\}$.

16. (i) tangent $(4,0)$, (ii) cuts circle at $(0,1)$ and $(-1,2)$. (iii) does not cut the circle. (iv) tangent $(0,0)$.

17. $\{(-2,2), (\frac{1}{7},\frac{4}{7})\}$.
18. $\{(8,6)\}$.

19. $\{(8,-6)\}$.
20. $\{(4,\frac{2}{3}), (-\frac{3}{2},\frac{5}{2})\}$.

21. $\{(4,2), (\frac{5}{2},\frac{7}{2})\}$.
22. $\{(-2,0), (-\frac{2}{5},3\frac{1}{5})\}$.

Page 84, Assignment 6.3

1. (i) $y^2 = 12x$. (ii) $x^2+y^2 = 9$. (iii) $2x+y = 7$. (iv) $4x^2+y^2 = 16$. (v) $y^2 = 4ax$. (vi) $xy = c^2$. (vii) $x^2+4y^2-6x-8y+9 = 0$. (viii) $\dfrac{x^2}{a^2}+\dfrac{y^2}{b^2} = 1$. (ix) $2y^2-9x = 9$. (x) $x+y^2-2y = 3$.
2. (i) $(2, -\frac{1}{2}), (2\sqrt{3}, \frac{1}{2})$, (ii) 45, 315; 0, 0. (iii) 30, 150, 210, 330; $2\sqrt{3}, -2\sqrt{3}, -2\sqrt{3}, 2\sqrt{3}$. (iv) $x^2-8y = 8$.

4. 2.
5. $\sqrt{2}$.

6. (i) $2y^2+x = 1$. (ii) $y = 12-12x+6x^2-x^3$. (iii) $x^2-2y = 2$. (iv) $x^2+y^2 = 2$.
7. (i) $(5,0), (1,0)$. (ii) $(4,\frac{3}{2})$.

(iii)

θ	0	$\frac{\pi}{6}$	$\frac{\pi}{4}$	$\frac{\pi}{3}$	$\frac{\pi}{2}$	$\frac{2\pi}{3}$	$\frac{3\pi}{4}$	$\frac{5\pi}{6}$	π
x	5	4·73	4·41	4	3	2	1·59	1·27	1
y	0	$\frac{1}{2}$	1	$\frac{3}{2}$	2	$\frac{3}{2}$	1	$\frac{1}{2}$	0

(iv) $-(\cos\alpha+\cos\beta)$. (v) $x^2-6x+2y+5 = 0$.

8. $-1\pm\sqrt{2}$.

Page 85, Assignment 6.4

1. E. 2. C. 3. E. 4. B. 5. D. 6. C.
7. E. 8. B. 9. A. 10. E. 11. A. 12. B.

Page 88, Assignment 7.1

1. a^{12}.
2. $a^{\frac{3}{4}}$.
3. $p^{\frac{1}{2}}$.
4. q^8.
5. x^2.
6. $a^{\frac{1}{8}}$.
7. 1.
8. $\dfrac{1}{2x}$.
9. $\dfrac{1}{64x^3}$.
10. $\dfrac{3}{x^2}$.
11. $12x^{\frac{3}{2}}$.
12. 24.
13. $2a^{\frac{1}{6}}$.
14. $\dfrac{1}{a^2}$.
15. $\dfrac{1}{s^{\frac{3}{2}}}$.
16. $x+x^2$.
17. $1-a$.
18. $x-2+\dfrac{1}{x}$.
19. $12x^2$.
20. $\dfrac{a}{b}$.
21. $\dfrac{b^4}{16a^4}$.
22. (i) 2. (ii) $\frac{1}{3}$. (iii) 9. (iv) $\frac{1}{8}$. (v) $\frac{4}{3}$. (vi) 5, 7. (vii) $\frac{3}{10}$. (viii) $\frac{1}{49}$.
23. $\frac{1}{16}$.
24. ± 3.
25. (i) $9xy^4$. (ii) $\frac{3}{2}ab$. (iii) $\dfrac{1}{2a^{\frac{1}{2}}b^{\frac{1}{2}}}$. (iv) $\dfrac{9}{4a^2b^2}$. (v) $\dfrac{1}{b^{2m+3n}}$. (vi) x^6y^9.
26. (i) $\frac{2}{3}$. (ii) 5. (iii) $\frac{8}{27}$. (iv) $\frac{1}{4}$. (v) 81.
27. $\frac{7}{57}$.
28. $\frac{1}{125}$.
29. (i) 6. (ii) $\frac{3}{4}$, (iii) $\frac{7}{3}$. (iv) $\frac{1}{5}$.
30. $\dfrac{2^{5k}}{5^{2k}}$.
32. $y = x^{p+1}$.
33. $\frac{1}{4}$.
34. $b = \dfrac{8c^{\frac{1}{2}}}{a^{\frac{3}{4}}}$, $\frac{8}{3}$.

Page 91, Assignment 7.2

1. 1·4, 1·73.
2. (a) 1·8, 0·56. (b) (i) 0·3, (ii) 1·7. (c) 0·78.
3.

x	-2	-1	0	1	2	3
$(2^x)^2$	$\frac{1}{16}$	$\frac{1}{4}$	1	4	16	64

4. (0, 1). 6. 0·71, 1·4. 7. $\frac{1}{5}$. 8. $g(x) = 4^x$.

Page 94, Assignment 7.3

2. (i) $\log_3 x$. (ii) $\log_{10} x$. (iii) $\log_4 y$. (iv) $\log_a z$. (v) 2^x. (vi) 4^x. (vii) b^a.
3. 2, 1, 0, -1, -2. 4. -2, -1, 0, 1, 2, 3.
5. (i) 4. (ii) 3. (iii) True. 6. -2, -1, 0, 1, 2, 3.
7. $\log_a x$; 0, 1, -1, 2, -2, 3.
8. (i) $\log_3 27 = 3$. (ii) $\log_4 64 = 3$. (iii) $\log_2 32 = 5$. (iv) $\log_3 \frac{1}{9} = -2$. (v) $\log_5 \frac{1}{125} = -3$. (vi) $\log_a 4 = b$. (vii) $\log_6 1 = 0$. (viii) $\log_a p = n$.
9. (i) $8 = 2^3$. (ii) $125 = 5^3$. (iii) $64 = 2^6$. (iv) $\frac{1}{128} = 2^{-7}$. (v) $1 = a^0$. (vi) $6 = 6^1$. (vii) $\frac{1}{27} = 3^{-3}$. (viii) $a = b^x$. (ix) $\sqrt{5} = 5^{\frac{1}{2}}$.
10. (i) 3. (ii) 1. (iii) $\frac{1}{2}$. (iv) 256. (v) -3. (vi) $\frac{1}{32}$. (vii) 3.
11. (i) 3. (ii) 0. (iii) 3. (iv) 1. (v) -4. (vi) 3. (vii) $\frac{1}{3}$.

Page 98, Assignment 7.4

1. (i) log 24. (ii) log 9. (iii) log 16. (iv) log 6. (v) $\log \frac{1}{27}$.
2. (i) 3. (ii) 5. (iii) 4. (iv) -1. (v) 0. (vi) 1.
3. (i) 16. (ii) $10^{\frac{1}{3}}$. (iii) $\dfrac{1}{7^{\frac{1}{2}}}$. (iv) $\frac{1}{8}$. (v) 1. (vi) $\frac{1}{4}$. (vii) $\frac{1}{3}$.
4. (i) $\log_3 16$. (ii) $\log_2 64$. (iii) $\log_5 35$. (iv) $\log_{10} 300$. (v) $\log_6 72$. (vi) $\log_{10} 3^{8/3}$.
5. (i) 3. (ii) -1. (iii) 0. (iv) $\frac{1}{4}$. (v) $\log a^2b$. (vi) $\log xy$. (vii) $\log \dfrac{y}{x}$. (viii) $\frac{1}{3}$.
6. 3. 7. 2. 8. 1.
9. 1. 10. 1. 11. 2.
12. 3. 13. $\frac{1}{9}$. 14. 8.
15. $\frac{4}{3}$. 16. $2^{\frac{1}{2}}$.
17. second is greater.
19. (i) $-2 \log x$. (ii) 0. (iii) 3.
20. (i) 6. (ii) $\frac{3}{5}$.
21. (i) 1·55. (ii) 1·67. (iii) 2·46. (iv) 0·423.
22. (i) $\sqrt{3}$. (ii) $x = 32$, $y = 8$.
23. (i) True. (ii) False. (iii) False. (iv) True.
24. $2^{\frac{1}{3}}$. 25. 0·631. 26. $-4·42$.
27. (i) $\frac{1}{2}$, (ii) -2. 28. $\frac{3}{2}$ or -1.

Page 103, Assignment 7.5

1. $a = 2·8$, $m = 1·2$. 2. $a = 2·04$, $n = 0·6$.
3. $a = 31·6$, $n = -3·7$, 4. $a = 8·47$, $m = -2·1$.
5. $y = 6·92x^{-4}$. 6. $w = 57500000d^{-3·13}$.
7. $A = 2·5t^{1·5}$. 8. $p = 1·86v^{0·33}$.

Page 107, Assignment 7.6

1. (i) 1·89. (ii) 3·17. (iii) 1·89.
2. (i) 0·912. (ii) 1·95. (iii) 1·65. (iv) 1·14.
3. 12·0. 4. 8 or $\frac{1}{8}$. 5. 3 or 27.
7. $\log_3 2$, 0·63. 9. 625.
10. 10,000. 11. 1·56.

Page 108, Assignment 7.7

1. 0·288.
2. (i) 335 mg. (ii) 34·7 years.
3. 2·57 secs.
5. (ii) 205 mg.
6. (i) 0·081. (ii) 8·57 cm.
7. (ii) 31 days.
8. 1·32.
9. $A = 45{\cdot}7$, $a = 0{\cdot}880$.

Page 109, Assignment 7.8

1. D. 2. E. 3. D. 4. D. 5. D. 6. D.
7. E. 8. C. 9. E. 10. E. 11. C. 12. D.

Page 112, Assignment 8.1

1. Miss Brown has studied botany.
2. ABCDE is not a quadrilateral.
3. Square ABCD is a kite and has two axes of symmetry.
4. No valid conclusion.
5. (i) Mr. Smith has passed his "O" levels.
   (ii) Nothing can be said.
   (iii) Nothing can be said.
   (iv) Miss James is not a civil servant.
6. (i) Nothing can be said.
   (ii) Nothing can be said.
   (iii) Miss Hill may vote.
   (iv) Nothing can be said.
7. Valid.
8. I, III, IV valid.
9. I valid.
10. II, III valid.

Page 114, Assignment 8.2

1. True. 2. True. 3. False. 4. True.
5. False. 6. True. 7. False. 8. True.
9. False. 10. True. 11. True.
12. Tom is not 16 years old.
13. $\triangle$ABC is an isosceles triangle.
14. James Scott is not a good footballer.
15. That hat is not brown.
16. Mary Jones wears spectacles.
17. It is true that 337 is a prime number.
18. James is not older than Paul.
19. $a^m \times a^n \neq a^{mn}$.
20. This boy has blue eyes.

Page 117, Assignment 8.3

1. Some four-legged animals enjoy milk.
2. Some periodic functions do not have a period of 2π.
3. Some sports fans like music.
4. Some flowering plants are evergreens.
5. (i) Some dogs do not hate cats.
   (ii) All dogs do not hate cats. *or* No dogs hate cats.
   (iii) All dogs hate cats.
   (iv) Some dogs hate cats.
6. (i) Some trees are not plants.
   (ii) All men are sports fans.
   (iii) All girls do not like hockey. *or* No girl likes hockey.
   (iv) Some men have blond hair.
   (v) Some scientists are not engineers.
   (vi) Some boys are good at mathematics.
7. (i) Some triangles are isosceles—True.
   (ii) All matrices are not commutative under multiplication—False.
   (iii) Some numbers of the form $6n-1$ are not prime—True.
   (iv) Some mappings are functions—True.
   (v) There is no value of θ such that $\cos\theta > 1$—True.
   (vi) Some functions have no inverses—True.

8. False. 9. True. 10. False.
11. False. 12. False.
13. No sensible quick tempered person has bad manners.
14. Some antiques are a joy for ever.
15. Some kites are not squares.
16. II and III valid.

Page 120, Assignment 8.4

1. Their corresponding sides are in proportion.
2. Is less than 11 units. 3. $f(\alpha) = 0$.
4. 3 equal angles. 5. Valid.
6. Not valid. 7. Not valid.
8. Not valid. 9. (i) valid, (ii) not valid.
10. If triangle ABC is isosceles it is equilateral—converse not true.
11. n divides $x^2 \Rightarrow n$ divides x—converse not true.
12. $-x < -y \Rightarrow x > y$—true; can be replaced by two-way implication.
13. $\log_2 p = n \Rightarrow 2^n = p$—true; can be replaced by two-way implication.
14. If a car is expensive then it is big—converse not true.
15. $a^2 > 9 \Rightarrow a > 3$ for all $a \in R$—converse not true.
16. $A \subset B \Rightarrow A \cap B = A$—true; can be replaced by two-way implication.
17. False $-2 > -3$, but $(-2)^2 < (-3)^2$.
18. False, February has not 31 days.
19. False, $2x - 4y = 7$ and $6x - 12y = -4$ have no solution.
20. False $1^2 \not> 1$.
21. False $n = 12$, $a = 6$, $b = 4$.
22. True. 23. True.

Page 121, Assignment 8.5

1. B. 2. C. 3. E. 4. B. 5. E. 6. D.
7. D. 8. A. 9. C. 10. B. 11. A. 12. B.

Page 123, Assignment 8.6

1. $p = 2$. 2. (b) (i) -1, (ii) -4.
3. $\dfrac{p^{15}}{r^2}$. 4. $x^2 - 3x - 5 = 0$.
5. $-\frac{1}{2}(5 + 4\sqrt{3})$. 6. 3·32.
7. $\pm\frac{1}{2}$, $30 \leqq \theta \leqq 90$. 8. 2, 3, 5.
9. (i) $\frac{1}{9}$, 1; no real roots since $9k^2 - 10k + 1 < 0$.
10. $p = a^2q^3$. 11. $p = 2\sqrt{3}$.
12. $\frac{1}{4}$. 13. $b^2 - 4ac < 0$.
14. 69·67 mins. 15. $x^2 - 10x + 13 = 0$.
16. 6, 30.
17. (i) $\log_3 9 = 2$. (ii) $\log_{10} 1000 = 3$. (iii) $\log_8 2 = \frac{1}{3}$. (iv) $\log_2 \frac{1}{4} = -2$. (v) $\log_{16} 2 = \frac{1}{4}$.
18. (i) $2^2 = 4$. (ii) $10^3 = 1000$. (iii) $2^{-2} = \frac{1}{4}$. (iv) $3^4 = 81$. (v) $x^n = p$.
19. (0, 1). 20. $\{(-5, 1, -3)\}$.
21. $\{(\frac{1}{2}, 2, -6)\}$. 22. $\{(-1, 3, 2)\}$.
23. $\{(2, 0), (6, 4)\}$. 24. $\{(0, 2), (3\frac{1}{5}, \frac{2}{5})\}$.
25. $\{(\frac{2}{3}, -4), (\frac{5}{2}, \frac{3}{2})\}$. 26. $\{(10, 2), (\frac{8}{5}, -\frac{4}{5})\}$.
27. (i) $y - 3x = 1$. (ii) $\log_2 xy = 2^{x+y}$. (iii) $x^2 = y + x$. (iv) $\frac{1}{2}(x - y) = \log_{10} \frac{1}{2}(x + y)$. (v) $xy = 1$.
28. $\dfrac{x^2}{16} + \dfrac{y^2}{4} = 1$. 29. 75 percent.
30. I nothing, II has been trained rigorously, III not a record breaker, IV nothing.
31. ABCD is not a square.
32. I not valid, II not valid, III not valid, IV not valid.

GEOMETRY

Page 134, Assignment 1.1

1. (i) $(7,4)$ (ii) $(-1,9)$ (iii) $(-18,-2)$ (iv) $(16,-3)$ (v) $(-2,3)$ (vi) $(-1,-18)$ (vii) $(2,-5)$ (viii) $(-2,13)$.
2. (i) $(-5,-7)$ (ii) $(-2,5)$ (iii) $(11,-4)$ (iv) $(1,14)$ (v) $(10,17)$ (vi) $(0,15)$ (vii) $(-2,-4)$ (viii) $(-7,-19)$.
3. $a_1 = 2s-a$, $b_1 = 2t-b$.
4. (i) $(11,2)$ (ii) $(-1,2)$ (iii) $(-11,-4)$ (iv) $(-10,-5)$ (v) $(2t-2s-9,4)$ (vi) $(7,-18)$ (vii) $(-7,18)$ (viii) $(5,8)$ (ix) $(-2,2b-2a+4)$ (x) $(a,2t-2s+b)$.
5. (i) $(3,13)$ (ii) $(-1,1)$ (iii) $(4,4)$ (iv) $(-10,3)$.
6. No. 7. No.
8. $y+2x = 2$, $y = 2x-10$; $\begin{pmatrix}6\\0\end{pmatrix}$.
9. $[2(x_2-x_1)+x, 2(y_2-y_1)+y]$; $\begin{pmatrix}2(x_2-x_1)\\2(y_2-y_1)\end{pmatrix}$; $\overrightarrow{PP'} = 2\overrightarrow{QR}$.
10. (i) $(-2,4)$ (ii) $(2,-4)$ (iii) $(2,-4)$ (iv) $(-2,4)$; $Y \circ M = M \circ X$, $X \circ M = M \circ Y$.
11. Yes.
12. (ii) a rotation about O through an angle twice the size of angle BOY.
14. $(5, 195°)$.

Page 139, Assignment 2.1

1. (i) $(-2,1)$, $(-3,-5)$, $(0,0)$, $(2,-6)$ (ii) $y = 2x$ (iii) Reflection in the y-axis.
2. (i) $a = \frac{1}{2}(a'-\frac{1}{2}b')$, $b = -\frac{1}{2}b'$ (ii) $(0,0)$, $(4,0)$, $(0,-8)$.
3. (i) $(0,0)$, $(2,0)$, $(8,2)$, $(6,2)$ (ii) Yes.
4. (i) A translation $\begin{pmatrix}2\\-2\end{pmatrix}$ followed by a reflection in the y-axis.
5. (ii) $(0,0)$, $(\sqrt{3},1)$, $(\sqrt{3}+1, 1-\sqrt{3})$, $(1,-\sqrt{3})$ (iii) Area of image is four times area of square OABC.

Page 146, Assignment 2.2

1. $\begin{pmatrix}1 & 2\\0 & 1\end{pmatrix}$.
2. $\begin{pmatrix}x_1\\y_1\end{pmatrix} = \begin{pmatrix}1 & 0\\2 & 1\end{pmatrix}\begin{pmatrix}x\\y\end{pmatrix}$; a shear parallel to the y-axis; $\{(0,0), (1,2), (-1,-3), (0,-1)\}$.
3. $\begin{pmatrix}0 & 1\\1 & 0\end{pmatrix}$; O′(0, 0), P′(0, 3), Q′(2, 3), R′(2, 0).
4. $\{(-2,-4), (-6,-12), (18,36), (a+3b, 2a+6b)\}$.
5. (i) $\begin{pmatrix}0 & -1\\1 & 0\end{pmatrix}$, $\begin{pmatrix}\frac{1}{\sqrt{2}} & -\frac{1}{\sqrt{2}}\\\frac{1}{\sqrt{2}} & \frac{1}{\sqrt{2}}\end{pmatrix}$, $\begin{pmatrix}\frac{\sqrt{3}}{2} & -\frac{1}{2}\\\frac{1}{2} & \frac{\sqrt{3}}{2}\end{pmatrix}$, $\begin{pmatrix}-\frac{1}{2} & -\frac{\sqrt{3}}{2}\\\frac{\sqrt{3}}{2} & -\frac{1}{2}\end{pmatrix}$.

(ii) $\begin{pmatrix} \frac{1}{2} & -\frac{\sqrt{3}}{2} \\ \frac{\sqrt{3}}{2} & \frac{1}{2} \end{pmatrix}, \begin{pmatrix} -1 & 0 \\ 0 & -1 \end{pmatrix}, \begin{pmatrix} -\frac{\sqrt{3}}{2} & -\frac{1}{2} \\ \frac{1}{2} & -\frac{\sqrt{3}}{2} \end{pmatrix}.$

(iii) $\begin{pmatrix} \frac{1}{2} & \frac{\sqrt{3}}{2} \\ -\frac{\sqrt{3}}{2} & \frac{1}{2} \end{pmatrix}, \begin{pmatrix} \frac{\sqrt{3}}{2} & \frac{1}{2} \\ -\frac{1}{2} & \frac{\sqrt{3}}{2} \end{pmatrix}, \begin{pmatrix} 0 & 1 \\ -1 & 0 \end{pmatrix}.$

(iv) $\begin{pmatrix} 1 & 0 \\ 0 & 1 \end{pmatrix}$; the identity or no-change matrix.

6. $\begin{pmatrix} -1 & 0 \\ 0 & 1 \end{pmatrix}$; matrix associated with a reflection in the y-axis.

7. $\begin{pmatrix} 0 & -1 \\ -1 & 0 \end{pmatrix}$; matrix associated with a reflection in the line $y = -x$.

8. (i) Reflection in the origin. (ii) $\begin{pmatrix} -1 & 0 \\ 0 & -1 \end{pmatrix}$ (iii) yes.

9. (i) $\begin{pmatrix} 0 & 1 \\ 1 & 0 \end{pmatrix}, \begin{pmatrix} 0 & -1 \\ -1 & 0 \end{pmatrix}$ (ii) reflection in the origin

(iii) $\begin{pmatrix} -1 & 0 \\ 0 & -1 \end{pmatrix}$ (iv) yes (v) yes.

10. (i) $\begin{pmatrix} -1 & 0 \\ 0 & 1 \end{pmatrix}, \begin{pmatrix} 0 & -1 \\ 1 & 0 \end{pmatrix}; \begin{pmatrix} 0 & 1 \\ 1 & 0 \end{pmatrix}, \begin{pmatrix} 0 & -1 \\ -1 & 0 \end{pmatrix}$

(iii) $\begin{pmatrix} -1 & 0 \\ 0 & -1 \end{pmatrix}.$

11. (i) $\begin{pmatrix} 1 & 0 \\ 0 & -1 \end{pmatrix}, \begin{pmatrix} 0 & -1 \\ 1 & 0 \end{pmatrix}; \begin{pmatrix} 0 & -1 \\ -1 & 0 \end{pmatrix}, \begin{pmatrix} 0 & 1 \\ 1 & 0 \end{pmatrix}$

(iii) $\begin{pmatrix} -1 & 0 \\ 0 & -1 \end{pmatrix}$; half-turn about the origin.

12. (i) a rotation of the plane through $+90°$.

(ii) a reflection in the y-axis; $X = \begin{pmatrix} 0 & -1 \\ -1 & 0 \end{pmatrix}$,

reflection in the line $y = -x$; $X^{-1} = \begin{pmatrix} 0 & -1 \\ -1 & 0 \end{pmatrix}$.

13. Matrix P is associated with a reflection in the line $y = x$.
Matrix Q is associated with a reflection in the line $y = -x$.

	I	P	Q	X
I	I	P	Q	X
P	P	I	X	Q
Q	Q	X	I	P
X	X	Q	P	I

14. $p = 2$, $q = -2$; $\begin{pmatrix} 4 & 0 \\ 2 & -2 \end{pmatrix}$; $X = \begin{pmatrix} 0 & -1 \\ 1 & 0 \end{pmatrix}$, a rotation of the plane through $+90°$; $y+3x = 0$.

15. $A = \begin{pmatrix} 1 & 0 \\ 0 & -1 \end{pmatrix}$, $B = \begin{pmatrix} 0 & 1 \\ 1 & 0 \end{pmatrix}$, $C = \begin{pmatrix} 0 & -1 \\ 1 & 0 \end{pmatrix}$;
a rotation of the plane through 90°.

16. $A = \begin{pmatrix} 0 & -1 \\ 1 & 0 \end{pmatrix}$, $B = \begin{pmatrix} 0 & 1 \\ 1 & 0 \end{pmatrix}$, $AB = \begin{pmatrix} -1 & 0 \\ 0 & 1 \end{pmatrix}$,

$BA = \begin{pmatrix} 1 & 0 \\ 0 & -1 \end{pmatrix}$;

Matrix AB is associated with a reflection in the y-axis,
Matrix BA is associated with a reflection in the x-axis;

$X = \begin{pmatrix} -1 & 0 \\ 0 & -1 \end{pmatrix}$, associated with a reflection in the origin; $y = 3x-2$.

17. $R = \begin{pmatrix} \sin 2\alpha & -\cos 2\alpha \\ \cos 2\alpha & \sin 2\alpha \end{pmatrix}$; $\begin{matrix} x' = x \sin 2\alpha - y \cos 2\alpha \\ y' = x \cos 2\alpha + y \sin 2\alpha \end{matrix}$.

18. $p = k$, $q = k-1$ where $k \in R$, $k \neq 0$ or 1.

19. $P = \begin{pmatrix} 1 & -1 \\ 1 & 1 \end{pmatrix}$; Q is the matrix associated with a dilation $[0, 2]$.

$R = \begin{pmatrix} \frac{1}{2} & -\frac{1}{2} \\ \frac{1}{2} & \frac{1}{2} \end{pmatrix}$.

20. (ii) $C = \begin{pmatrix} 0 & -1 \\ -1 & 0 \end{pmatrix}$, $B = \begin{pmatrix} -1 & -1 \\ 1 & -1 \end{pmatrix}$.

Page 150, Assignment 2.3

1. B. 2. E. 3. D. 4. C. 5. B. 6. B.
7. B. 8. E. 9. D. 10. D. 11. C. 12. A.

Page 155, Assignment 3.1

1. (i) $\boldsymbol{a}+\boldsymbol{b}$ (ii) $\boldsymbol{b}-\boldsymbol{c}$ (iii) $\boldsymbol{a}+\boldsymbol{b}-\boldsymbol{c}$.
2. $\boldsymbol{p}-\boldsymbol{q}+\frac{1}{3}\boldsymbol{r}$. 3. $\boldsymbol{a}+\boldsymbol{b}$, $\boldsymbol{a}-\boldsymbol{b}$.
4. $2\boldsymbol{t}+\frac{1}{3}\boldsymbol{s}$. 5. $\frac{2}{7}\boldsymbol{a}+\frac{5}{7}\boldsymbol{c}$.
6. $-\boldsymbol{a}$, $-\boldsymbol{b}$, $-\boldsymbol{c}$; parallelogram.
7. (i) $\frac{1}{2}\boldsymbol{u}$, $\boldsymbol{v}-\frac{1}{2}\boldsymbol{u}$, $\frac{1}{2}(\boldsymbol{u}+\boldsymbol{v})$ (ii) $\boldsymbol{u}-\boldsymbol{v}$.
8. (a) $\frac{1}{3}(\boldsymbol{s}+\boldsymbol{q})$, $\frac{1}{3}(\boldsymbol{q}-2\boldsymbol{s})$ (b) $\frac{1}{3}(2\boldsymbol{s}-\boldsymbol{q})$, $-\frac{1}{3}(\boldsymbol{s}+\boldsymbol{q})$.

Page 159, Assignment 3.2

1. $\boldsymbol{a}-\boldsymbol{b}$, $3(\boldsymbol{b}-\boldsymbol{a})$, $2(\boldsymbol{a}-\boldsymbol{b})$.
2. $\boldsymbol{q}-\boldsymbol{p}$, $3\boldsymbol{p}-2\boldsymbol{q}$, $-4\boldsymbol{p}-\boldsymbol{q}$, $3\boldsymbol{q}+\boldsymbol{p}$; $\frac{3}{2}\boldsymbol{p}$, $\boldsymbol{p}-\frac{3}{2}\boldsymbol{q}$, $-\frac{1}{2}(\boldsymbol{p}+\boldsymbol{q})$; $\frac{2}{3}(\boldsymbol{p}-\boldsymbol{q})$.
3. $\boldsymbol{a}-\boldsymbol{b}+\boldsymbol{c}$, $\frac{1}{2}(\boldsymbol{a}+\boldsymbol{c})$. 6. $p = 3$, $q = -2$.
7. (i) $\begin{pmatrix} -10 \\ 5 \end{pmatrix}$ (ii) $\begin{pmatrix} -1 \\ 9 \end{pmatrix}$ (iii) $\begin{pmatrix} 12 \\ 0 \end{pmatrix}$ (iv) $\begin{pmatrix} 5 \\ 16 \end{pmatrix}$.
8. $3\boldsymbol{v}-\boldsymbol{u}$, $3\boldsymbol{u}-\boldsymbol{v}$; $\frac{3}{4}(\boldsymbol{u}+\boldsymbol{v})$, $\frac{1}{4}(3\boldsymbol{u}-\boldsymbol{v})$.

Page 160, Assignment 3.3

1. E. 2. A. 3. C. 4. A. 5. C. 6. A.
7. E. 8. D. 9. A. 10. E. 11. B. 12. B.

Page 164, Assignment 4.1

1. (i) $\boldsymbol{p}+\boldsymbol{s}$ (ii) $\boldsymbol{p}+\boldsymbol{r}-\boldsymbol{s}$; $\boldsymbol{p}-\boldsymbol{r}+\boldsymbol{s}$; $\boldsymbol{r}+\boldsymbol{s}-\boldsymbol{p}$
   (iii) $|\overrightarrow{VP}|^2 = |\boldsymbol{p}|^2+|\boldsymbol{r}|^2+|\boldsymbol{s}|^2$.
2. (i) $\boldsymbol{w}-\boldsymbol{u}$; $\boldsymbol{w}-\boldsymbol{u}-\boldsymbol{v}$; $\frac{1}{2}(\boldsymbol{u}+\boldsymbol{v})$ (ii) $\frac{1}{2}(\boldsymbol{u}+\boldsymbol{v})-\boldsymbol{w}$
   (iii) 12 units.
3. (i) $\overrightarrow{AB}+\overrightarrow{BC}+\overrightarrow{CR}$, $\overrightarrow{AD}+\overrightarrow{DC}+\overrightarrow{CR}$, $\overrightarrow{AP}+\overrightarrow{PQ}+\overrightarrow{QR}$, $\overrightarrow{AP}+\overrightarrow{PS}+\overrightarrow{SR}$, $\overrightarrow{AD}+\overrightarrow{DS}+\overrightarrow{SR}$.
   (ii) $\overrightarrow{AB}$, $\overrightarrow{DC}$, $\overrightarrow{PQ}$, $\overrightarrow{SR}$; $\overrightarrow{AP}$, $\overrightarrow{BQ}$, $\overrightarrow{DS}$, $\overrightarrow{CR}$; $\overrightarrow{AD}$, $\overrightarrow{BC}$, $\overrightarrow{QR}$, $\overrightarrow{PS}$.
   (iii) $\boldsymbol{w}+\frac{1}{2}\boldsymbol{v}-\boldsymbol{u}$; $\boldsymbol{v}-\frac{1}{2}\boldsymbol{u}$; $\frac{1}{2}(\boldsymbol{v}+\boldsymbol{u})-\boldsymbol{w}$.
4. (b) $\frac{1}{2}(\boldsymbol{a}+\boldsymbol{b})-\boldsymbol{c}$.

Page 168, Assignment 4.2

1. (i) $\dfrac{3\boldsymbol{q}+5\boldsymbol{p}}{8}$ (ii) $\dfrac{5\boldsymbol{q}-2\boldsymbol{p}}{3}$ (iii) $\frac{31}{24}(\boldsymbol{q}-\boldsymbol{p})$; $\frac{31}{24}$.
2. (i) $\frac{1}{2}(\boldsymbol{a}+\boldsymbol{b})$ (ii) $\frac{1}{3}(\boldsymbol{a}+\boldsymbol{b}+\boldsymbol{d})$ (iii) $\boldsymbol{b}+\boldsymbol{d}-\boldsymbol{a}$.
3. (i) $\frac{1}{2}(\boldsymbol{q}+\boldsymbol{r})$ (ii) $\boldsymbol{p}+\boldsymbol{r}-\boldsymbol{q}$; $\frac{1}{3}(\boldsymbol{p}+\boldsymbol{q}+\boldsymbol{r})$.
4. (b) $\frac{1}{2}(\boldsymbol{a}+\boldsymbol{b})$
   (c) Yes, centroid of triangle PQR has position vector $\frac{1}{3}(\boldsymbol{a}+\boldsymbol{b}+\boldsymbol{c})$.
5. (i) $\frac{1}{2}(\boldsymbol{b}+\boldsymbol{c})$, $\frac{1}{3}(\boldsymbol{a}+\boldsymbol{b}+\boldsymbol{c})$ (ii) $\frac{1}{2}(\boldsymbol{c}+\boldsymbol{a})$, $\frac{1}{3}(\boldsymbol{a}+\boldsymbol{b}+\boldsymbol{c})$
   (iii) $\frac{1}{2}(\boldsymbol{a}+\boldsymbol{b})$, $\frac{1}{3}(\boldsymbol{a}+\boldsymbol{b}+\boldsymbol{c})$.
6. $\frac{1}{2}(\boldsymbol{a}+\boldsymbol{d})$, $\frac{1}{2}(\boldsymbol{b}+\boldsymbol{c})$; $\frac{1}{4}(\boldsymbol{a}+\boldsymbol{b}+\boldsymbol{c}+\boldsymbol{d})$.
7. (i) $2\boldsymbol{c}-\boldsymbol{b}$, $\dfrac{3\boldsymbol{a}+4\boldsymbol{c}}{7}$, $\dfrac{3\boldsymbol{a}+2\boldsymbol{b}}{5}$.

8. (i) $\dfrac{3\boldsymbol{b}+4\boldsymbol{a}}{7}, \dfrac{2\boldsymbol{a}+3\boldsymbol{c}}{5}$ (ii) $2\boldsymbol{c}-\boldsymbol{b}$ (iii) $\frac{3}{7}$.

Page 173, Assignment 4.3

1. $\begin{pmatrix}1\\-2\\4\end{pmatrix}, \begin{pmatrix}3\\6\\-4\end{pmatrix}, \begin{pmatrix}2\\8\\-8\end{pmatrix}; \begin{pmatrix}-2\\-8\\8\end{pmatrix}; 2\sqrt{33}$.
2. (i) 3 (ii) 9 (iii) $2\sqrt{2}$.
4. (i) $(4, -3, 0)$ (ii) $(4, -5\frac{1}{2}, 5)$ (iii) $(10, 15, -13)$ (iv) $(11, -18, 13)$.
5. (i) $(5, -1)$ (ii) $(0, 0)$ (iii) $(4, 22)$ (iii) $(2, 4)$.
6. (i) $(1, 2)$ (ii) $(-1, 0)$.
7. $-1:3$.
8. $-2:1, (-5, -2)$.
9. $1:2, (3, 1, 0)$.
10. $2:1, (0, -1, -2)$.
11. (a) $(4, 2, -3)$.
12. (i) $\begin{pmatrix}5\\8\\-8\end{pmatrix}$ (ii) $\begin{pmatrix}25\\-16\\9\end{pmatrix}$ (iii) $\begin{pmatrix}13\\12\\-11\end{pmatrix}$
13. (i) collinear (ii) not collinear (iii) collinear.
14. $(0, 3, -2)$.
15. $\begin{pmatrix}-4\\-5\\-18\end{pmatrix}$.
16. (i) $(4, \frac{8}{5}, -4)$ (ii) $(-8, -8, -16)$.
17. $A(0, -2, -1), B(5, 8, -6)$.
18. $k = -11, m = 9, n = 2$.
19. (i) $3\sqrt{5}$ (iii) $(13, -14, 18)$.

Page 180, Assignment 4.4

1. $\boldsymbol{a} = 3\boldsymbol{i}-\boldsymbol{j}+2\boldsymbol{k}, \boldsymbol{b} = -\boldsymbol{i}-2\boldsymbol{j}, \boldsymbol{c} = 2\boldsymbol{i}+2\boldsymbol{j}-\boldsymbol{k}$.
2. (i) $\boldsymbol{a} = 2\boldsymbol{i}-4\boldsymbol{j}, \boldsymbol{b} = -3\boldsymbol{i}-2\boldsymbol{j}+\boldsymbol{k}, \boldsymbol{c} = -2\boldsymbol{i}-\boldsymbol{j}+3\boldsymbol{k}$
   (ii) $7\boldsymbol{i}-6\boldsymbol{j}-\boldsymbol{k}; -2\boldsymbol{i}-7\boldsymbol{j}-\boldsymbol{k}; -8\boldsymbol{i}-4\boldsymbol{j}+12\boldsymbol{k}$.
3. (i) $\sqrt{14}$ (ii) $6\boldsymbol{i}-3\boldsymbol{j}+\boldsymbol{k}; \sqrt{46}$
   (iii) $-\boldsymbol{i}-5\boldsymbol{j}; -7\boldsymbol{i}-8\boldsymbol{j}+\boldsymbol{k}$.
4. (i) $(3, -2, 5)$ (ii) $\boldsymbol{a} = 3\boldsymbol{i}-2\boldsymbol{j}+5\boldsymbol{k}$.
   (iii) $\begin{pmatrix}3\\0\\0\end{pmatrix}, 3\boldsymbol{i}; \begin{pmatrix}0\\-2\\0\end{pmatrix}, -2\boldsymbol{j}; \begin{pmatrix}0\\0\\5\end{pmatrix}, 5\boldsymbol{k}; \begin{pmatrix}3\\-2\\0\end{pmatrix}, 3\boldsymbol{i}-2\boldsymbol{j}$.
5. $14\sqrt{2}$.
6. (i) No (ii) they are perpendicular.
7. (i) $\boldsymbol{p}.\boldsymbol{q}+\boldsymbol{p}.\boldsymbol{r}$ (ii) $\boldsymbol{p}.\boldsymbol{r}+\boldsymbol{p}.\boldsymbol{s}+\boldsymbol{q}.\boldsymbol{r}+\boldsymbol{q}.\boldsymbol{s}$
   (iii) $\boldsymbol{p}.\boldsymbol{r}-\boldsymbol{p}.\boldsymbol{s}+\boldsymbol{q}.\boldsymbol{r}-\boldsymbol{q}.\boldsymbol{s}$ (iv) $\boldsymbol{p}^2-\boldsymbol{q}^2$.
8. $2|\boldsymbol{a}|^2+2\boldsymbol{a}.\boldsymbol{b}$.
11. $\boldsymbol{a}+\boldsymbol{c}, \boldsymbol{a}-\boldsymbol{c}$.
12. $-16; -22$.
13. $150°$.
15. (iii) 9.
16. not collinear; $30°$, $3\sqrt{3}$ units^2.
17.

·	$\boldsymbol{i}$	$\boldsymbol{j}$	$\boldsymbol{k}$
$\boldsymbol{i}$	1	0	0
$\boldsymbol{j}$	0	1	0
$\boldsymbol{k}$	0	0	1

20. (i) $\dfrac{|\boldsymbol{q}|^2+|\overrightarrow{PQ}|^2-|\boldsymbol{p}|^2}{2|\boldsymbol{q}||\overrightarrow{PQ}|}$.
21. (i) $p_1q_1+p_2q_2+p_3q_3, |\boldsymbol{p}||\boldsymbol{q}|\cos\theta$ (ii) $48{\cdot}2°$
    (iii) $0; 90°$.
22. (a) $P(4, 2, 0), Q(3, 3, 6), R(1, 1, 6)$
    (b) $\begin{pmatrix}-2\\-2\\0\end{pmatrix}, \begin{pmatrix}3\\1\\-6\end{pmatrix}, \begin{pmatrix}-1\\1\\6\end{pmatrix}$.
23. $a-c, b-c$.
24. (i) $(2, -5, 3)$.

Page 183, Assignment 4.5

1. E. 2. E. 3. B. 4. E. 5. A. 6. D.
7. D. 8. A. 9. D. 10. C. 11. B. 12. C.